AF594988

Nutrition and Specific Diseases by Women during the Life Course

Editors

Birgit-Christiane Zyriax
Nataliya Makarova

Basel • Beijing • Wuhan • Barcelona • Belgrade • Novi Sad • Cluj • Manchester

Editors
Birgit-Christiane Zyriax
Preventive Medicine and Nutrition, Midwifery Science—Health Services Research and Prevention, Institute for Health Services Research in Dermatology and Nursing (IVDP), University Medical Center Hamburg-Eppendorf (UKE)
Hamburg, Germany

Nataliya Makarova
Preventive Medicine and Nutrition, Midwifery Science—Health Services Research and Prevention, Institute for Health Services Research in Dermatology and Nursing (IVDP), University Medical Center Hamburg-Eppendorf (UKE)
Hamburg, Germany

Editorial Office
MDPI
St. Alban-Anlage 66
4052 Basel, Switzerland

This is a reprint of articles from the Special Issue published online in the open access journal *Nutrients* (ISSN 2072-6643) (available at: https://www.mdpi.com/journal/nutrients/special_issues/Nutrition_Diseases_Women).

For citation purposes, cite each article independently as indicated on the article page online and as indicated below:

Lastname, A.A.; Lastname, B.B. Article Title. *Journal Name* **Year**, *Volume Number*, Page Range.

ISBN 978-3-0365-9584-9 (Hbk)
ISBN 978-3-0365-9585-6 (PDF)
doi.org/10.3390/books978-3-0365-9585-6

Contents

About the Editors . . . . **vii**

Nataliya Makarova and Birgit-Christiane Zyriax
Nutrition and Specific Diseases in Women during the Life Course
Reprinted from: *Nutrients* **2023**, *15*, 3401, doi:10.3390/nu15153401 . . . . **1**

Daniel Engler, Renate B. Schnabel, Felix Alexander Neumann, Birgit-Christiane Zyriax and Nataliya Makarova
Sex-Specific Dietary Patterns and Social Behaviour in Low-Risk Individuals
Reprinted from: *Nutrients* **2023**, *15*, 1832, doi:10.3390/nu15081832 . . . . **5**

Julia Maria Assies, Martje Dorothea Sältz, Frederik Peters, Christian-Alexander Behrendt, Annika Jagodzinski, Elina Larissa Petersen and et al.
Cross-Sectional Association of Dietary Patterns and Supplement Intake with Presence and Gray-Scale Median of Carotid Plaques—A Comparison between Women and Men in the Population-Based Hamburg City Health Study
Reprinted from: *Nutrients* **2023**, *15*, 1468, doi:10.3390/nu15061468 . . . . **21**

Anne Esser, Leonie Neirich, Sabine Grill, Stephan C. Bischoff, Martin Halle, Michael Siniatchkin and et al.
How Does Dietary Intake Relate to Dispositional Optimism and Health-Related Quality of Life in Germline *BRCA1/2* Mutation Carriers?
Reprinted from: *Nutrients* **2023**, *15*, 1396, doi:10.3390/nu15061396 . . . . **37**

Steven D. Hicks, Desirae Chandran, Alexandra Confair, Anna Ward and Shannon L. Kelleher
Human Milk-Derived Levels of let-7g-5p May Serve as a Diagnostic and Prognostic Marker of Low Milk Supply in Breastfeeding Women
Reprinted from: *Nutrients* **2023**, *15*, 567, doi:10.3390/nu15030567 . . . . **53**

Kaylee Slater, Tracy L. Schumacher, Ker Nee Ding, Rachael M. Taylor, Vanessa A. Shrewsbury and Melinda J. Hutchesson
Modifiable Risk Factors for Cardiovascular Disease among Women with and without a History of Hypertensive Disorders of Pregnancy
Reprinted from: *Nutrients* **2023**, *15*, 410, doi:10.3390/nu15020410 . . . . **69**

Catrin Herpich, Stephanie Lehmann, Bastian Kochlik, Maximilian Kleinert, Susanne Klaus, Ursula Müller-Werdan and et al.
The Effect of Dextrose or Protein Ingestion on Circulating Growth Differentiation Factor 15 and Appetite in Older Compared to Younger Women
Reprinted from: *Nutrients* **2022**, *14*, 4066, doi:10.3390/nu14194066 . . . . **83**

Kaja Michalczyk, Patrycja Kapczuk, Grzegorz Witczak, Mateusz Bosiacki, Mateusz Kurzawski, Dariusz Chlubek and et al.
The Associations between Metalloestrogens, GSTP1, and SLC11A2 Polymorphism and the Risk of Endometrial Cancer
Reprinted from: *Nutrients* **2022**, *14*, 3079, doi:10.3390/nu14153079 . . . . **93**

Mingling Chen, Siew Lim and Cheryce L. Harrison
Limiting Postpartum Weight Retention in Culturally and Linguistically Diverse Women: Secondary Analysis of the HeLP-her Randomized Controlled Trial
Reprinted from: *Nutrients* **2022**, *14*, 2988, doi:10.3390/nu14142988 . . . . **107**

Sabrina Aliné, Chien-Yeh Hsu, Hsiu-An Lee, Rathi Paramastri and Jane C.-J. Chao
Association of Dietary Pattern with Cardiovascular Risk Factors among Postmenopausal Women in Taiwan: A Cross-Sectional Study from 2001 to 2015
Reprinted from: *Nutrients* **2022**, *14*, 2911, doi:10.3390/nu14142911 . . . . **119**

Alexandra Tijerina, Yamile Barrera, Elizabeth Solis-Pérez, Rogelio Salas, José L. Jasso, Verónica López and et al.
Nutritional Risk Factors Associated with Vasomotor Symptoms in Women Aged 40–65 Years
Reprinted from: *Nutrients* **2022**, *14*, 2587, doi:10.3390/nu14132587 . . . . **133**

Joanna Maria Pieczyńska, Ewa Pruszyńska-Oszmałek, Paweł Antoni Kołodziejski, Anna Łukomska and Joanna Bajerska
The Role of a High-Fat, High-Fructose Diet on Letrozole-Induced Polycystic Ovarian Syndrome in Prepubertal Mice
Reprinted from: *Nutrients* **2022**, *14*, 2478, doi:10.3390/nu14122478 . . . . **147**

Iwona Szydłowska, Jolanta Nawrocka-Rutkowska, Agnieszka Brodowska, Aleksandra Marciniak, Andrzej Starczewski and Małgorzata Szczuko
Dietary Natural Compounds and Vitamins as Potential Cofactors in Uterine Fibroids Growth and Development
Reprinted from: *Nutrients* **2022**, *14*, 734, doi:10.3390/nu14040734 . . . . **159**

About the Editors

Birgit-Christiane Zyriax

Birgit-Christiane Zyriax is a professor of midwifery sciences with a focus on health services research and prevention at the University Medical Center Hamburg-Eppendorf. Her research and training focus on nutrition and lifestyle particularly with regard to women's health including the phase of pregnancy as well as menopause. She is a member of a broad multidisciplinary network of scientists with research activities in the field of lifestyle and chronic diseases. Her further research addresses health care needs of patients and health literacy. She has many years of experience with the implementation of observational and intervention studies aiming to develop primary, secondary and tertiary prevention programs. At the Medical University Birgit-Christiane oversees the academic training of midwifery and medical students and is a member of both curricula.

Nataliya Makarova

Nataliya Makarova has a postdoc position within the research group health services research, preventive medicine and nutrition. She is currently involved in different projects researching risk factors and biomarkers within women's health with a specific focus on life course approach and developing of prevention strategies.

Editorial

Nutrition and Specific Diseases in Women during the Life Course

Nataliya Makarova [1,2] and Birgit-Christiane Zyriax [1,2,*]

[1] German Center for Cardiovascular Research (DZHK), Partner Site Hamburg/Kiel/Lübeck, 20246 Hamburg, Germany
[2] Preventive Medicine and Nutrition, Midwifery Science—Health Services Research and Prevention, Institute for Health Services Research in Dermatology and Nursing (IVDP), University Medical Center Hamburg-Eppendorf (UKE), 20246 Hamburg, Germany
* Correspondence: b.zyriax@uke.de

In Western countries, the prevalence rates of risk factors for premature mortality and early non-communicable diseases are growing due to the increasing prevalence of poor nutrition habits, increasing levels of stress, and sedentary lifestyles.

The life course approach, developed by Diana Kuh, Yoav Ben-Shlomo and colleagues, offers an integrative approach which guides research on health, human development, and ageing [1]. The translation of this life course approach to women's health, the study of reproductive health, is indispensable. It comprises the investigation of factors across one's life that influence the timing of menarche, fertility, pregnancy outcomes, gynaecological disorders, and age at menopause. It also recognises the important influence of reproductive health on non-communicable disease risks in later life [2]. This integrative approach considers the continuity of reproductive health and the inter-relationship between different biomarkers and risk factors [2]. Within women's life courses, lifestyle plays an essential role throughout different phases, ranging from young age to pregnancy, menopause, and healthy aging. Factors such as pregnancy outcomes, long-term health, quality of life, and disease risk are consequently influenced by women's nutrition, levels of stress, physical activity, and over-riding lifestyle choices.

Correlations between nutrition and many diseases have been observed for many years. However, the underlying mechanisms and the effect sizes are only partly known due to the often multifactorial disease processes. The link between lifestyle and the growing rates of different diseases (e.g., cervical, ovarian carcinoma, breast cancer, cardiovascular diseases, or diabetes mellitus type 2) needs to be investigated further. New scientific approaches are being used to try to relate individual biomarkers to dietary patterns and changes in the microbiome in order to make risk potentials visible earlier. However, what are the key players in nutrition or physical activity that dominate lifestyle and need to be highlighted in prevention programs? More population-based studies and especially RCTs investigating the association between dietary factors and the occurrence of diseases are needed, leading to the central question: How can we prevent multimorbidity and reduced life quality in elderly women?

Within this Special Issue on nutrition and women's health, we share twelve evidence-based research papers with the scientific community.

Currently, the average number of views within this Special Issue is 1649.

The first contribution within this Special Issue performed an analysis of the literature generated over the past 20 years regarding risks of uterine fibroids' occurrence and dietary factors [3].

One study investigated the role of a high-fat, high-fructose diet in mice [4], whereas the remainder of the studies in this Special Issue investigated humans. The results from this animal model study showed that high-fat and high-fructose feeding given around puberty may directly affect reproductive and metabolic symptoms [4].

Citation: Makarova, N.; Zyriax, B.-C. Nutrition and Specific Diseases in Women during the Life Course. *Nutrients* **2023**, *15*, 3401. https://doi.org/10.3390/nu15153401

Received: 25 June 2023
Accepted: 12 July 2023
Published: 31 July 2023

Four other research papers investigated cardiovascular disease risk factors within the large, population-based, longitudinal Hamburg City Health Study (HCHS) [5,6], the Australian Longitudinal Study of Women's Health (ALSWH) [7], and a cross-sectional study from Taiwan [8]. In addition to conventional modifiable risk factors such as overweight, hypertension, hyperlipidaemia, and smoking or the consumption of alcohol, nutrition-related risk factors were highlighted as being more important. These included the intake of dietary supplements as well as psychologic risk factors such as depressive symptoms and social behaviour, including personality traits, political preferences, and altruism. The understanding of mechanisms of action between these risk factors allows clinicians to provide appropriate and tailored interventions in and disease prevention and management for women. A greater adherence to dietary quality scores, such as the Mediterranean dietary score or the DASH (Dietary Approaches to Stop Hypertension) score, was associated with a lower risk for and lower severity of risk factors. However, scores based on the Mediterranean diet were developed specifically for the southern European population and thus show weak adherence to diet, for example, in the German population.

A total of four studies highlighted the relationship between biomarkers such as GDF-15 [9], let-7g-5p [10] and germline BRCA1/2 mutation carriers [11]; metalloestrogens, GSTP1, and SLC11A2 polymorphisms [12] and dietary patterns and outcomes.

Finally, in a secondary analysis of the HeLP-her randomised controlled trial, postpartum weight retention (PPWR) was examined in non-Australian-born culturally and linguistically diverse women compared with Australian-born women [7].

Specific diseases in women such as letrozole-induced polycystic ovarian syndrome [4], uterine fibroids [3], vasomotor symptoms [13], and endometrial cancer [12] were examined within this Special Issue.

From the life course perspective, the phases of prepuberty [4], pregnancy [14], and postpartum [7] as well as breastfeeding [10], middle-aged [5,6], and postmenopausal [8] women were investigated. In one study, older women were compared to younger women [9]. Two studies compared men with women [5,6].

Despite the important role of lifestyle and the influence of nutrition on the development and course of diseases, there is little research considering nutrition in the preconceptional phase, in new-borns, and in young girls. Questions on how nutrition in these phases influence the life courses of women remain unanswered.

Further research with additional nutrition-related short-screening tools needs to be introduced to support clinical management and health services to identify patients with unhealthy diets. Additionally, the development of such tools can offer a simple aid for patients to improve their eating habits in a self-effective way.

Author Contributions: Conceptualization, N.M. and B.-C.Z.; methodology, N.M.; writing—original draft preparation, N.M.; writing—review and editing, N.M. and B.-C.Z.; supervision, B.-C.Z.; project administration, N.M. and B.-C.Z. All authors have read and agreed to the published version of the manuscript.

Conflicts of Interest: The authors declare no conflict of interest.

References

1. Kuh, D.; Ben-Shlomo, Y.; Lynch, J.; Hallqvist, J.; Power, C. Life course epidemiology. *J. Epidemiol. Community Health* **2003**, *57*, 778. [CrossRef] [PubMed]
2. Mishra, G.D.; Cooper, R.; Kuh, D. A life course approach to reproductive health: Theory and methods. *Maturitas* **2010**, *65*, 92–97. [CrossRef] [PubMed]
3. Szydłowska, I.; Nawrocka-Rutkowska, J.; Brodowska, A.; Marciniak, A.; Starczewski, A.; Szczuko, M. Dietary natural compounds and vitamins as potential cofactors in uterine fibroids growth and development. *Nutrients* **2022**, *14*, 734. [CrossRef] [PubMed]
4. Pieczyńska, J.M.; Pruszyńska-Oszmałek, E.; Kołodziejski, P.A.; Łukomska, A.; Bajerska, J. The Role of a High-Fat, High-Fructose Diet on Letrozole-Induced Polycystic Ovarian Syndrome in Prepubertal Mice. *Nutrients* **2022**, *14*, 2478. [CrossRef] [PubMed]
5. Engler, D.; Schnabel, R.B.; Neumann, F.A.; Zyriax, B.C.; Makarova, N. Sex-Specific Dietary Patterns and Social Behaviour in Low-Risk Individuals. *Nutrients* **2023**, *15*, 1832. [CrossRef] [PubMed]

6. Assies, J.M.; Sältz, M.D.; Peters, F.; Behrendt, C.A.; Jagodzinski, A.; Petersen, E.L.; Schäfer, I.; Twerenbold, R.; Blankenberg, S.; Rimmele, D.L.; et al. Cross-Sectional Association of Dietary Patterns and Supplement Intake with Presence and Gray-Scale Median of Carotid Plaques—A Comparison between Women and Men in the Population-Based Hamburg City Health Study. *Nutrients* **2023**, *15*, 1468. [CrossRef] [PubMed]
7. Chen, M.; Lim, S.; Harrison, C.L. Limiting Postpartum Weight Retention in Culturally and Linguistically Diverse Women: Secondary Analysis of the HeLP-her Randomized Controlled Trial. *Nutrients* **2022**, *14*, 2988. [CrossRef] [PubMed]
8. Aliné, S.; Hsu, C.Y.; Lee, H.A.; Paramastri, R.; Chao, J.C.J. Association of dietary pattern with cardiovascular risk factors among postmenopausal women in Taiwan: A cross-sectional study from 2001 to 2015. *Nutrients* **2022**, *14*, 2911. [CrossRef] [PubMed]
9. Herpich, C.; Lehmann, S.; Kochlik, B.; Kleinert, M.; Klaus, S.; Müller-Werdan, U.; Norman, K. The Effect of Dextrose or Protein Ingestion on Circulating Growth Differentiation Factor 15 and Appetite in Older Compared to Younger Women. *Nutrients* **2022**, *14*, 4066. [CrossRef] [PubMed]
10. Hicks, S.D.; Chandran, D.; Confair, A.; Ward, A.; Kelleher, S.L. Human Milk-Derived Levels of let-7g-5p May Serve as a Diagnostic and Prognostic Marker of Low Milk Supply in Breastfeeding Women. *Nutrients* **2023**, *15*, 567. [CrossRef] [PubMed]
11. Esser, A.; Neirich, L.; Grill, S.; Bischoff, S.C.; Halle, M.; Siniatchkin, M.; Yahiaoui, M.; Kiechle, M.; Lammert, J. How Does Dietary Intake Relate to Dispositional Optimism and Health-Related Quality of Life in Germline BRCA1/2 Mutation Carriers? *Nutrients* **2023**, *15*, 1396. [CrossRef] [PubMed]
12. Michalczyk, K.; Kapczuk, P.; Witczak, G.; Bosiacki, M.; Kurzawski, M.; Chlubek, D.; Cymbaluk-Płoska, A. The associations between metalloestrogens, GSTP1, and SLC11A2 polymorphism and the risk of endometrial cancer. *Nutrients* **2022**, *14*, 3079. [CrossRef] [PubMed]
13. Tijerina, A.; Barrera, Y.; Solis-Pérez, E.; Salas, R.; Jasso, J.L.; López, V.; Erik Ramírez, E.; Pastor, R.; Tur, J.A.; Bouzas, C. Nutritional Risk Factors Associated with Vasomotor Symptoms in Women Aged 40–65 Years. *Nutrients* **2022**, *14*, 2587. [CrossRef] [PubMed]
14. Slater, K.; Schumacher, T.L.; Ding, K.N.; Taylor, R.M.; Shrewsbury, V.A.; Hutchesson, M.J. Modifiable Risk Factors for Cardiovascular Disease among Women with and without a History of Hypertensive Disorders of Pregnancy. *Nutrients* **2023**, *15*, 410. [CrossRef] [PubMed]

Article

Sex-Specific Dietary Patterns and Social Behaviour in Low-Risk Individuals

Daniel Engler [1,*], Renate B. Schnabel [1,2], Felix Alexander Neumann [3], Birgit-Christiane Zyriax [2,3] and Nataliya Makarova [2,3]

1 Department of Cardiology, University Heart & Vascular Center Hamburg-Eppendorf (UHZ), University Medical Center Hamburg-Eppendorf (UKE), 20246 Hamburg, Germany
2 German Center for Cardiovascular Research (DZHK), Partner Site Hamburg/Kiel/Lübeck, 20246 Hamburg, Germany
3 Preventive Medicine and Nutrition, Midwifery Science—Health Services Research and Prevention, Institute for Health Services Research in Dermatology and Nursing (IVDP), University Medical Center Hamburg-Eppendorf (UKE), 20246 Hamburg, Germany
* Correspondence: d.engler@uke.de

Abstract: Dietary and social behaviour are non-medical factors that influence health outcomes. Non-communicable diseases are related to dietary patterns. To date, little is known about how social behaviour is associated with health-related dietary patterns, and, in particular, we lack information about the role of sex within this possible relation. Our cross-sectional study investigated associations between dietary patterns and social behaviour including personality traits (self-control, risk taking), political preferences (conservative, liberal, ecological, social) and altruism (willingness to donate, club membership, time discounting) in men and women. We performed sex-specific correlation analyses to investigate relationships between dietary patterns based on self-reported protocols from the Mediterranean Diet Adherence Screener (MEDAS) and the validated Healthy Eating Index (HEI) from the EPIC Study and a self-reported social behaviour questionnaire. In linear regression models, we analysed associations between dietary and social behaviour patterns. Sex differences were measured by interaction analysis for each social behaviour item. The study sample consisted of N = 102 low-risk individuals. The median age of the study participants was 62.4 (25th/75th percentile 53.6, 69.1) years, and 26.5% were women. Analyses showed that a lower HEI score was correlated with a higher BMI in both women and men. MEDAS and HEI showed a positive correlation with each other in men. In men, a higher MEDAS showed a positive correlation when they estimated their ability as high, with the same for self-control and preference for ecological politics and MEDAS. A weak negative correlation has been shown between men with a preference for conservative politics and MEDAS. HEI showed a positive significant correlation with age in men. Male participants without club membership scored significantly higher in the HEI compared to non-members. A negative correlation was shown for time discounting in men. Linear regression models showed positive associations between preferences for ecological-oriented politics and nutrition for both HEI and MEDAS. No sex interactions were observed. We faced a few limitations, such as a small sample size, particularly for women, and a limited age spectrum in a European cohort. However, assuming that individuals with a preference for ecological-oriented politics act ecologically responsibly, our findings indicate that ecological behaviour in low-risk individuals might determine, at least in part, a healthy diet. Furthermore, we observed dietary patterns such as higher alcohol consumption in men or higher intake of butter, margarine and cream in women that indicate that women and men may have different needs for nutritional improvement. Thus, further investigations are needed to better understand how social behaviour affects nutrition, which could help to improve health. Our findings have the potential to inform researchers and practitioners who investigate the nature of the relationship between social behaviour and dietary patterns to implement strategies to create first-stage changes in health behaviour for individuals with a low cardiovascular risk profile.

Keywords: social behaviour and preferences; dietary patterns; lifestyle; atrial fibrillation; cardiovascular disease; sex-specific differences

Citation: Engler, D.; Schnabel, R.B.; Neumann, F.A.; Zyriax, B.-C.; Makarova, N. Sex-Specific Dietary Patterns and Social Behaviour in Low-Risk Individuals. *Nutrients* **2023**, *15*, 1832. https://doi.org/10.3390/nu15081832

Academic Editor: Lutz Schomburg

Received: 11 March 2023
Revised: 5 April 2023
Accepted: 7 April 2023
Published: 11 April 2023

1. Introduction

1.1. Dietary Patterns and Cardiovascular Disease

Cardiovascular disease (CVD) continues to be the leading cause of mortality in developed countries and is one of the primary leading causes worldwide [1]. The burden of CVD is attributable to multiple risk factors for disease development [1]. While understanding the underlying medical causes of death, it is important to investigate lifestyle and social drivers of the disease. Dietary patterns have been shown to have a fundamental role in the prevention of CVD [2]. A meta-analysis showed a 22% CVD risk reduction for individuals scoring high in diet quality assessments [3], 19–28% in women and 14–26% in men. Current indices such as the Mediterranean Diet Adherence Screener (MEDAS) showed associations with dietary patterns and several health benefits, including a reduction in total mortality [4] and a decrease in metabolic syndrome risk [5]. Poor diet is a known risk factor for overweight and obesity and is associated with the development of CVD. It has been proposed that personality may be linked to dietary patterns [1,6]. Findings from different studies show a positive association between openness to experience and the consumption of fruits and vegetables and between healthy eating habits [7,8].

1.2. Relation between Social Behaviour, Dietary Pattern and Cardiovascular Health

It has been proposed that personality may be linked to dietary patterns [1,6]. Findings from different studies show a positive association between openness to experience and the consumption of fruits and vegetables and between healthy eating habits [7,8]. Individual social circumstances can influence the type and variety of food consumed in multiple pathways and thereby impact health. Eating behaviour is influenced by social context and, for instance, it is more likely to follow an eating norm if it is perceived to be relevant based on social comparison [9]. Thus, norms of healthy eating are set by the behaviour of other individuals. These social norms have a powerful effect on both food choices and intake [10]. Furthermore, psychosocial mechanisms such as social support or isolation, social influence or independency, social engagement or separation and the access to resources and information are involved in food choices [11]. However, little is known about social behaviour such as personality traits (self-control, risk taking), political preferences (conservative, liberal, ecological, social) and altruism (willingness to donate, club membership, time discounting) in relation to adherence to dietary recommendations. Previous studies have shown that prosocial factors such as conscientiousness are associated with a number of health-promoting behaviours that include a reduced BMI, avoiding alcohol-related harm, binge drinking and smoking and adherence to medication regimens [12–14]. Moreover, an individual participant meta-analysis showed that conscientiousness appears to be related to mortality risk across populations [15].

Evidence suggests that risk factors such as conscientiousness, self-control and risk taking influence health and illness by shaping barriers and facilitators to access to care and health-related behaviours [16].

1.3. The Role of Sex-Specific Dietary Patterns for Social Behaviour and Health

Women and men differ in their different types of social relationships. Women, older individuals and more educated individuals consider health aspects more important than other factors, whilst men consider the taste of food and eating habits as the main determinants of food choices [17]. In a survey of attitudes to food, nutrition and health, results indicated that factors such as quality/freshness, price, taste, trying to eat healthily and the eating habits of family members have an important influence on food choices [18].

However, social attitudes and beliefs vary in individuals; to date, little is known about how social behaviour is associated with health-related dietary patterns. In particular, we lack information about the role of sex within this possible interrelation. This evidence might be applicable for social behaviour-based prevention strategies, e.g., concerning sex-specific dietary habits, the evidence is largely lacking.

1.4. Aim of the Study

Unhealthy dietary patterns are important drivers of an increased cardiovascular risk. They may vary between men and women and social determinates. To date, research examining the relationship of social behaviour with healthy dietary patterns has been limited. Therefore, we investigated individuals with a low risk profile to identify possible relationships between social behaviour factors and sex-specific dietary patterns.

2. Materials and Methods

2.1. Design, Setting and Participants

2.1.1. AHRI Study

The AFHRI cohort is a prospective, monocentric, clinical cohort study, to improve the prediction of personal risk for AF. In the current study, a sub-sample of atrial fibrillation (AF) patients with a low cardiovascular risk factor burden—in particular, no prevalent cardiovascular disease, thyroid dysfunction or cancer—was invited for participation (AFHRI-C). The AFHRI-C was planned as a case–control study. Case participants of the study needed to have AF without other cardiovascular diseases. The population-based controls from the Hamburg City Health Study (HCHS) had to fulfil the following criteria: no cardiovascular disease and limited or no risk factors that are related to AF and no known AF. All participants personally signed informed consent forms.

2.1.2. HCHS Study

The Hamburg City Health Study (HCHS) is a single-centre, prospective, epidemiologic cohort study with an emphasis on imaging to improve the identification of individuals at risk for major chronic diseases, to improve early diagnosis and survival. The enrolled participants were selected from a statistical sample provided by the local residents' registration office. Participants between 45 and 74 years of age from the general population of Hamburg, Germany are included in the study.

For our analysis, we combined the participants from the AFHRI cohort matched with a sub-sample from the population-based Hamburg City Health Study (HCHS). Matching of individuals was based on age, sex and risk factors [19]. Detailed information on both cohorts, their design and the matching approach was published previously [20].

2.2. Variables, Measurements and Processes

Questionnaire data and peripheral venous blood were acquired on the day of enrolment. The participants' characteristics were collected through a questionnaire administered by a healthcare professional, and from patient records.

A three-day dietary record was assessed before their first study participation appointment. A standardised procedure during data collection, with precise questions administered by a study nurse, was applied. The food data were analysed in terms of energy and nutrient intake and adherence to the dietary patterns of the Mediterranean Diet Adherence Screener (MEDAS) [21] and the Healthy Eating Index (HEI) [21,22]. The 14-item MEDAS includes the frequency of food consumption (olive oil, vegetables, fruits, red meat, animal fats, carbonated drinks, red wine, fish/seafood, legumes, nuts, commercial foods and traditional Mediterranean dishes with tomato sauce) as well as the preferred cooking fat and meat consumed. Each item was scored zero or one depending on whether the item-specific criteria were met, resulting in a score between 0 and 14. For the HEI, we used the edition validated in German. The HEI scores five food groups from 0 to 10 (cereal and potatoes; dairy; meat, sausages, fish and eggs; fat or oil; sweets and foods high in fat) and three food groups from 0 to 20 (vegetables; fruits; beverages), allowing total scores between 0 and 110 points. Detailed information of the clinical visits, ECG registration and collection of the nutrition data have been published previously [20]. The following behaviour and preference items were used for analysis: personality traits of risk taking by survey questions "Are you generally a risk-taker, or do you try to avoid risk?"; self-control "When I set my mind to something, I follow through"; religiousness "How strongly religious do you consider yourself to be?"; political preferences (conservative, liberal, ecological, social) "I identify myself with ... -oriented politics"; club membership "Are you an active and/or passive member in a club?"; and time discounting. Time discounting means the willingness to give up something at present in order to benefit from something in the future. Each participant received an expense allowance of EUR 50 for study participation. To assess altruistic behaviour, we asked the participants if and how much they were willing to donate to the United Nations Children's Fund (UNICEF). For the analysis, we used the continuous variable of the willingness to donate scaled from EUR 0 (no funding—less altruistic behaviour) to EUR 50 (maximum funding—more altruistic behaviour). All variables of the present analysis and detailed definitions of the five behaviour preferences are explained in Supplementary Table S1.

2.3. Data Handling

Of N = 104, two participants were excluded from the analysis due to missing values of more than 50% in the behaviour questionnaire. The variables were tested using Shapiro–Wilk tests for normal distribution and visualisations such as scatter plots and box plots for outliers. Behaviour preferences and the willingness to donate showed a normal distribution. After checking whether missing values were distributed randomly, using Little's MCAR Test (chi^2 (69) = 69.91, $p = 0.245$), we performed a multiple data imputation by single value regression analysis with five iterations for the imputation [23] and aggregated these into a pooled value using the Bar procedure [24]. Missing values were imputed for income (14), resting heart rate (11), NT-proBNP (8) and weight (2). No impact of the imputation on our results was detected. Details of the iteration steps for the imputation were published perilously [20]. We used SPSS Statistics (version 27.0; IBM Corp., Armonk, NY, USA) for all the analyses.

2.4. Statistical Analysis

Categorical variables of participant characteristics are presented by their absolute and relative frequencies, and Pearson's chi-square and Fisher's exact tests were used to determine subgroup comparisons. Shapiro–Wilk tests indicated that most of the continuous variables significantly varied from a normal distribution. Therefore, continuous variables are reported as the median and the first and third quartiles. The Mann–Whitney U test was applied for group comparisons. Furthermore, bivariate correlations of variables were assessed using Spearman's rho test. To analyse sex-specific dietary patterns, we applied the standardised test statistics of Pearson's Chi.

To analyse the association between behaviour preferences and the indices (MEDAS and HEI), we performed multiple linear regression analysis. We analysed outliers by cook test and tested variables for autocorrelation by Durban–Watson. Multi-collinearity was tested by VIF. Additionally, we performed interaction analysis to investigate a possible difference or similaritiy with sex. All possible confounders were tested for significance within the model. The variable of club membership was significantly correlated with social politics; thus, we excluded membership from the regression analysis. Results of the regressions are presented as beta and their 95% lower and upper confidence intervals (CI). R^2 was used to describe the regression model's fit. Linear regression models with beta coefficients and interaction analyses were performed to show the association between social behaviour and both HEI and MEDAS. The significance level was set to $a < 0.05$. All statistical tests were computed using SPSS Statistics (version 27.0; IBM Corp., Armonk, NY, USA) and Microsoft Excel 2013 for the visualisation in Figure 1 (Microsoft, Redmond, WA, USA).

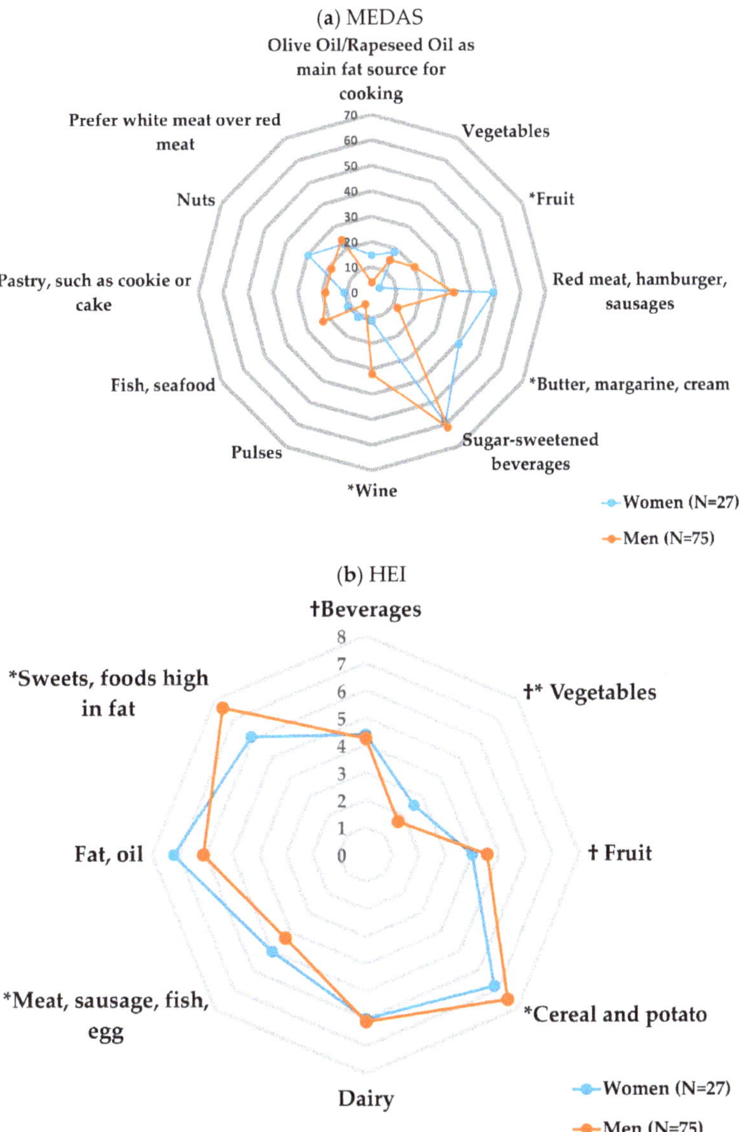

Figure 1. Sex differences in the food items of the Mediterranean Diet Adherence Screener (MEDAS) and the Healthy Eating index (HEI). Dietary patterns of men (N = 75) and women (N = 27) were determined according to average scores achieved for items in the Mediterranean Diet Adherence Screener (MEDAS) and items in the Healthy Eating Index (HEI). For MEDAS, the category "olive oil, rapeseed oil" was excluded as no participant scored it. Scores are presented in percentages for MEDAS and as the average median for items in the Healthy Eating Index (HEI). The scores ranged within 0–10 or 0–20 points depending on the item. † maximum points downscaled from 20 to 10 points. * $p < 0.05$ for significance. Detailed numerical results are presented in Supplementary Table S2 for MEDAS and Supplementary Table S3 for HEI.

3. Results

3.1. Characteristics of Study Participants

Overall, N = 102 participants with a median age of 62.4 (26.5% women) were included in the analysis. Significant sex differences were shown for income (z = 10.3, $p = 0.008$). Men had a significantly higher income compared to women. Dyslipidaemia was more frequent in women (z = 6.1, $p = 0.014$). Women reported significantly less alcohol intake (z = 4.3, $p = 0.069$). The social behaviour assessment showed significant results for active club membership in men (z = 10.7, $p = 0.004$) and a higher preference for conservative politics (z = 17.8, $p = 0.032$) than in women. More detailed information on study participants is provided in Table 1.

Table 1. Characteristics of study participants.

Variables	Total (N = 102)	Women (n = 27)	Men (n = 75)	Standardised Test Statistic	p-Value
Sociodemographic and clinical data					
Age, years	62.4 (53.6, 69.1)	62.4 (56.6, 70.3)	63.5 (53.3, 69.0)	−0.0618 [c]	0.536
Income categories					
Low (<EUR 2500)	16 (15.5%)	**7 (25.9%)**	**9 (12.0%)**	**10.268 [b]**	**0.008**
Middle (EUR 2500 < EUR 5000)	56 (54.9%)	**18 (66.7%)**	**38(50.7%)**		
High (>EUR 5000)	30 (29.4%)	**2 (7.4%)**	**28 (37.3%)**		
Body mass index, kg/m^2	25.5 (23.1, 27.8)	25.0 (23.1, 27.3)	25.4 (23.1, 27.8)	0.155 [c]	0.876
Body mass index categories					
Underweight	1 (1.0%)	-	1 (1.4%)	0.985 [b]	0.875
Normal weight	48 (47.1%)	14 (51.9%)	33 (44.6%)		
Overweight	46 (45.1%)	11 (40.7%)	36 (47.3%)		
Obesity	7 (7.9%)	2 (7.4%)	5 (6.8%)		
Prevalent diseases					
Type 2 diabetes mellitus	2 (2.0%)	-	2 (2.7%)	1.421 [a]	0.153
Arterial hypertension	22 (21.6%)	8 (29.6%)	14 (18.7%)	1.410 [a]	0.235
atrial fibrillation	52 (51%)	14 (51.0%)	38 (50.7%)	0.011 [a]	0.916
Dyslipidaemia	21 (20.6%)	**10 (37.0%)**	**11 (14.7%)**	**6.077 [a]**	**0.014**
Lifestyle factors					
MEDAS, points	3 (1, 4)	3 (1, 5)	2 (2, 4)	−0.320 [c]	0.749
Healthy Eating Index, points	54.9 (47.3, 60.3)	55.6 (48.6, 61,8)	54.8 (67.2, 59.3)	−0.804 [c]	0.421
Healthy Eating Index, categories					
Poor (≤40 pts.)	10 (9.8%)	1 (13.7%)	9 (12.0%)	1.355	0.556
Improvable (>40–64 pts.)	77 (75.5%)	22 (80.4%)	55 (73.3%)		
Good (>64 pts.)	15 (14.7)	4 (5.9%)	11 (14.7%)		
Diet change past 12 months					

Table 1. *Cont.*

Variables	Total (N = 102)	Women (n = 27)	Men (n = 75)	Standardised Test Statistic	p-Value
No	82 (80.4%)	22 (78.4%)	60 (80.0%)	0.028 [a]	0.868
Yes, partially	20 (19.6%)	5 (18.5%)	15 (20.0%)		
Energy intake, kcal/day	2187 (1904, 2504)	**2138 (1224, 2418)**	**2253 (1974, 2583)**	**3.243**	**0.001**
Physical activity, MET-h/day (in Log10)	3.3 (3.1, 3.6)	3.5 (3.1, 3.8)	3.3 (3.1, 3.5)	1.737 [c]	0.082
Alcohol consumption	82 (80.4)	**18 (66.7%)**	**63 (84.0%)**	**4.249 [a]**	**0.039**
Smoking					
Current	62 (60.8%)	6 (22.2%)	10 (13.3%)	2.018 [b]	0.379
Former		9 (33.3%)	41 (54.7%)		
Social behaviour					
Personality traits					
Risk taking	6 (5.7)	6 (5.6)	6 (5.8)	10.563 [b]	0.250
Self-control	9 (8.10)	8 (7.8)	9 (8.10)	12.425 [b]	0.139
Religiousness	4 (1.6)	4 (1.5)	4 (1.7)	10.468 [b]	0.276
Political preferences					
Conservative	5 (2.7)	**4 (1.5)**	**6 (4.8)**	**17.628 [b]**	**0.032**
Liberal	6 (2.7)	5 (2.6)	6 (4.8)	14.993 [b]	0.088
Social	8 (6.9)	8 (7.9)	8 (6.9)	6.444 [b]	0.826
Ecological	8 (7.9)	8 (7.9)	8 (6.9)	6.670 [b]	0.783
Altruism					
Willingness to donate in EUR (EUR 0 to 50)	EUR 20 ± 22	25 (20.50)	20 (20.50)	−0.580 [c]	0.562
Low willingness to donate (EUR 0–25)	61 (59.8%)	16 (59.3%)	45 (60.0%)	0.005 [a]	0.946
High willingness to donation (EUR 25–50)	41 (41.2%)	11 (40.7%)	30 (40.0%)		
Club membership					
No	37 (36.3%)	17 (63.3%)	20 (26.7%)	10.655 [b]	**0.004**
Yes, passive	12 (11.8%)	2 (7.4%)	10 (13.3%)		
Yes, active	53 (52.0%)	8 (29.6)	45 (60.0%)		
Time discounting	4 (2.6)	3 (2.4)	4 (3.5)	3.175 [b]	0.957

Note. Data presented as percentages for categorical variables and as median (1st quartile, 3rd quartile) for continuous variables. [a] Pearson's Chi-square test. [b] Fisher's exact test. [c] Mann–Whitney U test. Multiple data imputation with five imputations was performed to fill in missing values of income (n = 14), and based on N = 102. Significant correlations ($p < 0.05$) marked in bold.

3.2. *Sex-Specific Dietary Patterns*

For MEDAS, women scored significantly lower for the food groups "fruit" (z = 4.0, p = 0.046) and "wine" (z = 4.6, p = 0.035) compared to men. The intake of butter, margarine and cream was higher in women compared to men (z = 10.3, p = 0.001). For the HEI, only the food intake of vegetables was significantly higher when reported by women compared with the male participants (z = 2.8, p = 0.005). The overall HEI score did not show differences between the groups Figure 1.

3.3. Correlation Analysis

For MEDAS, no significant correlations were shown in women. The HEI showed a significant positive correlation for energy intake (r = 0.446, p = 0.020) in women. For both women and men, the HEI was negatively corrected with BMI (women: r = −0.391, p = 0.044; men: r = −0.406, p = 0.001). MEDAS and HEI showed a positive correlation with each other (r = 0.352, p = 0.002) in men. A positive correlation was found in men for MEDAS with self-control (r = 0.293, p = 0.011) and ecological (r = 0.479, p = 0.001) and a negative correlation for conservative party preferences (r = −0.230, p = 0.047). Meanwhile, HEI showed a positive correlation with age (r = 0.301, p = 0.009) and a negative correlation with club membership (r = −0.228, p = 0.049) and time discounting (r = −0.250, p = 0.030) Table 2.

Table 2. Correlations of MEDAS and HEI with independent variables by sex.

Independent Variables	MEDAS				HEI			
	Women		Men		Women		Men	
	R	*p*-Value	R	*p*-Value	R	*p*-Value	R	*p*-Value
Sociodemographic and clinical data								
Age, years	−0.042	0.836	0.104	0.375	0.037	0.854	**0.301**	**0.009**
Income	0.078	0.698	−0.009	0.942	0.014	0.889	−0.069	0.555
Body mass index, kg/m^2	−0.077	0.702	**−0.432**	**0.001**	**−0.391**	**0.044**	**−0.406**	**0.001**
Lifestyle factors								
MEDAS, points	-	-	-	-	0.307	0.119	**0.352**	**0.002**
Healthy Eating Index, points	0.307	0.119	**−0.352**	**0.002**	-	-	-	-
Diet change past year	0.257	0.196	0.184	0.114	−0.037	0.856	0.050	0.670
Energy intake, kcal/day	0.003	0.986	−0.010	0.932	**0.446**	**0.020**	0.165	0.158
Physical activity, MET-h/day	0.233	0.242	0.089	0.448	−0.255	0.199	−0.065	0.577
Social behaviour								
Altruism								
Willingness to donate	0.112	0.577	−0.032	0.783	0.252	0.206	−0.007	0.955
Club membership	0.359	0.066	0.038	0.745	0.192	0.337	**−0.228**	**0.049**
Time discounting	−0.290	0.142	−0.218	0.060	0.191	0.341	**−0.250**	**0.030**
Personality traits								
Risk taking	−0.221	0.268	0.103	0.379	0.188	0.348	0.014	0.905
Self-control	0.026	0.898	**0.293**	**0.011**	0.093	0.646	0.153	0.326
Religiousness	−0.025	0.900	0.029	0.804	0.140	0.487	0.075	0.521
Political preferences								
Conservative	0.026	0.899	**−0.230**	**0.047**	0.146	0.486	0.45	0.700
Liberal	0.035	0.862	0.029	0.802	0.074	0.714	0.145	0.214
Social	0.110	0.584	0.165	0.158	0.173	0.387	0.097	0.407
Ecological	0.155	0.441	**0.479**	**0.001**	0.048	0.812	0.244	0.053

Spearman correlation coefficients of MEDAS and HEI with independent variables (n = 102). BMI, body mass index; MET, metabolic equivalents. Significant correlations ($p < 0.05$) marked in bold.

3.4. Association of Social Behaviour and Nutrition Patterns and the Effect of Moderation by Sex

For MEDAS, the social behaviour model was statistically significant; it explained 20% of the variation in the score. The results of the linear regression in Figure 2a showed that the preference for ecological behaviour (beta = 0.348, 95% CI 0.151–0.546, p = 0.001) was significantly associated with higher scoring in the MEDAS. A preference for an ecological party explained 13.9% of the model variation. A statistically borderline negative association was shown for the preference of social politics preferences (Beta = −0.196, 95% CI −0.399–0.007, p = 0.059).

(**a**) MEDAS

Association of social- behaviour and nutrition patterns for the MEDAS

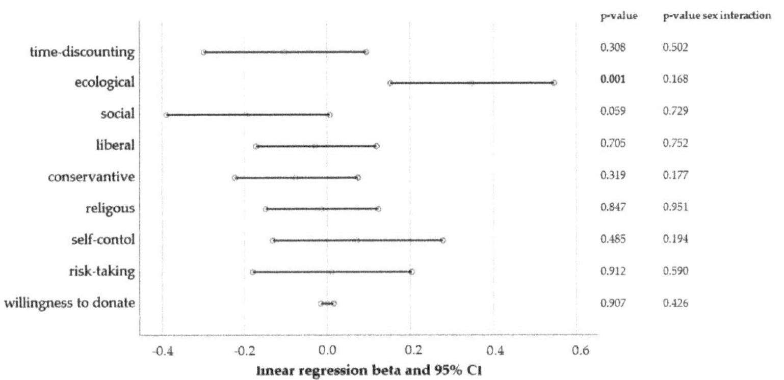

(**b**) HEI

Association of social- behaviour and nutrition patterns for the HEI

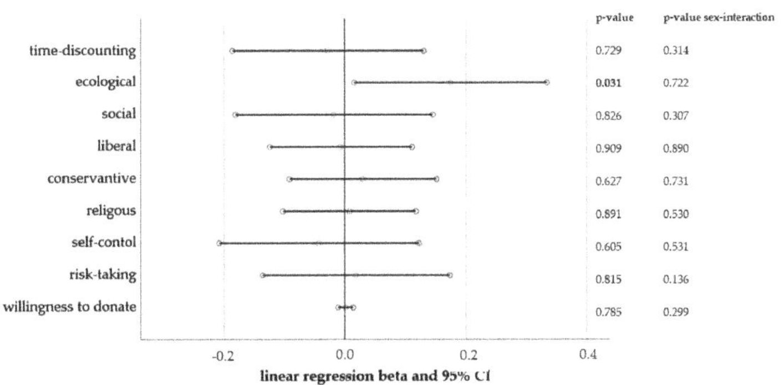

Figure 2. (**a**) Association of social behaviour items with the MEDAS. Linear regression model: R2 = 0.200; p = 0.011. Detailed numerical results are presented in Supplementary Tables S6 and S8. (**b**) Association of social behaviour items with the HEI. Significant associations (p < 0.05) marked in bold. Linear regression model: R2 = 0.082; p = 0.515. Detailed numerical results are presented in Supplementary Tables S5 and S7. Significant associations (p < 0.05) marked in bold.

Similarly, ecological politics preferences were significantly associated with the HEI (beta = 0.175, 95% CI 0.02–0.33, p = 0.031). However, the overall social behaviour model was not significant and only explained 8.2% of the variation in HEI (F (9.92) = 0.917, p = 0.515)

Figure 2b. We did not observe any sex interactions for social behaviour items. Detailed results are shown in Supplementary Tables S5 and S6.

4. Discussion

In our cross-sectional study based on a contemporary cohort of individuals with a low cardiovascular risk factor burden, we investigated the associations between dietary patterns and social behaviour including personality traits (self-control, risk taking), political preferences (conservative, liberal, ecological, social) and altruism (willingness to donate, club membership, time discounting) in men and women.

4.1. Main Findings and Implications for Future Investigations

Overall, we could demonstrate some sex-specific differences in dietary patterns. Further, we showed that social behaviour such as personality traits, political preferences and altruism explain between 8.2 and 20% of the variability in dietary patterns. In the association analyses, an ecological party preference was consistently related to MEDAS and HEI. The findings provide behavioural insights into the field of (un)healthy food choices that can support the development of behavioural primary prevention strategies.

4.2. Dietary Patterns and Sex

Our findings support the assumption that dietary patterns differ between women and men, and we showed that different social behaviour patterns and personal traits are associated with dietary patterns for both sexes. The results underpin the importance of sex differences in dietary patterns as a modifiable risk factor for possible disease development [25]. Recent studies reported more detailed results concerning sex differences for adherence to dietary recommendations using a large sample size of 210,106 women and men [26] and for food preferences and their implications for promoting sustainable dietary patterns in a systematic review [27]. Our study provided detailed information of sex-specific dietary patterns by using two well-established scores, MEDAS and HEI.

Both indices, MEDAS and HEI, are comprehensive assessments of diet quality for this study population. In particular, we observed a lower HEI score for participants with a higher BMI for both sexes, and for MEDAS, we showed that a higher BMI also decreased the scoring in men, but not in women. Although HEI is primarily a measure of overall diet quality, it may also be a predictor of obesity. Other studies have also demonstrated that dietary consumption that follows the HEI is associated with a lower risk for obesity [28]. Sex differences exist in the regulation of energy homeostasis, with a greater intake of energy in men [26,29]. We estimated that these factors were less revenant for the differences in energy intake between women and men given that almost all female participants were in menopause. However, a relationship with higher energy intake in men compared to women has been identified in this study. Factors such as the higher energy needs of men due to a larger body surface, more muscle mass and usually more sports compared to women might have a influence on this result. Despite other study results that showed that lower energy intake is associated with healthy eating behaviours [30], in our cohort, we observed that women with a higher energy intake had higher HEI scores. It is possible that this observation could be explained by the low sample size of women within our study. Consistent with previous studies, men reported higher consumption of alcohol compared to women [31]. In the latest global status report on alcohol and health of the WHO in 2018, 54% men (1.46 billion) and 32% of women (0.88 billion) aged 15 and older worldwide consumed alcohol. Men experience an estimated 2.3 million deaths and 106.5 million DALYs attributable to alcohol consumption, while women experience 0.7 million deaths and 26.1 million DALYs attributable to alcohol consumption [32]. Thus, the assessment of alcohol as an important indicator for dietary patterns shows potential to detect and prevent unhealthy eating habits in consideration of sex-specific patterns of consumption in low-risk individuals. According to average scores achieved for items in the MEDAS and items in the HEI, women scored significantly lower for fruits and wine, but significantly higher for vegetables, compared to men. The low fruit consumption in women is in contrast to other studies, where women compared to men had a higher intake of fruits [33,34]. However, the tendency of healthy dietary patterns such as less or no alcohol consumption and higher intake of vegetables in women within this cohort is mainly in line with the results of other established population studies [26,33]. Thus, the low consumption of fruit could possibly be explained by, again, the low sample size of women and the generally lower overall food consumption in women rather than in men.

In our study, women had a higher burden of dyslipidaemia compared to men and also scored significantly higher for butter, margarine and cream consumption, which is known to increase total and LDL cholesterol [35]. As opposed to our findings, few studies show that women tend to have more favourable levels of blood cholesterol compared to men [36,37]. The onset of dyslipidaemia occurs later in women, and often is more poorly controlled compared to men [38].

4.3. Social Behaviour, Personality Traits and Political Preferences

In general, we showed that men were frequently more active members in a club, self-control correlated with an increase in the MEDAS scoring, and we identified significance in the preference for conservative politics as compared to women.

We showed that, in men, active club membership correlated with a higher MEDAS and HEI score. An active role in the community is a strong indicator of social support and amplifies integration into society. In a few studies, social factors such as informal networks have been identified to have an influence on food-related behaviours [39–41]. If we assume that active participation in a club helps to avoid social isolation, our findings are in line with previous studies.

Self-control, as a major personality characteristic, explains several health-related behaviours [42]. We did not observe an association between self-control and dietary patterns in the regression analysis. However, self-control was correlated with MEDAS in men. Other studies support the importance of self-control for healthy eating attitudes and its role in the maintenance of weight/shape concerns and disordered eating for both women and men [43,44].

We identified a significant association between the preference for ecological-oriented politics and the tendency towards healthy dietary patterns for both indices. Ecological behaviour is a pro-environmental attitude and it refers to the human relationship with the natural environment and is a complex, diverse and dynamic phenomenon, and it seems that ecological behaviour is a significant factor in lifestyle management and food choices. The literature suggests that environment and exposure can predict food-related health risk behaviours and health outcomes [45]. Altruism, tested by the willingness to donate, showed no sex-specific differences, despite a significant difference in income between men and women. There have been a few investigations with the objective to identify associations of altruistic behaviour for recruitment and enrolment optimisations in RCTs [46,47]. Weissberger and colleagues identified, for example, that increased financial altruism is associated with disease occurrence in older adults [48]. Conversely, Shim and colleagues incorporated altruism into a game-theoretic epidemiological model to determine how altruistic behaviour impacts the disease burden. They recommended promoting altruism to improve public health outcomes [49]. In the association analysis, we could not find an interaction effect of sex between dietary patterns and social behaviour items. However, it should be considered that a number of gender-based stereotypes about food exist in every human culture. Although the causes of this are far from being fully elucidated, the consequences for food choices and dietary habits might be relevant because both men and women tend to adhere to those expectations most likely to reinforce their own gender identities [50,51].

4.4. Recommendations for Primary Prevention

- The association of ecological preferences and healthy dietary patterns inform preventive agendas to focus on different behavioural strategies to promote environmentally sustainable food consumption in high-income countries.
- Health aspects are not the only determinants of food choices. People may have various food-related goals, such as to save money or maintain a sustainable lifestyle, which are often more salient and compete with the importance of health considerations.
- Social characteristics such as self-control and ecological preferences could support the ability to reflect on the influence of environmental factors such as marketing and peers. To educate individuals about this, behavioural insights that could support healthy dietary patterns could be integrated into behavioural primary prevention strategies.

The importance of the determination of human health behaviours, investigation of personality traits and preferences and consequently the development of appropriate prevention programs has been underlined by prior research [52].

4.5. Strengths and Limitations

Our results are limited by the cohort approach, which does not permit statements on causality. Bias may exist in the form of recall bias for food items and social desirability bias. The relationship between dietary patterns and social behaviour is complex and we identified weak associations. Therefore, we accounted for possible sex differences and incomes, but may have missed other relevant factors, such as education levels. We had to deal with a relatively small sample size (especially for women). Despite careful adjustment, residual confounding may exist. Additionally, the present research was restricted to the metropolitan region of Hamburg in Germany, and the results may not be generalisable to other populations. The main strength of our study is a well-characterised and consciously selected cohort with low-risk participants, where we were able to identify at least weak differences for social behaviours in men and women. The other strength is the advantageous comprehensive assessment of dietary patterns with different items based on protocols from both the Mediterranean Diet Adherence Screener and the validated Healthy Eating Index and the provision of a broader picture of food consumption.

5. Conclusions

In our cohort of low-risk individuals, we could demonstrate that social behaviour such as personality traits (self-control, risk taking), political preferences (conservative, liberal, ecological, social) and altruism (willingness to donate, club membership, time discounting) may be related to dietary patterns. In particular, ecological preferences showed a significant association with healthy dietary habits.

However, the observed associations were weak. There were no sex differences observed between social behaviour and dietary patterns. Based on our analyses, primary prevention might address behavioural aspects in order to improve dietary habits and thus health in both women and men.

Supplementary Materials: The following supporting information can be downloaded at: https://www.mdpi.com/article/10.3390/nu15081832/s1, Table S1: Assessment of social behaviour, Table S2: Sex differences in food items of the Mediterranean Diet Adherence Screener, Table S3: Sex differences in the food items of the Healthy Eating Index (HEI), Table S4: Differences individuals with low and high willingness to donate in the food items of the Mediterranean Diet Adherence Screener, Table S5: Linear regressions of social behaviour variables with HEI, Table S6: Linear regressions of social behaviour variables with MEDAS, Table S7 Moderation effect of sex for social behaviour with HEI, Table S8: Moderation effect of sex for social behaviour with MEDAS, Figure S1: Dietary patterns of individuals with high (N = 61) and low (N = 41) willingness to donate according to average scores achieved for items in the Mediterranean Diet Adherence Screener (MEDAS).

Author Contributions: Methodology, D.E.; Formal analysis, D.E.; Writing—original draft, D.E. and N.M.; Writing—review & editing, R.B.S., F.A.N. and B.-C.Z.; Supervision, R.B.S. and B.-C.Z.; Project administration, N.M.; Funding acquisition, R.B.S. All authors have read and agreed to the published version of the manuscript.

Funding: R.B.S. received funding from the European Research Council (ERC) under the European Union's Horizon 2020 research and innovation programme under the grant agreement No. 648131, from the European Union's Horizon 2020 research and innovation programme under the grant agreement No. 847770 (AFFECT-EU) and from the German Center for Cardiovascular Research (DZHK e.V.) (81Z1710103), as well as the German Ministry of Research and Education (BMBF 01ZX1408A) and ERACoSysMed3 (031L0239). R.B.S. received conference lecture fees from BMS/Pfizer outside this work. N.M. received conference lecture fees from Abbott Laboratories outside this work.

Institutional Review Board Statement: This study was approved by the Local Psychological Ethics Committee of the Hamburg Medical Association (ethics code: PV5705).

Informed Consent Statement: Informed consent was obtained from all subjects involved in the study.

Data Availability Statement: The data analysed during the current study are not publicly available due to the German National Data Protection Regulation. They are available on reasonable request from the corresponding author.

Acknowledgments: We thank all study participants for their generous participation in the cohort study.

Conflicts of Interest: R.B.S. has received lecture fees and advisory board fees from BMS/Pfizer outside this work. N.M. received conference lecture fees from Abbott Laboratories outside this work. D.E., F.A.N. and B.-C.Z. declare no conflicts of interest.

References

1. Vaduganathan, M.; Mensah, G.A.; Turco, J.V.; Fuster, V.; Roth, G.A. The Global Burden of Cardiovascular Diseases and Risk. *J. Am. Coll. Cardiol.* **2022**, *80*, 2361–2371. [CrossRef] [PubMed]
2. Zampelas, A.; Magriplis, E. Dietary patterns and risk of cardiovascular diseases: A review of the evidence. *Proc. Nutr. Soc.* **2020**, *79*, 68–75. [CrossRef]
3. Schwingshackl, L.; Bogensberger, B.; Hoffmann, G. Diet Quality as Assessed by the Healthy Eating Index, Alternate Healthy Eating Index, Dietary Approaches to Stop Hypertension Score, and Health Outcomes: An Updated Systematic Review and Meta-Analysis of Cohort Studies. *J. Acad. Nutr. Diet.* **2018**, *118*, 74–100. [CrossRef] [PubMed]
4. Eleftheriou, D.; Benetou, V.; Trichopoulou, A.; La Vecchia, C.; Bamia, C. Mediterranean diet and its components in relation to all-cause mortality: Meta-analysis. *Br. J. Nutr.* **2018**, *120*, 1081–1097. [CrossRef]
5. Kastorini, C.-M.; Milionis, H.J.; Esposito, K.; Giugliano, D.; Goudevenos, J.A.; Panagiotakos, D.B. The effect of Mediterranean diet on metabolic syndrome and its components: A meta-analysis of 50 studies and 534,906 individuals. *J. Am. Coll. Cardiol.* **2011**, *57*, 1299–1313. [CrossRef]
6. Lunn, T.E.; Nowson, C.A.; Worsley, A.; Torres, S.J. Does personality affect dietary intake? *Nutrition* **2014**, *30*, 403–409. [CrossRef]
7. Conner, T.S.; Thompson, L.M.; Knight, R.L.; Flett, J.A.M.; Richardson, A.C.; Brookie, K.L. The Role of Personality Traits in Young Adult Fruit and Vegetable Consumption. *Front. Psychol.* **2017**, *8*, 119. [CrossRef] [PubMed]
8. Plante, C.N.; Rosenfeld, D.L.; Plante, M.; Reysen, S. The role of social identity motivation in dietary attitudes and behaviors among vegetarians. *Appetite* **2019**, *141*, 104307. [CrossRef] [PubMed]
9. Levy, D.E.; Pachucki, M.C.; O'Malley, A.J.; Porneala, B.; Yaqubi, A.; Thorndike, A.N. Social connections and the healthfulness of food choices in an employee population. *Nat. Hum. Behav.* **2021**, *5*, 1349–1357. [CrossRef]
10. Higgs, S. Social norms and their influence on eating behaviours. *Appetite* **2015**, *86*, 38–44. [CrossRef] [PubMed]
11. Conklin, A.I.; Forouhi, N.G.; Surtees, P.; Khaw, K.T.; Wareham, N.J.; Monsivais, P. Social relationships and healthful dietary behaviour: Evidence from over-50s in the EPIC cohort, UK. *Soc. Sci. Med.* **2014**, *100*, 167–175. [CrossRef]
12. Provencher, V.; Bégin, C.; Gagnon-Girouard, M.P.; Tremblay, A.; Boivin, S.; Lemieux, S. Personality traits in overweight and obese women: Associations with BMI and eating behaviors. *Eat. Behav.* **2008**, *9*, 294–302. [CrossRef] [PubMed]
13. Molloy, G.J.; O'Carroll, R.E.; Ferguson, E. Conscientiousness and Medication Adherence: A Meta-analysis. *Ann. Behav. Med.* **2013**, *47*, 92–101. [CrossRef] [PubMed]
14. Hakulinen, C.; Jokela, M. Chapter 3—Personality as Determinant of Smoking, Alcohol Consumption, Physical Activity, and Diet Preferences. In *Personality and Disease*; Johansen, C., Ed.; Academic Press: San Diego, CA, USA, 2018; pp. 33–48. [CrossRef]
15. Jokela, M.; Batty, G.D.; Nyberg, S.T.; Virtanen, M.; Nabi, H.; Singh-Manoux, A.; Kivimäki, M. Personality and All-Cause Mortality: Individual-Participant Meta-Analysis of 3,947 Deaths in 76,150 Adults. *Am. J. Epidemiol.* **2013**, *178*, 667–675. [CrossRef]
16. Essien, U.R.; Kornej, J.; Johnson, A.E.; Schulson, L.B.; Benjamin, E.J.; Magnani, J.W. Social determinants of atrial fibrillation. *Nat. Rev. Cardiol.* **2021**, *18*, 763–773. [CrossRef]

17. Shepherd, R. Social determinants of food choice. *Proc. Nutr. Soc.* **1999**, *58*, 807–812. [CrossRef]
18. Lappalainen, R.; Kearney, J.; Gibney, M. A pan EU survey of consumer attitudes to food, nutrition and health: An overview. *Food Qual. Prefer.* **1998**, *9*, 467–478. [CrossRef] [PubMed]
19. Jagodzinski, A.; Johansen, C.; Koch-Gromus, U.; Aarabi, G.; Adam, G.; Anders, S.; Augustin, M.; der Kellen, R.B.; Beikler, T.; Behrendt, C.A.; et al. Rationale and Design of the Hamburg City Health Study. *Eur. J. Epidemiol.* **2020**, *35*, 169–181. [CrossRef] [PubMed]
20. Neumann, F.A.; Jagemann, B.; Makarova, N.; Börschel, C.S.; Aarabi, G.; Gutmann, F.; Schnabel, R.B.; Zyriax, B.-C. Mediterranean Diet and Atrial Fibrillation: Lessons Learned from the AFHRI Case–Control Study. *Nutrients* **2022**, *14*, 3615. [CrossRef]
21. Schröder, H.; Fitó, M.; Estruch, R.; Martínez-González, M.A.; Corella, D.; Salas-Salvadó, J.; Lamuela-Raventós, R.; Ros, E.; Salaverría, I.; Fiol, M. A short screener is valid for assessing Mediterranean diet adherence among older Spanish men and women. *J. Nutr.* **2011**, *141*, 1140–1145. [CrossRef]
22. von Rusten, A.; Illner, A.-K.; Boeing, H.; Flothkotter, M. Begutachtetes Original-Die Bewertung der Lebensmittelaufnahme mittels eines' Healthy Eating Index'(HEI-EPIC). *Ernahr.-Umsch.* **2009**, *56*, 450.
23. Sterne, J.A.C.; White, I.R.; Carlin, J.B.; Spratt, M.; Royston, P.; Kenward, M.G.; Wood, A.M.; Carpenter, J.R. Multiple imputation for missing data in epidemiological and clinical research: Potential and pitfalls. *BMJ* **2009**, *338*, b2393. [CrossRef] [PubMed]
24. Baranzini, D. SPSS Single Dataframe Aggregating SPSS Multiply Imputed Split Files. 2018. Available online: https://www.researchgate.net/profile/Daniele_Baranzini/publication/328887514_SPSS_Single_dataframe_aggregating_SPSS_Multiply_Imputed_split_files/data/5be9a1cf299bf1124fce0d62/The-Bar-Procedure.docx (accessed on 22 January 2023).
25. Northstone, K. Dietary patterns: The importance of sex differences. *Br. J. Nutr.* **2012**, *108*, 393–394. [CrossRef] [PubMed]
26. Bennett, E.; Peters, S.A.; Woodward, M. Sex differences in macronutrient intake and adherence to dietary recommendations: Findings from the UK Biobank. *BMJ Open* **2018**, *8*, e020017. [CrossRef]
27. Modlinska, K.; Adamczyk, D.; Maison, D.; Pisula, W. Gender differences in attitudes to vegans/vegetarians and their food preferences, and their implications for promoting sustainable dietary patterns—A systematic review. *Sustainability* **2020**, *12*, 6292. [CrossRef]
28. Tande, D.L.; Magel, R.; Strand, B.N. Healthy Eating Index and abdominal obesity. *Public Health Nutr.* **2010**, *13*, 208–214. [CrossRef]
29. Wang, C.; Xu, Y. Mechanisms for sex differences in energy homeostasis. *J. Mol. Endocrinol.* **2019**, *62*, R129. [CrossRef] [PubMed]
30. Vahid, F.; Jalili, M.; Rahmani, W.; Nasiri, Z.; Bohn, T. A Higher Healthy Eating Index Is Associated with Decreased Markers of Inflammation and Lower Odds for Being Overweight/Obese Based on a Case-Control Study. *Nutrients* **2022**, *14*, 5127. [CrossRef] [PubMed]
31. Colpani, V.; Baena, C.P.; Jaspers, L.; Van Dijk, G.M.; Farajzadegan, Z.; Dhana, K.; Tielemans, M.J.; Voortman, T.; Freak-Poli, R.; Veloso, G.G. Lifestyle factors, cardiovascular disease and all-cause mortality in middle-aged and elderly women: A systematic review and meta-analysis. *Eur. J. Epidemiol.* **2018**, *33*, 831–845. [CrossRef]
32. World Health Organization. *Global Status Report on Alcohol and Health 2018*; World Health Organization: Geneva, Switzerland, 2019.
33. Prättälä, R.; Paalanen, L.; Grinberga, D.; Helasoja, V.; Kasmel, A.; Petkeviciene, J. Gender differences in the consumption of meat, fruit and vegetables are similar in Finland and the Baltic countries. *Eur. J. Public Health* **2007**, *17*, 520–525. [CrossRef] [PubMed]
34. Martín-María, N.; Lara, E.; Cabello, M.; Olaya, B.; Haro, J.M.; Miret, M.; Ayuso-Mateos, J.L. To be happy and behave in a healthier way. A longitudinal study about gender differences in the older population. *Psychol. Health* **2023**, *38*, 307–323. [CrossRef] [PubMed]
35. Engel, S.; Tholstrup, T. Butter increased total and LDL cholesterol compared with olive oil but resulted in higher HDL cholesterol compared with a habitual diet. *Am. J. Clin. Nutr.* **2015**, *102*, 309–315. [CrossRef]
36. Wang, X.; Magkos, F.; Mittendorfer, B. Sex Differences in Lipid and Lipoprotein Metabolism: It's Not Just about Sex Hormones. *J. Clin. Endocrinol. Metab.* **2011**, *96*, 885–893. [CrossRef]
37. Peters, S.A.E.; Muntner, P.; Woodward, M. Sex Differences in the Prevalence of, and Trends in, Cardiovascular Risk Factors, Treatment, and Control in the United States, 2001 to 2016. *Circulation* **2019**, *139*, 1025–1035. [CrossRef]
38. Soriano-Maldonado, C.; Lopez-Pineda, A.; Orozco-Beltran, D.; Quesada, J.A.; Alfonso-Sanchez, J.L.; Pallarés-Carratalá, V.; Navarro-Perez, J.; Gil-Guillen, V.F.; Martin-Moreno, J.M.; Carratala-Munuera, C. Gender Differences in the Diagnosis of Dyslipidemia: ESCARVAL-GENERO. *Int. J. Environ. Res. Public Health* **2021**, *18*, 2419. [CrossRef] [PubMed]
39. Turrini, A.; D'Addezio, L.; Maccati, F.; Davy, B.M.; Arber, S.; Davidson, K.; Grunert, K.; Schuhmacher, B.; Pfau, C.; Kozłowska, K.; et al. The Informal Networks in Food Procurement by Older People—A Cross European Comparison. *Ageing Int.* **2010**, *35*, 253–275. [CrossRef]
40. Alshehri, M.; Kruse-Diehr, A.J.; McDaniel, J.T.; Partridge, J.; Null, D.B. Impact of social support on the dietary behaviors of international college students in the United States. *J. Am. Coll. Health* **2021**, 1–9. [CrossRef]
41. Ellis, A.; Jung, S.E.; Palmer, F.; Shahan, M. Individual and interpersonal factors affecting dietary intake of community-dwelling older adults during the COVID-19 pandemic. *Public Health Nutr.* **2022**, *25*, 1667–1677. [CrossRef] [PubMed]
42. Junger, M.; van Kampen, M. Cognitive ability and self-control in relation to dietary habits, physical activity and bodyweight in adolescents. *Int. J. Behav. Nutr. Phys. Act.* **2010**, *7*, 22. [CrossRef]
43. Stutts, L.A.; Blomquist, K.K. A longitudinal study of weight and shape concerns and disordered eating groups by gender and their relationship to self-control. *Eat. Weight Disord.–Stud. Anorex. Bulim. Obes.* **2021**, *26*, 227–237. [CrossRef]

44. Szabo, K.; Piko, B.F.; Fitzpatrick, K.M. Adolescents' attitudes towards healthy eating: The role of self-control, motives and self-risk perception. *Appetite* **2019**, *143*, 104416. [CrossRef]
45. Van Dooren, C.; Marinussen, M.; Blonk, H.; Aiking, H.; Vellinga, P. Exploring dietary guidelines based on ecological and nutritional values: A comparison of six dietary patterns. *Food Policy* **2014**, *44*, 36–46. [CrossRef]
46. Baca-Motes, K.; Edwards, A.M.; Waalen, J.; Edmonds, S.; Mehta, R.R.; Ariniello, L.; Ebner, G.S.; Talantov, D.; Fastenau, J.M.; Carter, C.T. Digital recruitment and enrollment in a remote nationwide trial of screening for undiagnosed atrial fibrillation: Lessons from the randomized, controlled mSToPS trial. *Contemp. Clin. Trials Commun.* **2019**, *14*, 100318. [CrossRef] [PubMed]
47. Schreier, H.M.; Schonert-Reichl, K.A.; Chen, E. Effect of volunteering on risk factors for cardiovascular disease in adolescents: A randomized controlled trial. *JAMA Pediatr.* **2013**, *167*, 327–332. [CrossRef]
48. Weissberger, G.H.; Samek, A.; Mosqueda, L.; Nguyen, A.L.; Lim, A.C.; Fenton, L.; Han, S.D. Increased Financial Altruism is Associated with Alzheimer's Disease Neurocognitive Profile in Older Adults. *J. Alzheimer's Dis.* **2022**, *88*, 995–1005. [CrossRef]
49. Shim, E.; Chapman, G.B.; Townsend, J.P.; Galvani, A.P. The influence of altruism on influenza vaccination decisions. *J. R. Soc. Interface* **2012**, *9*, 2234–2243. [CrossRef] [PubMed]
50. Cavazza, N.; Guidetti, M.; Butera, F. Ingredients of gender-based stereotypes about food. Indirect influence of food type, portion size and presentation on gendered intentions to eat. *Appetite* **2015**, *91*, 266–272. [CrossRef] [PubMed]
51. Çınar, Ç.; Wesseldijk, L.W.; Karinen, A.K.; Jern, P.; Tybur, J.M. Sex differences in the genetic and environmental underpinnings of meat and plant preferences. *Food Qual. Prefer.* **2022**, *98*, 104421. [CrossRef]
52. Obara-Gołębiowska, M.; Michałek-Kwiecień, J. Personality traits, dieting self-efficacy and health behaviors in emerging adult women: Implications for health promotion and education. *Health Promot. Perspect.* **2020**, *10*, 230. [CrossRef]

 *nutrients*

Article

Cross-Sectional Association of Dietary Patterns and Supplement Intake with Presence and Gray-Scale Median of Carotid Plaques—A Comparison between Women and Men in the Population-Based Hamburg City Health Study

Julia Maria Assies [1,†], Martje Dorothea Sältz [1,†], Frederik Peters [2], Christian-Alexander Behrendt [3], Annika Jagodzinski [4], Elina Larissa Petersen [3,5], Ines Schäfer [3,5], Raphael Twerenbold [3,5,6], Stefan Blankenberg [3,5,6], David Leander Rimmele [7], Götz Thomalla [7], Nataliya Makarova [1,6,*,‡] and Birgit-Christiane Zyriax [1,6,‡]

1 Midwifery Science—Health Care Research and Prevention, Research Group Preventive Medicine and Nutrition, Institute for Health Service Research in Dermatology and Nursing (IVDP), University Medical Center Hamburg-Eppendorf, Martinistraße 52, W26, 20246 Hamburg, Germany; julia.assies@googlemail.com (J.M.A.); martje@saeltz.de (M.D.S.); b.zyriax@uke.de (B.-C.Z.)
2 Hamburg Cancer Registry, 20097 Hamburg, Germany
3 Population Health Research Department, University Heart and Vascular Center, 20246 Hamburg, Germany
4 Lohfert & Lohfert Working Group, 20148 Hamburg, Germany
5 Department of Cardiology, University Heart and Vascular Center, 20246 Hamburg, Germany
6 German Center for Cardiovascular Research (DZHK), Partner Site Hamburg/Kiel/Luebeck, 20246 Hamburg, Germany
7 Department of Neurology, University Medical Centre Hamburg-Eppendorf, Martinistraße 52, 20246 Hamburg, Germany
* Correspondence: n.makarova@uke.de
† These authors contributed equally to this work.
‡ These authors contributed equally to this work.

Citation: Assies, J.M.; Sältz, M.D.; Peters, F.; Behrendt, C.-A.; Jagodzinski, A.; Petersen, E.L.; Schäfer, I.; Twerenbold, R.; Blankenberg, S.; Rimmele, D.L.; et al. Cross-Sectional Association of Dietary Patterns and Supplement Intake with Presence and Gray-Scale Median of Carotid Plaques—A Comparison between Women and Men in the Population-Based Hamburg City Health Study. *Nutrients* **2023**, *15*, 1468. https://doi.org/10.3390/nu15061468

Academic Editor: Hayato Tada

Received: 16 February 2023
Revised: 15 March 2023
Accepted: 16 March 2023
Published: 18 March 2023

Abstract: This population-based cross-sectional cohort study investigated the association of the Mediterranean and DASH (Dietary Approach to Stop Hypertension) diet as well as supplement intake with gray-scale median (GSM) and the presence of carotid plaques comparing women and men. Low GSM is associated with plaque vulnerability. Ten thousand participants of the Hamburg City Health Study aged 45–74 underwent carotid ultrasound examination. We analyzed plaque presence in all participants plus GSM in those having plaques (n = 2163). Dietary patterns and supplement intake were assessed via a food frequency questionnaire. Multiple linear and logistic regression models were used to assess associations between dietary patterns, supplement intake and GSM plus plaque presence. Linear regressions showed an association between higher GSM and folate intake only in men (+9.12, 95% CI (1.37, 16.86), p = 0.021). High compared to intermediate adherence to the DASH diet was associated with higher odds for carotid plaques (OR = 1.18, 95% CI (1.02, 1.36), p = 0.027, adjusted). Odds for plaque presence were higher for men, older age, low education, hypertension, hyperlipidemia and smoking. In this study, the intake of most supplements, as well as DASH or Mediterranean diet, was not significantly associated with GSM for women or men. Future research is needed to clarify the influence, especially of the folate intake and DASH diet, on the presence and vulnerability of plaques.

Keywords: dietary patterns; supplements; carotid artery disease; cardiovascular disease; peripheral artery disease; carotid plaques; GSM; prevention

1. Introduction

Atherosclerotic cardiovascular disease (CVD) is widespread and is a leading cause of morbidity and mortality worldwide [1,2]. Atherosclerosis refers to a slowly progressive process of plaque formation in the vessel wall. Plaque rupture, platelet activation and, consequently, secondary thrombosis may occur during the progression of the disease [3–6]. Thereby, the risk of cardio- and cerebrovascular events is increased. Ischemic strokes are caused by the rupture of plaques in the carotids in about 15% of cases [7,8]. The stability and vulnerability of plaques have a major impact on the risk of plaque rupture [9]. Stable plaques consist of a high amount of fibrous tissue and calcification. Unstable plaques, in contrast, are rupture-prone due to high lipid content, an oftentimes necrotic core and intra-plaque hemorrhage [10,11].

Measurement of carotid intima-media thickness (cIMT) and its progression is an established and widely used prognostic biomarker for future CVD events [12,13]. However, it does not provide any information about plaque composition and, thus, the vulnerability of plaques. Therefore, the measurement of plaque gray-scale median (GSM) may improve the detection of vulnerable plaques. GSM provides additional information on plaque morphology due to the measurement of densitometry of the plaque [14,15].

The previous literature has demonstrated that GSM is a suitable measurement to quantify and assess the vulnerability of carotid plaques based on their echogenicity on B-mode ultrasonography [14,16,17]. GSM correlates with histopathological findings in patients after carotid endarterectomy and thus reflects the composition of plaques [10,18–22]. More precisely, high GSM values correlate with predominantly echogenic, stable plaques with a higher grade of calcification and fibrosis, whereas low GSM values are associated with echolucent, vulnerable plaques [11,14]. Low GSM values in carotid plaques are associated with an increased risk for CVD events, especially ischemic strokes [23–25].

GSM and cIMT are associated with different risk factors. While cIMT correlates with traditional risk factors such as hypertension and smoking status, GSM correlates with other traditional risk factors like dyslipidemia as well as with markers of inflammation and oxidative stress [26,27]. This suggests that cIMT and GSM may depict different aspects of atherosclerosis, with GSM relating more to metabolic aspects [28]. In addition to GSM, the presence of carotid plaque is associated with the incidence of CVD events [29,30] and is further an established ultrasound surrogate of CVD [31,32]. Nutritional aspects are known to play a relevant role in the development of atherosclerosis and in the formation of plaques [33]. Hence, it is of great interest how both can be influenced through diet.

CVD may be prevented in up to 90% of cases by a healthy lifestyle [34]. Numerous studies investigating the association between dietary patterns and CVD have included Dietary Approaches to Stop Hypertension (DASH) diet and the Mediterranean diet. Both adherence to the DASH diet and the Mediterranean diet are usually higher in women than in men [35–37]. Mediterranean diet has shown a primary prevention effect on CVD events as well as a tendency to slow down carotid plaque progression [38–44]. However, data on the association with GSM has been missing until now, and even the association between the Mediterranean diet and cIMT remains to be confirmed [45]. DASH diet is associated with fewer CVD events and lower cIMT values, while data regarding the association with GSM or plaque presence is not available [35,46,47].

Furthermore, dietary supplement intake is widespread in the general population [48]; for example, more than half of US adults take at least one supplement daily [49,50]. For Germany, the EPIC-Heidelberg cohort has shown an increasing prevalence of up to about 45% for vitamin/mineral supplement intake in a follow-up reassessment (2004–2006) [51]. The EPIC-Heidelberg cohort and many other studies have also revealed that women, in particular, are more likely to take supplements compared to men [49,51–54]. The main reasons for intake are general health and well-being and filling nutrient gaps [55].

According to previous studies, the associations between dietary supplements and CVD or cIMT remain unclear. Some data on B vitamins exist, especially for folic acid supplementation, which appears to be associated with benefits for CVD and, in particular, stroke risk [56–59]. Studies investigating associations between dietary supplement intake with GSM or plaque presence are lacking.

This study, therefore, aimed to examine associations between the dietary patterns Mediterranean diet and the DASH diet as well as dietary supplements (specifically multivitamins, multiminerals, calcium, magnesium, vitamin B and folate) and (a) the presence or (b) GSM of carotid plaques as predictors of CVD in women and men.

2. Materials and Methods

2.1. Study Population and Study Design

This study is part of the Hamburg City Health Study (HCHS). HCHS is a prospective, single-center, population-based cohort study. It aims to identify risk and prognostic factors of main chronic diseases. Participants must be inhabitants of Hamburg, Germany, at the time of enrollment, aged 45–74 years and must provide sufficient language skills for participating in the study. Participants are chosen randomly via the registration office. They sign an informed consent and undergo an extensive baseline evaluation. Detailed information on the HCHS has been published separately [60]. For this study, data from the first sub-cohort (n = 10,000) was used. Data acquisition took place between 8 February 2016–30 November 2018.

2.2. Ultrasound Images

B-mode duplex sonography was performed by trained study assistants using a Siemens SC2000® Ultrasound System and a 7.5 Mhz broadband linear transducer. Measurement of the cIMT was performed three times. The carotid bulb, common carotid artery and internal and external carotid artery were then scanned for plaques using the longitudinal view of carotid artery. A plaque was defined as a local cIMT $\geq$ 1.5 mm.

2.3. Gray-Scale Median

Carotid ultrasound scans were saved in DICOM (digital imaging and communications in medicine) format after performing the sonography. In the next step, echogenicity of carotid plaques was analyzed using software that was specifically written for this project's purpose, based on the open-source project JS Paint [61,62]. Plaques were segmented manually by outlining the plaques using the computer mouse. One additional marker was drawn in the vessel lumen, and a second in the adventitia. Each plaque was segmented twice by different operators to minimize interobserver reliability. Interobserver reliability was determined based on a random sub-sample of 135 (5%) participants that were evaluated by all observers. Remeasurements of outliers were performed. Images were saved as portable network graphics (PNG) files after segmentation. Next, image brightness was normalized using the vessel lumen as the reference structure for darkness (GSM = 0) and the adventitia as the reference structure for brightness (GSM = 190). Both grayscale values were chosen based on the existing literature [63]. In general, GSM values range from 0, indicating total black, to 255, indicating total white. Noise reduction and cropping of the images were performed automatically. Finally, minimum, maximum, mean and median grayscale values were calculated and output in a comma-separated values (CSV) file. Primary outcome of the present study was the mean value over all individual echogenicity measurements as numerical variable.

2.4. Questionnaires and Dietary Scores

Dietary habits and intake of nutrition supplements were assessed in questionnaires. For dietary intake, the food frequency questionnaire (version 2, FFQ2) developed for the European Perspective Investigation into Cancer and Nutrition (EPIC) study was used [64]. It samples information on frequency and portion size of 102 food items consumed during the previous year. Information was collected and analyzed in terms of energy intake, food groups and nutrients.

The validated German translation of the Mediterranean Diet Adherence Score (MEDAS) was used for evaluating adherence to a Mediterranean diet [65]. It contains twelve questions on food items and two questions on food habits (Supplementary Material Table S1). For each item, a score of 0 indicates a non-adherence, whereas a score of 1 indicates adherence. Finally, the score was grouped by quantiles into the categories 0–3, 4, 5 and 6+.

Adherence to the Dietary Approaches to Stop Hypertension (DASH) diet was assessed using a scoring system adapted from Folsom et al. [66]. The score includes ten items on consumption of grains, vegetables, fruits, dairy, meat/poultry/fish, nuts/seeds/legumes and sweets (obtained from raw data) and average daily intake of nutrients (saturated fat, fat, sodium) (Supplementary Material Table S2). Each item was scored from 0 to 1. Finally, the score was grouped by quantiles into the categories 0–3.5, 3.6–4.5, 4.6–5.0 and 5.1+.

The FFQ2 continued to ask about the use of dietary supplements for at least one month in the last twelve months, specifically multipreparations (multivitamin or multimineral preparations or both) or 14 single and simple combination preparations, as well as nine natural health products. For this study, data on multivitamin and multimineral preparations as well as calcium, magnesium, vitamin B complex and folic acid, were included.

2.5. Statistical Analysis

In the descriptive analysis, continuous data are presented as the median and interquartile range (IQR), and categorical data as absolute numbers and percentages.

Multiple linear regressions were used to assess the association between echogenicity and dietary and supplement intake, i.e., nutritional supplements, DASH diet, MEDAS within GSM-sub-cohort (n = 2163). All models were estimated separately for males and females and adjusted for not performing any sports (examined as 'never performing sports except for cycling or walking'), age, socioeconomic status index (including education, profession, salary), body mass index (BMI), smoking status, energy intake (kcal), dyslipidemia, hypertension, diabetes mellitus, myocardial infarction, heart failure, atrial fibrillation, history of stroke or transient ischemic attack (TIA), peripheral arterial disease, estimated glomerular filtration rate (eGFR), lipid-lowering drugs, antihypertensive medication, antidiabetic medication, use of antiplatelets. Central results were presented as betas with 95% confidence intervals. We did not adjust for multiple comparisons. We imputed missing values by multivariate imputation by chained equations separately for twenty copies of the data with ten iterations. Subsequently, estimates were averaged, and standard errors were adjusted using Rubin's rules [67].

We performed additional analysis regarding the presence of at least one carotid plaque using multiple logistic regressions within a full cohort of 10,000 participants. For the full-adjusted model, age, sex, education, body-mass index, diabetes mellitus, arterial hypertension, hyperlipidemia, smoking status, heart failure, atrial fibrillation, myocardial infarction, stroke and sports were used for adjustment. Education was divided into three categories (low, medium and high) based on the International Standard Classification of Education (ISCED 1011).

Statistical significance was defined as an $\alpha = 0.05$. We adhered to the Strengthening the Reporting of Observational Studies in Epidemiology (STROBE) statement [68]. All analyses were performed in R version 4.0.3.

3. Results

3.1. Baseline Characteristics of GSM-Sub-Cohort

From the HCHS cohort of 10,000 participants, GSM was assessed for 2163 participants having at least one carotid plaque (Figure 1). The baseline characteristics of these participants, consisting of 921 (42.6%) women and 1242 (57.4%) men, are shown in Table 1. Here, the median age of women and men at recruitment was 68 (IQR (62, 73)) years. Obesity was found in 272 (21.9%) men and 187 (20.3%) women. Overall, 486 (22.5%) were current smokers. Of men, 397 (32.0%) were not performing any sports, whereas 249 (27.0%) women were not exercising.

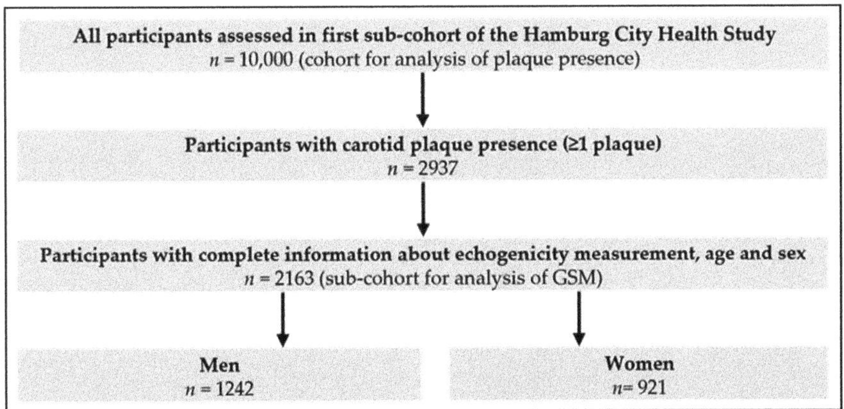

Figure 1. Flow Chart for assessment of echogenicity measurement in participants of HCHS.

Table 1. Baseline characteristics of the study population of GSM-sub-cohort.

	Overall	Men	Women	*p*-Value
n (%)	2163 (100)	1242 (57, 4)	921 (42, 6)	
Age in years (median [IQR])	68 (62, 73)	68 (62, 73)	68 (62, 72)	0.44
SES index (median [IQR])	12.30 (9.97, 16.00)	13.30 (10.30, 16.70)	11.40 (9.40, 14.20)	<0.001
NA	1307 (60.4)	653 (52.6)	654 (71)	
BMI (kg/m^2) (%)				<0.001
NA	130 (6.0)	67 (5.4)	63 (6.8)	
Normal weight (BMI 18.5–24.9 kg/m^2)	677 (31.3)	322 (25.9)	355 (38.5)	
Obesity (BMI $\geq$ 30 kg/m^2)	459 (21.2)	272 (21.9)	187 (20.3)	
Overweight (BMI 25–29.9 kg/m^2)	883 (40.8)	576 (46.4)	307 (33.3)	
Underweight (BMI < 18.5 kg/m^2)	14 (0.6)	5 (0.4)	9 (1.0)	
Smoking status (%)	486 (22.5)	282 (22.7)	204 (22.1)	0.697
NA	11 (0.5)	5 (0.4)	6 (0.7)	
Not performing any sports (%)	646 (29.9)	397 (32.0)	249 (27.0)	0.047
NA	180 (8.3)	100 (8.1)	80 (8.7)	
Total energy intake, kcal (median [IQR])	2009.48 (1613.75, 2564.23)	2311.52 (1853.48, 2837.52)	1729.92 (1419.01, 2101.23)	<0.001
NA	242 (11.2)	141 (11.4)	101 (11)	
MEDAS score (%)				<0.001
NA	244 (11.3)	142 (11.4)	102 (11.1)	
0–3 points	576 (26.6)	453 (36.5)	123 (13.4)	
4 points	439 (20.3)	272 (21.9)	167 (18.1)	
5 points	356 (16.5)	175 (14.1)	181 (19.7)	
6+ points	548 (25.3)	200 (16.1)	348 (37.8)	
DASH score (%)				<0.001
NA	244 (11.3)	142 (11.4)	102 (11.1)	
0–3.5 points	489 (22.6)	395 (31.8)	94 (10.2)	
3.6–4.5 points	643 (29.7)	377 (30.4)	266 (28.9)	
4.6–5.0 points	325 (15.0)	152 (12.2)	173 (18.8)	
5.1+ points	462 (21.4)	176 (14.2)	286 (31.1)	
Any supplement intake (%)	755 (34.9)	352 (28.3)	403 (43.8)	<0.001
NA	177 (8.2)	100 (8.1)	77 (8.4)	
Multivitamins (%)	167 (7.7)	96 (7.7)	71 (7.7)	1
Multiminerals (%)	180 (8.3)	78 (6.3)	102 (11.1)	<0.001
Calcium (%)	127 (5.9)	68 (5.5)	59 (6.4)	0.413
Magnesium (%)	382 (17.7)	179 (14.4)	203 (22.0)	<0.001
Vitamin B (%)	110 (5.1)	45 (3.6)	65 (7.1)	<0.001

Table 1. *Cont.*

	Overall	Men	Women	*p*-Value
Folate (%)	76 (3.5)	30 (2.4)	46 (5.0)	0.002
Hyperlipidemia (%)	744 (34.4)	498 (40.1)	246 (26.7)	<0.001
NA	100 (4.6)	53 (4.3)	47 (5.1)	
Arterial hypertension (%)	1644 (76.0)	974 (78.4)	670 (72.7)	<0.001
NA	62 (2.9)	44 (3.5)	18 (2.0)	
Diabetes mellitus (%)	256 (11.8)	171 (13.8)	85 (9.2)	0.002
NA	118 (5.5)	59 (4.8)	59 (6.4)	
Prior MI (%)	119 (5.5)	103 (8.3)	16 (1.7)	<0.001
NA	17 (0.8)	10 (0.8)	7 (0.8)	
Heart failure (%)	174 (8.0)	115 (9.3)	59 (6.4)	0.014
NA	20 (0.9)	15 (1.2)	5 (0.5)	
Atrial fibrillation (%)	183 (8.5)	120 (9.7)	63 (6.8)	0.014
NA	199 (9.2)	101 (8.1)	98 (10.6)	
Prior stroke (%)	100 (4.6)	65 (5.2)	35 (3.8)	0.121
NA	17 (0.8)	7 (0.6)	10 (1.1)	
PAD (ABI < 0.9) (%)	255 (11.8)	140 (11.3)	115 (12.5)	0.308
NA	1140 (52.7)	645 (51.9)	495 (53.7)	
GFR (median [IQR])	87.20 (78.40, 93.20)	88.90 (81.20, 94.60)	84.80 (76.30, 90.30)	<0.001
NA	211 (9.8)	105 (8.5)	106 (11.5)	
Lipid-lowering drugs (%)	617 (28.5)	413 (33.3)	204 (22.1)	<0.001
NA	60 (2.8)	39 (3.1)	21 (2.3)	
Antihypertensives (%)	1000 (46.2)	598 (48.1)	402 (43.6)	0.035
NA	60 (2.8)	39 (3.1)	21 (2.3)	
Antidiabetics (%)	171 (7.9)	119 (9.6)	52 (5.6)	0.001
NA	60 (2.8)	39 (3.1)	21 (2.3)	
Antiplatelets (%)	600 (27.7)	409 (32.9)	191 (20.7)	<0.001
NA	60 (2.8)	39 (3.1)	21 (2.3)	

This table shows baseline characteristics related to participants with assessed GSM (n = 2163). Abbreviations: IQR, interquartile range; SES, socioeconomic status; NA, not available (missings with respect to line above); BMI, body mass index (calculated as weight in kilograms divided by height in meters squared); MEDAS, Mediterranean Diet Adherence Score; DASH, Dietary Approach to Stop Hypertension; MI, myocardial infarction; PAD, peripheral artery disease; ABI, ankle-brachial-pressure-Index; GFR, glomerular filtration rate.

Women reached higher MEDAS scores more often than men; women reached a score of 6+ points in 37.8% of the cases, whereas men reached a score of 6+ points in 16.1%. A similar trend holds true for the DASH score: 31.1% of women and 14.2% of men achieved a score of 5.1+ points. Men reached the largest distribution range at 0–3.5 points (31.8%) and 3.6–4.5 points (30.4%). In comparison, fewer women had low score values.

A total of 755 (34.9%) participants had an intake of any supplement. Intake was higher among women (43.8%) than men (23.8%). As Table 1 shows, for each of the examined supplements, intake was higher in women than in men, with the exception of multivitamins. Here, an equal supplementation distribution of 7.7% each for women and men was assessed.

Figure 2 shows the distribution of GSM levels separately for men (shown in blue) and women (shown in red). The median GSM was 56.50 with IQR between 46.00 and 68.50 for men and 55.80 with IQR between 44.25 and 70.33 for women.

3.2. Linear Regression of Nutrition Parameters and Examined Supplements with GSM in Women and Men

Table 2 shows the results of multivariate linear regression models of nutrition parameters, examined supplements and GSM in men and women of the GSM-sub-cohort (n = 2163). A significant correlation could only be found for folic acid intake in men (GSM 9.12 (95% CI (1.37, 16.86), p = 0.021).

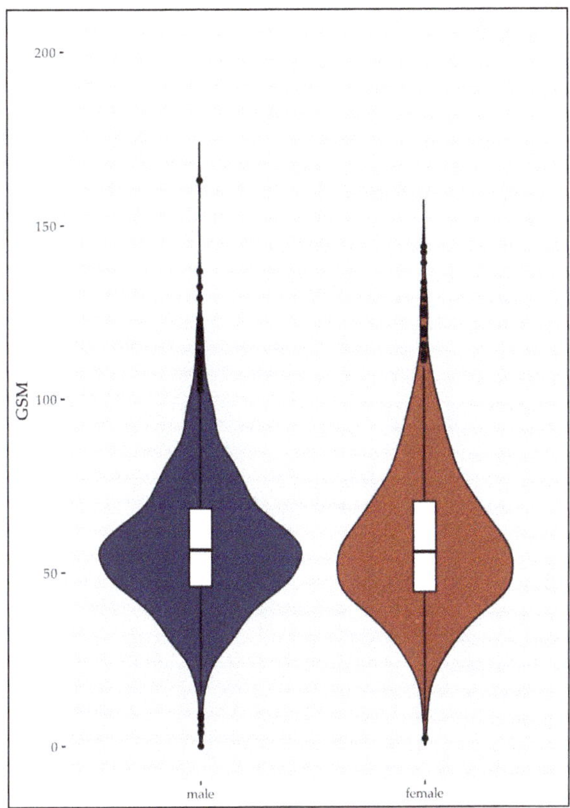

Figure 2. Distribution of gray-scale median (GSM) in women (red) and men (blue). n = 2163. Abbreviations: GSM, gray-scale median.

Table 2. Linear regression models for Outcome GSM (0 to 255) in men and women.

Model Parameters	Men		Women	
	GSM [95% CI] [1]	p-Value	GSM [95% CI] [1]	p-Value
MEDAS 4 points [2]	−0.14 (−3.06, 2.79)	0.927	−2.82 (−7.78, 2.15)	0.267
MEDAS 5 points [2]	−1.10 (−4.45, 2.25)	0.521	−2.02 (−6.78, 2.74)	0.406
MEDAS 6+ points [2]	−1.53 (−4.71, 1.66)	0.347	−1.88 (−6.19, 2.43)	0.393
DASH 3.6–4.5 points [3]	−1.27 (−4.09, 1.55)	0.378	−1.69 (−6.57, 3.19)	0.497
DASH 4.6–5.0 points [3]	−0.95 (−4.66, 2.75)	0.614	−3.76 (−9.06, 1.53)	0.164
DASH 5.1+ points [3]	−1.45 (−4.86, 1.96)	0.406	−0.33 (−5.22, 4.57)	0.896
Any supplement intake [4]	−0.69 (−4.12, 2.74)	0.693	−1.06 (−4.78, 2.67)	0.578
Multivitamins [4]	−2.67 (−7.94, 2.61)	0.322	−2.03 (−8.53, 4.47)	0.540
Multiminerals [4]	1.52 (−4.11, 7.15)	0.597	−0.87 (−6.53, 4.79)	0.763
Calcium [4]	−0.14 (−5.57, 5.29)	0.960	2.57 (−3.61, 8.76)	0.415
Magnesium [4]	−0.01 (−4.20, 4.18)	0.995	−0.19 (−4.43, 4.05)	0.930
Vitamin B [4]	−2.19 (−8.59, 4.21)	0.502	3.93 (−1.93, 9.79)	0.189
Folate [4]	9.12 (1.37, 16.86)	0.021	−2.50 (−9.31, 4.31)	0.472

This table shows results of linear regression models regarding GSM related to participants with assessed GSM presented as betas (n = 2163). [1] Model adjusted for: age, socioeconomic status, not doing sport, BMI, smoking status, energy intake (kcal), dyslipidemia, hypertension, diabetes mellitus, myocardial infarction, heart failure, atrial fibrillation, history of stroke or TIA, peripheral arterial disease, eGFR, lipid-lowering drugs, antihypertensive medication, antidiabetic medication, antiplatelets. [2] Reference category: <4 points. [3] Reference category: <3.6 points. [4] Reference category: No supplement intake. Abbreviations: GSM, gray-scale median; CI, confidence interval; MEDAS, Mediterranean Diet Adherence Score; DASH, Dietary Approach to Stop Hypertension; BMI, body-mass-index; TIA, transient ischemic attack; eGFR, estimated glomerular filtration rate.

A non-significant opposing trend was found in women with GSM of −2.50 (95% CI −9.31, 4.31), $p = 0.472$. No significant associations could be found between dietary patterns or intake of the other examined supplements and GSM.

3.3. Results of Logistic Regression Regarding the Presence of Carotid Plaques

The results of logistic regressions with multivariable adjustments as OR related to the reference category for the presence of at least one carotid plaque in full HCHS-sub-cohort, including 10,000 participants, are shown in Table 3. In all three adjusted logistic regression models, the odds for the presence of at least one carotid plaque were significantly higher among the categories men, older age, low education, arterial hypertension, hyperlipidemia and smoking status (Supplementary Material Table S3).

Table 3. Logistic regression models regarding the presence of carotid plaques.

Characteristics	OR (95% CI)	*p*-Value
High MEDAS vs. medium MEDAS	1.07 (0.92, 1.24)	0.367
Low MEDAS vs. medium MEDAS	0.86 (0.75, 1.00)	0.052
High DASH vs. medium DASH	1.18 (1.02, 1.36)	0.027
Low DASH vs. medium DASH	0.95 (0.82, 1.10)	0.469
Supplement intake yes vs. no	0.96 (0.85, 1.08)	0.490

This table shows results of additional analyses of logistic regression models regarding the presence of carotid plaques in full HCHS-sub-cohort, including 10,000 participants. All models are adjusted for age, sex, education, body-mass index, diabetes mellitus, arterial hypertension, hyperlipidemia, smoking status, heart failure, atrial fibrillation, myocardial infarction, stroke and sports. Abbreviations: OR, odds ratio; CI, confidence interval; MEDAS, Mediterranean Diet Adherence Score; DASH, Dietary Approach to Stop Hypertension.

A high DASH score showed significantly increased odds for the presence of at least one plaque compared to intermediate score values in adjusted models (OR = 1.18, 95% CI (1.02, 1.36), $p = 0.027$).

In adjusted models, no significant association between MEDAS or any supplement intake and the presence of carotid plaque was found.

4. Discussion

GSM was not associated with Mediterranean or DASH nutritional patterns and most supplements in an elderly German population. Folic acid intake was significantly associated with higher GSM only in men. A high DASH score was significantly associated with increased odds for the presence of carotid plaques compared to intermediate score values. However, in all other fully adjusted analyses, no significant associations were found between DASH/Mediterranean diet and plaque presence.

This study is the first to investigate associations between GSM and the Mediterranean Diet or DASH diet as well as the supplements examined in this study, plus the relation between the presence of carotid plaques with the DASH diet or supplement intake. There are only a few studies that have investigated plaque prevalence and MEDAS.

The study's baseline data fit with the demographics of previous studies, which have also shown that both following healthy dietary patterns—measured by high adherence scores—and taking supplements are more prevalent among women [35–37,49,53,69,70].

The significantly increased GSM in men taking folic acid should be considered with caution because only 30 men (2.4%) supplemented folic acid. Future studies should investigate the effect of folic acid on plaque vulnerability. In addition to that, the clinical implication should be mentioned. If the observed evidence of a 9.12 increased GSM by folic acid intake (95% CI (1.37, 16.86), $p = 0.021$) is not coincidental, this positive effect, however, is not necessarily clinically relevant. However, three reviews revealed a reduced stroke risk for folic acid supplementation and, thus, beneficial effects for stroke prevention [57–59]. Again, further studies are necessary to determine which GSM changes are clinically relevant to outcomes related to CVD, e.g., ischemic stroke. Thus, the findings probably exist due to confounders like traditional cardiovascular risk factors considering that supplement users tend to have more healthy habits than non-users [55].

Several studies have shown that the presence of carotid plaques is particularly associated with older age, male sex [71] and smoking [72], but also linked to diseases such as hypercholesterolemia [31], hypertension, diabetes mellitus [73,74] and cardiac disease [75]. Our findings are in line with previous studies that revealed the following associations: In adjusted regression models, the odds of having at least one plaque significantly increased in men, older age, low education, arterial hypertension, hyperlipidemia and smoking status. Evidence for correlations between supplement intake or DASH diet with plaque presence is missing in the existing literature. For any supplement intake, the odds of carotid plaque presence were lower, although no significant trend was observed after adjustment.

Contrary to our expectations, we have found a significant association between high DASH scores and a more frequent occurrence of carotid plaques in adjusted models. In contrast, Fung et al. showed that adherence to the DASH diet is associated with a reduced risk of CVD events such as stroke [46]. The reason for our findings could be that people having cardiovascular diseases are more willing to follow healthy nutrition recommendations. Likewise, individuals who have received nutritional counseling cause of their CVD are more likely to report healthy nutrition in questionnaires (recall/reporting bias).

We found an absence of proof regarding the association between MEDAS or supplement intake and the presence of carotid plaque. Previous studies confirm that there may be no association between MEDAS and the presence of carotid plaques. For example, neither Gardener et al. in the Northern Manhattan Study (NOMAS) [38] nor Mateo-Gallego et al. in the Aragon Workers' Health Study (AWHS) [41] observed an association between the Mediterranean diet and plaque presence. Jimenez-Torres et al. also did not find any effect of the Mediterranean diet on the number of carotid plaques [39]. In contrast, a Croatian study in a population of HIV-infected patients found that lower adherence to the Mediterranean diet was associated with increased odds of subclinical atherosclerosis defined as cIMT $\geq$ 0.9 mm or $\geq$1 carotid plaque [76].

Although no clinically relevant association between the Mediterranean/DASH diet or supplement intake and GSM has been found, some studies have shown associations between these lifestyle adjustments and the CVD predictor cIMT. For example, Maddock et al. describe significantly lower cIMT for greater adherence to the DASH diet [35].

Because GSM and cIMT may be associated with different risk factors [26,27] and represent different aspects of atherosclerosis [28], it is worth doubting whether GSM is an appropriate parameter for detecting associations with dietary adjustments. Perhaps other methods are more useful for investigating associations and, finally, causal influences on clinical outcomes related to diet or supplements. For example, using a juxtaluminal black area (JBA) instead of GSM could provide even more information [22]. While the GSM value is based on the echolucency measurement of the whole plaque, JBA focuses on a low GSM plaque area near the vessel lumen. Salem et al. found a stronger association between histological findings and JBA than with GSM [21].

In summary, further research regarding the relationship between GSM and the presence of carotid artery plaques with nutrition patterns or supplement intake is needed.

Strengths and Limitations

The present study consists of an exceptionally large sample size of 2163 participants within the GSM-sub-cohort and 10,000 participants in an additional analysis with the presence of at least one plaque. Almost no exclusion criteria (only insufficient German language skills and incapability to travel to the study center and to cooperate in the investigations) and random invitations via the registration office are used for the selection of study participants for HCHS. Still, selection bias cannot be excluded for certain. HCHS participants tend to be more health-conscious and educated, showing fewer cardiovascular risk factors than the general German population [77]. Furthermore, the HCHS study population consists of middle-aged individuals living in Hamburg, so generalizations to other age groups and individuals living in rural areas should not be made without careful consideration.

Being a cross-sectional analysis, no causal conclusions can be made. Data on dietary parameters were collected by self-reporting in questionnaires, so there is a risk of reporting and recall bias. In addition, no data were collected on the dose of the supplements nor on the continuity or duration of intake.

Furthermore, adjustments for multiple comparisons were not performed. This could lead to dismissing the null hypothesis hastily, especially in consideration of the wide variety of supplements.

Another limitation could be our grouping of the dietary scores in the GSM regression models (MEDAS 4/5/6+ points, DASH 3.6–4.5, 4.6–5, 5.1+) since differences in adherence between the groups are small. Comparison of, for instance, the highest tertial of adherence vs. the lowest tertial of adherence might have been more informative. In addition, adherence to the Mediterranean diet was low in our northern German participants. Another dietary pattern, e.g., an anti-inflammatory or Nordic diet, could have shown higher prevalence rates and thus more information.

Additionally, in the present study, GSM measurement was performed based on 2D ultrasound scans and thus cannot present information on the whole plaque as 3D files may have done.

Lastly, multiple trained operators drew in the plaques for the GSM determination. This leaves room for intra-observer and inter-observer variability. Plus, the reference values for normalization of the image brightness also had to be drawn in. This can lead to a bias in true GSM values if, for instance, an expert draws in an area that is too dark for the adventitia, the normalization thus becoming incorrect [78].

5. Conclusions

The current study found no clinically relevant significant associations between adherence to the DASH/Mediterranean diet or supplement intake and the GSM of carotid plaques. There may be an association between higher GSM and folate intake in men, but further studies are needed to confirm this association and clinical relevance.

High compared to intermediate adherence to the DASH diet was associated with higher odds for carotid plaque presence.

Further research is needed to examine whether nutrition patterns or supplement intake—particularly DASH diet and folate intake—are associated with plaque presence or GSM.

Supplementary Materials: The following supporting information can be downloaded at: https://www.mdpi.com/article/10.3390/nu15061468/s1, regarding the criteria for Mediterranean Dietary Score (Table S1) and scoring system for Dietary Approaches to Stop Hypertension (DASH) diet (Table S2) and additional results from logistic regression models regarding the presence of carotid plaques (Table S3).

Author Contributions: J.M.A. collected data, worked on methodology, conceptualized the paper, wrote the original draft, developed the discussion part, reviewed and edited the paper; M.D.S. collected data, worked on methodology, conceptualized the paper, wrote the original draft, developed the discussion part, reviewed and edited the paper; F.P. analyzed data statistically, worked on visualization, reviewed and edited the paper, C.-A.B. administrated and supervised this study, reviewed and edited the paper; S.B. acted as the expert from the cardiological field, contributed to the discussion part and supervised the paper as PI of the HCHS; D.L.R. controlled the quality of assessed data, contributed to the discussion, reviewed and edited the paper; G.T. controlled the quality of assessed data, contributed to the discussion, reviewed and edited the paper; A.J. was responsible for the data management and administration, quality control, reviewed and edited the paper; E.L.P. analyzed data statistically, worked on visualization, reviewed and edited the paper; I.S. was responsible for the data management and administration, reviewed and edited the paper; R.T. acted as the expert from the cardiological field, contributed to the discussion part, reviewed and edited the paper; N.M. curated data: developed statistical analysis plan for this manuscript, administrated and supervised this study, worked on discussion, reviewed and edited the paper; B.-C.Z. developed methods concerning the nutritional part of this study, administrated and supervised this study and reviewed and edited the paper. All authors have read and agreed to the published version of the manuscript.

Funding: This research received no external funding.

Institutional Review Board Statement: The local ethics committee of the Landesärztekammer Hamburg (State of Hamburg Chamber of Medical Practitioners, PV5131) was consulted, and its approval for the study protocol as well as the approval by the Data Protection Commissioner of the University Medical Center of the University Hamburg-Eppendorf and the Data Protection Commissioner of the Free and Hanseatic City of Hamburg were obtained. The study has been registered at https://clinicaltrials.gov/ (NCT03934957). The procedures set out in this study, pertaining to the conduct, evaluation, and documentation, are designed to ensure that all persons involved in the study abide by Good Clinical Practice (GCP), Good Epidemiological Practice (GEP), and the ethical principles described in the current revision of the Declaration of Helsinki. The study will be carried out in keeping with local legal and regulatory requirements. The requirements of the GCP and GEP regulation will be adhered to. In order to be admitted to HCHS, all participants are to consent to participate only after the nature and scope of the study have been explained to and understood by them. Written informed consent is obtained from all participants. The examinations were chosen because of the non-invasive nature of acquisition and standardized testing to assess intermediate phenotypes of the different diseases.

Informed Consent Statement: Informed consent was obtained from all subjects involved in the Hamburg City Health Study.

Data Availability Statement: Data for this study as well as statistical analyses and R code are available for purposes of review, transparency and comprehensibility.

Acknowledgments: The authors acknowledge the participants of the Hamburg City Health Study, the staff at the Epidemiological Study Center, the Hamburg City Health Study research consortium and steering board committee, as well as its cooperation partners and patrons.

Conflicts of Interest: S.B. reports institution funding from Bayer, Siemens, Novartis, and the City of Hamburg.

References

1. Kim, M.J.; Jung, S.K. Nutraceuticals for prevention of atherosclerosis: Targeting monocyte infiltration to the vascular endothelium. *J. Food Biochem.* **2020**, *44*, e13200. [CrossRef] [PubMed]
2. Torres, N.; Guevara-Cruz, M.; Velázquez-Villegas, L.A.; Tovar, A.R. Nutrition and Atherosclerosis. *Arch. Med. Res.* **2015**, *46*, 408–426. [CrossRef] [PubMed]
3. Badimon, L.; Vilahur, G. Thrombosis formation on atherosclerotic lesions and plaque rupture. *J. Intern. Med.* **2014**, *276*, 618–632. [CrossRef] [PubMed]
4. Penson, P.E.; Banach, M. The Role of Nutraceuticals in the Optimization of Lipid-Lowering Therapy in High-Risk Patients with Dyslipidaemia. *Curr. Atheroscler. Rep.* **2020**, *22*, 67. [CrossRef] [PubMed]
5. Scheen, A.J. From atherosclerosis to atherothrombosis: From a silent chronic pathology to an acute critical event. *Rev. Med. Liege* **2018**, *73*, 224–228.
6. Saba, L.; Sanagala, S.S.; Gupta, S.K.; Koppula, V.K.; Johri, A.M.; Sharma, A.M.; Kolluri, R.; Bhatt, D.L.; Nicolaides, A.; Suri, J.S. Ultrasound-based internal carotid artery plaque characterization using deep learning paradigm on a supercomputer: A cardiovascular disease/stroke risk assessment system. *Int. J. Cardiovasc. Imaging* **2021**, *37*, 1511–1528. [CrossRef]
7. Baud, J.M.; Stanciu, D.; Yeung, J.; Maurizot, A.; Chabay, S.; de Malherbe, M.; Chadenat, M.L.; Bachelet, D.; Pico, F. Contrast enhanced ultrasound of carotid plaque in acute ischemic stroke (CUSCAS study). *Rev. Neurol.* **2021**, *177*, 115–123. [CrossRef]
8. Flaherty, M.L.; Kissela, B.; Khoury, J.C.; Alwell, K.; Moomaw, C.J.; Woo, D.; Khatri, P.; Ferioli, S.; Adeoye, O.; Broderick, J.P.; et al. Carotid artery stenosis as a cause of stroke. *Neuroepidemiology* **2013**, *40*, 36–41. [CrossRef]
9. Gong, H.Y.; Shi, X.K.; Zhu, H.Q.; Chen, X.Z.; Zhu, J.; Zhao, B.W. Evaluation of carotid atherosclerosis and related risk factors using ultrasonic B-Flow technology in elderly patients. *J. Int. Med. Res.* **2020**, *48*, 300060520961224. [CrossRef]
10. Spanos, K.; Tzorbatzoglou, I.; Lazari, P.; Maras, D.; Giannoukas, A.D. Carotid artery plaque echomorphology and its association with histopathologic characteristics. *J. Vasc. Surg.* **2018**, *68*, 1772–1780. [CrossRef]
11. Karim, R.; Xu, W.; Kono, N.; Li, Y.; Yan, M.; Stanczyk, F.Z.; Hodis, H.N.; Mack, W.J. Comparison of Cardiovascular Disease Risk Factors Between 2 Subclinical Atherosclerosis Measures in Healthy Postmenopausal Women: Carotid Artery Wall Thickness and Echogenicity: Carotid Artery Wall Thickness and Echogenicity. *J. Ultrasound Med.* **2022**, *42*, 35–44. [CrossRef]

12. Lorenz, M.W.; Markus, H.S.; Bots, M.L.; Rosvall, M.; Sitzer, M. Prediction of clinical cardiovascular events with carotid intima-media thickness: A systematic review and meta-analysis. *Circulation* **2007**, *115*, 459–467. [CrossRef]
13. Willeit, P.; Tschiderer, L.; Allara, E.; Reuber, K.; Seekircher, L.; Gao, L.; Liao, X.; Lonn, E.; Gerstein, H.C.; Yusuf, S.; et al. Carotid Intima-Media Thickness Progression as Surrogate Marker for Cardiovascular Risk: Meta-Analysis of 119 Clinical Trials Involving 100,667 Patients. *Circulation* **2020**, *142*, 621–642. [CrossRef] [PubMed]
14. Della-Morte, D.; Dong, C.; Crisby, M.; Gardener, H.; Cabral, D.; Elkind, M.S.V.; Gutierrez, J.; Sacco, R.L.; Rundek, T. Association of Carotid Plaque Morphology and Glycemic and Lipid Parameters in the Northern Manhattan Study. *Front. Cardiovasc. Med.* **2022**, *9*, 793755. [CrossRef] [PubMed]
15. Naylor, R.; Rantner, B.; Ancetti, S.; de Borst, G.J.; De Carlo, M.; Halliday, A.; Kakkos, S.K.; Markus, H.S.; McCabe, D.J.H.; Sillesen, H.; et al. European Society for Vascular Surgery (ESVS) 2023 Clinical Practice Guidelines on the Management of Atherosclerotic Carotid and Vertebral Artery Disease. *Eur. J. Vasc. Endovasc. Surg.* **2022**, *65.*, 7–111. [CrossRef]
16. Kadoglou, N.P.E.; Moulakakis, K.G.; Mantas, G.; Kakisis, J.D.; Mylonas, S.N.; Valsami, G.; Liapis, C.D. The Association of Arterial Stiffness With Significant Carotid Atherosclerosis and Carotid Plaque Vulnerability. *Angiology* **2022**, *73*, 668–674. [CrossRef] [PubMed]
17. Li, Y.; Kwong, D.L.; Wu, V.W.; Yip, S.P.; Law, H.K.; Lee, S.W.; Ying, M.T. Computer-assisted ultrasound assessment of plaque characteristics in radiation-induced and non-radiation-induced carotid atherosclerosis. *Quant. Imaging Med. Surg.* **2021**, *11*, 2292–2306. [CrossRef] [PubMed]
18. Doonan, R.J.; Gorgui, J.; Veinot, J.P.; Lai, C.; Kyriacou, E.; Corriveau, M.M.; Steinmetz, O.K.; Daskalopoulou, S.S. Plaque echodensity and textural features are associated with histologic carotid plaque instability. *J. Vasc. Surg.* **2016**, *64*, 671–677.e678. [CrossRef]
19. Grogan, J.K.; Shaalan, W.E.; Cheng, H.; Gewertz, B.; Desai, T.; Schwarze, G.; Glagov, S.; Lozanski, L.; Griffin, A.; Castilla, M.; et al. B-mode ultrasonographic characterization of carotid atherosclerotic plaques in symptomatic and asymptomatic patients. *J. Vasc. Surg.* **2005**, *42*, 435–441. [CrossRef]
20. Mitchell, C.C.; Stein, J.H.; Cook, T.D.; Salamat, S.; Wang, X.; Varghese, T.; Jackson, D.C.; Sandoval Garcia, C.; Wilbrand, S.M.; Dempsey, R.J. Histopathologic Validation of Grayscale Carotid Plaque Characteristics Related to Plaque Vulnerability. *Ultrasound Med. Biol.* **2017**, *43*, 129–137. [CrossRef]
21. Salem, M.K.; Bown, M.J.; Sayers, R.D.; West, K.; Moore, D.; Nicolaides, A.; Robinson, T.G.; Naylor, A.R. Identification of patients with a histologically unstable carotid plaque using ultrasonic plaque image analysis. *Eur. J. Vasc. Endovasc. Surg.* **2014**, *48*, 118–125. [CrossRef]
22. Sztajzel, R.; Momjian, S.; Momjian-Mayor, I.; Murith, N.; Djebaili, K.; Boissard, G.; Comelli, M.; Pizolatto, G. Stratified gray-scale median analysis and color mapping of the carotid plaque: Correlation with endarterectomy specimen histology of 28 patients. *Stroke* **2005**, *36*, 741–745. [CrossRef]
23. Ariyoshi, K.; Okuya, S.; Kunitsugu, I.; Matsunaga, K.; Nagao, Y.; Nomiyama, R.; Takeda, K.; Tanizawa, Y. Ultrasound analysis of gray-scale median value of carotid plaques is a useful reference index for cerebro-cardiovascular events in patients with type 2 diabetes. *J. Diabetes Investig.* **2015**, *6*, 91–97. [CrossRef] [PubMed]
24. el-Barghouty, N.; Nicolaides, A.; Bahal, V.; Geroulakos, G.; Androulakis, A. The identification of the high risk carotid plaque. *Eur. J. Vasc. Endovasc. Surg.* **1996**, *11*, 470–478. [CrossRef] [PubMed]
25. Nicolaides, A.N.; Kakkos, S.K.; Kyriacou, E.; Griffin, M.; Sabetai, M.; Thomas, D.J.; Tegos, T.; Geroulakos, G.; Labropoulos, N.; Doré, C.J.; et al. Asymptomatic internal carotid artery stenosis and cerebrovascular risk stratification. *J. Vasc. Surg.* **2010**, *52*, 1481–1485.e5. [CrossRef] [PubMed]
26. Andersson, J.; Sundstrom, J.; Gustavsson, T.; Hulthe, J.; Elmgren, A.; Zilmer, K.; Zilmer, M.; Lind, L. Echogenecity of the carotid intima-media complex is related to cardiovascular risk factors, dyslipidemia, oxidative stress and inflammation: The Prospective Investigation of the Vasculature in Uppsala Seniors (PIVUS) study. *Atherosclerosis* **2009**, *204*, 612–618. [CrossRef]
27. Jung, M.; Parrinello, C.M.; Xue, X.; Mack, W.J.; Anastos, K.; Lazar, J.M.; Selzer, R.H.; Shircore, A.M.; Plankey, M.; Tien, P.; et al. Echolucency of the carotid artery intima-media complex and intima-media thickness have different cardiovascular risk factor relationships: The Women's Interagency HIV Study. *J. Am. Heart Assoc.* **2015**, *4*, e001405. [CrossRef]
28. Spence, J.D. Technology Insight: Ultrasound measurement of carotid plaque–patient management, genetic research, and therapy evaluation. *Nat. Clin. Pract. Neurol.* **2006**, *2*, 611–619. [CrossRef]
29. Gepner, A.D.; Young, R.; Delaney, J.A.; Budoff, M.J.; Polak, J.F.; Blaha, M.J.; Post, W.S.; Michos, E.D.; Kaufman, J.; Stein, J.H. Comparison of Carotid Plaque Score and Coronary Artery Calcium Score for Predicting Cardiovascular Disease Events: The Multi-Ethnic Study of Atherosclerosis. *J. Am. Heart Assoc.* **2017**, *6*, e005179. [CrossRef]
30. Hollander, M.; Bots, M.L.; Del Sol, A.I.; Koudstaal, P.J.; Witteman, J.C.; Grobbee, D.E.; Hofman, A.; Breteler, M.M. Carotid plaques increase the risk of stroke and subtypes of cerebral infarction in asymptomatic elderly: The Rotterdam study. *Circulation* **2002**, *105*, 2872–2877. [CrossRef]
31. Vilanova, M.B.; Franch-Nadal, J.; Falguera, M.; Marsal, J.R.; Canivell, S.; Rubinat, E.; Miró, N.; Molló, À.; Mata-Cases, M.; Gratacòs, M.; et al. Prediabetes Is Independently Associated with Subclinical Carotid Atherosclerosis: An Observational Study in a Non-Urban Mediterranean Population. *J. Clin. Med.* **2020**, *9*, 2139. [CrossRef]

32. Wang, C.; Fang, X.; Wu, X.; Hua, Y.; Zhang, Z.; Gu, X.; Tang, Z.; Guan, S.; Liu, H.; Liu, B.; et al. Metabolic syndrome and risks of carotid atherosclerosis and cardiovascular events in community-based older adults in China. *Asia Pac. J. Clin. Nutr.* **2019**, *28*, 870–878. [CrossRef] [PubMed]
33. Munjral, S.; Ahluwalia, P.; Jamthikar, A.D.; Puvvula, A.; Saba, L.; Faa, G.; Singh, I.M.; Chadha, P.S.; Turk, M.; Johri, A.M.; et al. Nutrition, atherosclerosis, arterial imaging, cardiovascular risk stratification, and manifestations in COVID-19 framework: A narrative review. *Front. Biosci.* **2021**, *26*, 1312–1339. [CrossRef]
34. McClintock, T.R.; Parvez, F.; Wu, F.; Islam, T.; Ahmed, A.; Rani Paul, R.; Shaheen, I.; Sarwar, G.; Rundek, T.; Demmer, R.T.; et al. Major dietary patterns and carotid intima-media thickness in Bangladesh. *Public. Health Nutr.* **2016**, *19*, 218–229. [CrossRef]
35. Maddock, J.; Ziauddeen, N.; Ambrosini, G.L.; Wong, A.; Hardy, R.; Ray, S. Adherence to a Dietary Approaches to Stop Hypertension (DASH)-type diet over the life course and associated vascular function: A study based on the MRC 1946 British birth cohort. *Br. J. Nutr.* **2018**, *119*, 581–589. [CrossRef]
36. GómezSánchez, M.; Gómez Sánchez, L.; Patino-Alonso, M.C.; Alonso-Domínguez, R.; Sánchez-Aguadero, N.; Lugones-Sánchez, C.; Rodríguez Sánchez, E.; García Ortiz, L.; Gómez-Marcos, M.A. Adherence to the Mediterranean Diet in Spanish Population and Its Relationship with Early Vascular Aging according to Sex and Age: EVA Study. *Nutrients* **2020**, *12*, 1025. [CrossRef] [PubMed]
37. Koutsonida, M.; Kanellopoulou, A.; Markozannes, G.; Gousia, S.; Doumas, M.T.; Sigounas, D.E.; Tzovaras, V.T.; Vakalis, K.; Tzoulaki, I.; Evangelou, E.; et al. Adherence to Mediterranean Diet and Cognitive Abilities in the Greek Cohort of Epirus Health Study. *Nutrients* **2021**, *13*, 3363. [CrossRef]
38. Gardener, H.; Wright, C.B.; Cabral, D.; Scarmeas, N.; Gu, Y.; Cheung, K.; Elkind, M.S.; Sacco, R.L.; Rundek, T. Mediterranean diet and carotid atherosclerosis in the Northern Manhattan Study. *Atherosclerosis* **2014**, *234*, 303–310. [CrossRef]
39. Jimenez-Torres, J.; Alcala-Diaz, J.F.; Torres-Pena, J.D.; Gutierrez-Mariscal, F.M.; Leon-Acuna, A.; Gomez-Luna, P.; Fernandez-Gandara, C.; Quintana-Navarro, G.M.; Fernandez-Garcia, J.C.; Perez-Martinez, P.; et al. Mediterranean Diet Reduces Atherosclerosis Progression in Coronary Heart Disease: An Analysis of the CORDIOPREV Randomized Controlled Trial. *Stroke* **2021**, *52*, 3440–3449. [CrossRef]
40. Martinez-Gonzalez, M.A.; Salas-Salvado, J.; Estruch, R.; Corella, D.; Fito, M.; Ros, E.; Predimed, I. Benefits of the Mediterranean Diet: Insights From the PREDIMED Study. *Prog. Cardiovasc. Dis.* **2015**, *58*, 50–60. [CrossRef]
41. Mateo-Gallego, R.; Uzhova, I.; Moreno-Franco, B.; Leon-Latre, M.; Casasnovas, J.A.; Laclaustra, M.; Penalvo, J.L.; Civeira, F. Adherence to a Mediterranean diet is associated with the presence and extension of atherosclerotic plaques in middle-aged asymptomatic adults: The Aragon Workers' Health Study. *J. Clin. Lipidol.* **2017**, *11*, 1372–1382. [CrossRef] [PubMed]
42. Penalvo, J.L.; Fernandez-Friera, L.; Lopez-Melgar, B.; Uzhova, I.; Oliva, B.; Fernandez-Alvira, J.M.; Laclaustra, M.; Pocock, S.; Mocoroa, A.; Mendiguren, J.M.; et al. Association Between a Social-Business Eating Pattern and Early Asymptomatic Atherosclerosis. *J. Am. Coll. Cardiol.* **2016**, *68*, 805–814. [CrossRef]
43. Petersen, K.S.; Clifton, P.M.; Keogh, J.B. The association between carotid intima media thickness and individual dietary components and patterns. *Nutr. Metab. Cardiovasc. Dis.* **2014**, *24*, 495–502. [CrossRef] [PubMed]
44. Sala-Vila, A.; Romero-Mamani, E.S.; Gilabert, R.; Nunez, I.; de la Torre, R.; Corella, D.; Ruiz-Gutierrez, V.; Lopez-Sabater, M.C.; Pinto, X.; Rekondo, J.; et al. Changes in ultrasound-assessed carotid intima-media thickness and plaque with a Mediterranean diet: A substudy of the PREDIMED trial. *Arter. Thromb. Vasc. Biol.* **2014**, *34*, 439–445. [CrossRef]
45. Bhat, S.; Mocciaro, G.; Ray, S. The association of dietary patterns and carotid intima-media thickness: A synthesis of current evidence. *Nutr. Metab. Cardiovasc. Dis.* **2019**, *29*, 1273–1287. [CrossRef]
46. Fung, T.T.; Chiuve, S.E.; McCullough, M.L.; Rexrode, K.M.; Logroscino, G.; Hu, F.B. Adherence to a DASH-style diet and risk of coronary heart disease and stroke in women. *Arch. Intern. Med.* **2008**, *168*, 713–720. [CrossRef] [PubMed]
47. Levitan, E.B.; Wolk, A.; Mittleman, M.A. Consistency with the DASH diet and incidence of heart failure. *Arch. Intern. Med.* **2009**, *169*, 851–857. [CrossRef]
48. Rautiainen, S.; Manson, J.E.; Lichtenstein, A.H.; Sesso, H.D. Dietary supplements and disease prevention—A global overview. *Nat. Rev. Endocrinol.* **2016**, *12*, 407–420. [CrossRef]
49. Rontogianni, M.O.; Kanellopoulou, A.; Markozannes, G.; Bouras, E.; Derdemezis, C.; Doumas, M.T.; Sigounas, D.E.; Tzovaras, V.T.; Vakalis, K.; Panagiotakos, D.B.; et al. Prevalence and Determinants of Sex-Specific Dietary Supplement Use in a Greek Cohort. *Nutrients* **2021**, *13*, 2857. [CrossRef]
50. Chen, F.; Du, M.; Blumberg, J.B.; Ho Chui, K.K.; Ruan, M.; Rogers, G.; Shan, Z.; Zeng, L.; Zhang, F.F. Association Among Dietary Supplement Use, Nutrient Intake, and Mortality Among U.S. Adults: A Cohort Study. *Ann. Intern. Med.* **2019**, *170*, 604–613. [CrossRef]
51. Li, K.; Kaaks, R.; Linseisen, J.; Rohrmann, S. Vitamin/mineral supplementation and cancer, cardiovascular, and all-cause mortality in a German prospective cohort (EPIC-Heidelberg). *Eur. J. Nutr.* **2012**, *51*, 407–413. [CrossRef] [PubMed]
52. Marques-Vidal, P.; Pécoud, A.; Hayoz, D.; Paccaud, F.; Mooser, V.; Waeber, G.; Vollenweider, P. Prevalence and characteristics of vitamin or dietary supplement users in Lausanne, Switzerland: The CoLaus study. *Eur. J. Clin. Nutr.* **2009**, *63*, 273–281. [CrossRef] [PubMed]
53. Rock, C.L. Multivitamin-multimineral supplements: Who uses them? *Am. J. Clin. Nutr.* **2007**, *85*, 277s–279s. [CrossRef] [PubMed]
54. Schwarzpaul, S.; Strassburg, A.; Luhrmann, P.M.; Neuhauser-Berthold, M. Intake of vitamin and mineral supplements in an elderly german population. *Ann. Nutr. Metab.* **2006**, *50*, 155–162. [CrossRef] [PubMed]

55. Dickinson, A.; Blatman, J.; El-Dash, N.; Franco, J.C. Consumer usage and reasons for using dietary supplements: Report of a series of surveys. *J. Am. Coll. Nutr.* **2014**, *33*, 176–182. [CrossRef]
56. Jenkins, D.J.A.; Spence, J.D.; Giovannucci, E.L.; Kim, Y.I.; Josse, R.G.; Vieth, R.; Sahye-Pudaruth, S.; Paquette, M.; Patel, D.; Blanco Mejia, S.; et al. Supplemental Vitamins and Minerals for Cardiovascular Disease Prevention and Treatment: JACC Focus Seminar. *J. Am. Coll. Cardiol.* **2021**, *77*, 423–436. [CrossRef]
57. Li, Y.; Huang, T.; Zheng, Y.; Muka, T.; Troup, J.; Hu, F.B. Folic Acid Supplementation and the Risk of Cardiovascular Diseases: A Meta-Analysis of Randomized Controlled Trials. *J. Am. Heart Assoc.* **2016**, *5*, e003768. [CrossRef]
58. Khan, S.U.; Khan, M.U.; Riaz, H.; Valavoor, S.; Zhao, D.; Vaughan, L.; Okunrintemi, V.; Riaz, I.B.; Khan, M.S.; Kaluski, E.; et al. Effects of Nutritional Supplements and Dietary Interventions on Cardiovascular Outcomes: An Umbrella Review and Evidence Map. *Ann. Intern. Med.* **2019**, *171*, 190–198. [CrossRef]
59. Tian, T.; Yang, K.Q.; Cui, J.G.; Zhou, L.L.; Zhou, X.L. Folic Acid Supplementation for Stroke Prevention in Patients with Cardiovascular Disease. *Am. J. Med. Sci.* **2017**, *354*, 379–387. [CrossRef]
60. Jagodzinski, A.; Johansen, C.; Koch-Gromus, U.; Aarabi, G.; Adam, G.; Anders, S.; Augustin, M.; der Kellen, R.B.; Beikler, T.; Behrendt, C.A.; et al. Rationale and Design of the Hamburg City Health Study. *Eur. J. Epidemiol.* **2020**, *35*, 169–181. [CrossRef]
61. Mathiesen, E.B.; Bonaa, K.H.; Joakimsen, O. Echolucent plaques are associated with high risk of ischemic cerebrovascular events in carotid stenosis: The tromso study. *Circulation* **2001**, *103*, 2171–2175. [CrossRef] [PubMed]
62. Odhner, I. JS Paint [software]. Available online: https://github.com/1j01/jspaint (accessed on 15 March 2023).
63. Petroudi, S. Segmentation of the common carotid intima-media complex in ultrasound images using active contours. *IEEE Trans. Biomed. Eng.* **2012**, *59*, 3060–3069. [CrossRef] [PubMed]
64. Nothlings, U.; Hoffmann, K.; Bergmann, M.M.; Boeing, H. Fitting portion sizes in a self-administered food frequency questionnaire. *J. Nutr.* **2007**, *137*, 2781–2786. [CrossRef] [PubMed]
65. Hebestreit, K.; Yahiaoui-Doktor, M.; Engel, C.; Vetter, W.; Siniatchkin, M.; Erickson, N.; Halle, M.; Kiechle, M.; Bischoff, S.C. Validation of the German version of the Mediterranean Diet Adherence Screener (MEDAS) questionnaire. *BMC Cancer* **2017**, *17*, 341. [CrossRef] [PubMed]
66. Folsom, A.R.; Parker, E.D.; Harnack, L.J. Degree of concordance with DASH diet guidelines and incidence of hypertension and fatal cardiovascular disease. *Am. J. Hypertens.* **2007**, *20*, 225–232. [CrossRef]
67. van Buuren, S.; Groothuis-Oudshoorn, C. MICE: Multivariate Imputation by Chained Equations in R. *J. Stat. Softw.* **2011**, *45*, 1–67. [CrossRef]
68. von Elm, E.; Altman, D.G.; Egger, M.; Pocock, S.J.; Gotzsche, P.C.; Vandenbroucke, J.P.; Initiative, S. The Strengthening the Reporting of Observational Studies in Epidemiology (STROBE) Statement: Guidelines for reporting observational studies. *Int. J. Surg.* **2014**, *12*, 1495–1499. [CrossRef]
69. Li, K.; Kaaks, R.; Linseisen, J.; Rohrmann, S. Consistency of vitamin and/or mineral supplement use and demographic, lifestyle and health-status predictors: Findings from the European Prospective Investigation into Cancer and Nutrition (EPIC)-Heidelberg cohort. *Br. J. Nutr.* **2010**, *104*, 1058–1064. [CrossRef]
70. Oliván-Blázquez, B.; Aguilar-Latorre, A.; Motrico, E.; Gómez-Gómez, I.; Zabaleta-Del-Olmo, E.; Couso-Viana, S.; Clavería, A.; Maderuelo-Fernandez, J.A.; Recio-Rodríguez, J.I.; Moreno-Peral, P.; et al. The Relationship between Adherence to the Mediterranean Diet, Intake of Specific Foods and Depression in an Adult Population (45–75 Years) in Primary Health Care. A Cross-Sectional Descriptive Study. *Nutrients* **2021**, *13*, 2724. [CrossRef]
71. Catalan, M.; Herreras, Z.; Pinyol, M.; Sala-Vila, A.; Amor, A.J.; de Groot, E.; Gilabert, R.; Ros, E.; Ortega, E. Prevalence by sex of preclinical carotid atherosclerosis in newly diagnosed type 2 diabetes. *Nutr. Metab. Cardiovasc. Dis.* **2015**, *25*, 742–748. [CrossRef]
72. de Kreutzenberg, S.V.; Coracina, A.; Volpi, A.; Fadini, G.P.; Frigo, A.C.; Guarneri, G.; Tiengo, A.; Avogaro, A. Microangiopathy is independently associated with presence, severity and composition of carotid atherosclerosis in type 2 diabetes. *Nutr. Metab. Cardiovasc. Dis.* **2011**, *21*, 286–293. [CrossRef] [PubMed]
73. Harari, F.; Barregard, L.; Östling, G.; Sallsten, G.; Hedblad, B.; Forsgard, N.; Borné, Y.; Fagerberg, B.; Engström, G. Blood Lead Levels and Risk of Atherosclerosis in the Carotid Artery: Results from a Swedish Cohort. *Environ. Health Perspect.* **2019**, *127*, 127002. [CrossRef] [PubMed]
74. Li, C.; Engström, G.; Berglund, G.; Janzon, L.; Hedblad, B. Incidence of ischemic stroke in relation to asymptomatic carotid artery atherosclerosis in subjects with normal blood pressure. A prospective cohort study. *Cereb. Dis.* **2008**, *26*, 297–303. [CrossRef] [PubMed]
75. Rundek, T.; Arif, H.; Boden-Albala, B.; Elkind, M.S.; Paik, M.C.; Sacco, R.L. Carotid plaque, a subclinical precursor of vascular events: The Northern Manhattan Study. *Neurology* **2008**, *70*, 1200–1207. [CrossRef]
76. Višković, K.; Rutherford, G.W.; Sudario, G.; Stemberger, L.; Brnić, Z.; Begovac, J. Ultrasound measurements of carotid intima-media thickness and plaque in HIV-infected patients on the Mediterranean diet. *Croat. Med. J.* **2013**, *54*, 330–338. [CrossRef] [PubMed]

77. Terschuren, C.; Damerau, L.; Petersen, E.L.; Harth, V.; Augustin, M.; Zyriax, B.C. Association of Dietary Pattern, Lifestyle and Chronotype with Metabolic Syndrome in Elderly-Lessons from the Population-Based Hamburg City Health Study. *Int. J. Environ. Res. Public. Health* **2021**, *19*, 377. [CrossRef]
78. Spahl, L. Automatische Dichtebestimmung von Plaques in der Arteria carotis. In *Projektpraktikumsbericht im Rahmen des Studienganges Medizinische Informatik der Universität zu Lübeck durchgeführt*; Universitätsklinikum Hamburg-Eppendorf: Hamburg, Germany, 2019.

 *nutrients* 

Article

How Does Dietary Intake Relate to Dispositional Optimism and Health-Related Quality of Life in Germline *BRCA1/2* Mutation Carriers?

Anne Esser [1,*], Leonie Neirich [1], Sabine Grill [1], Stephan C. Bischoff [2], Martin Halle [3], Michael Siniatchkin [4], Maryam Yahiaoui-Doktor [5], Marion Kiechle [1] and Jacqueline Lammert [1]

1 Department of Gynecology and Center for Hereditary Breast and Ovarian Cancer, University Hospital Rechts der Isar, Technical University of Munich (TUM), 81675 Munich, Germany
2 Institute of Nutritional Medicine, University of Hohenheim, 70599 Stuttgart, Germany
3 Department of Prevention and Sports Medicine, University Hospital Klinikum Rechts der Isar, Technical University of Munich (TUM), 81675 Munich, Germany
4 Clinic for Child and Adolescent Psychiatry and Psychotherapy, Medical Center Bethel, University of Bielefeld, 33617 Bielefeld, Germany
5 Institute for Medical Informatics, Statistics and Epidemiology, University of Leipzig, 04107 Leipzig, Germany
* Correspondence: anne.esser@mri.tum.de

Abstract: *Background*: The Mediterranean diet (MD) is an anti-inflammatory diet linked to improved health-related quality of life (HRQoL). Germline (g)*BRCA1/2* mutation carriers have an increased risk of developing breast cancer and are often exposed to severe cancer treatments, thus the improvement of HRQoL is important. Little is known about the associations between dietary intake and HRQoL in this population. *Methods*: We included 312 g*BRCA1/2* mutation carriers from an ongoing prospective randomized controlled lifestyle intervention trial. Baseline data from the EPIC food frequency questionnaire was used to calculate the dietary inflammatory index (DII), and adherence to MD was captured by the 14-item PREDIMED questionnaire. HRQoL was measured by the EORTC QLQ-C30 and LOT-R questionnaires. The presence of metabolic syndrome (MetS) was determined using anthropometric measurements, blood samples and vital parameters. Linear and logistic regression models were performed to assess the possible impact of diet and metabolic syndrome on HRQoL. *Results*: Women with a prior history of cancer (59.6%) reported lower DIIs than women without it ($p = 0.011$). A greater adherence to MD was associated with lower DII scores ($p < 0.001$) and reduced odds for metabolic syndrome (MetS) ($p = 0.024$). Women with a more optimistic outlook on life reported greater adherence to MD ($p < 0.001$), whereas a more pessimistic outlook on life increased the odds for MetS (OR = 1.15; $p = 0.023$). *Conclusions*: This is the first study in g*BRCA1/2* mutation carriers that has linked MD, DII, and MetS to HRQoL. The long-term clinical implications of these findings are yet to be determined.

Keywords: *BRCA1*; *BRCA2*; DII; Mediterranean diet; HRQoL; metabolic syndrome

Citation: Esser, A.; Neirich, L.; Grill, S.; Bischoff, S.C.; Halle, M.; Siniatchkin, M.; Yahiaoui-Doktor, M.; Kiechle, M.; Lammert, J. How Does Dietary Intake Relate to Dispositional Optimism and Health-Related Quality of Life in Germline *BRCA1/2* Mutation Carriers? *Nutrients* **2023**, *15*, 1396. https://doi.org/10.3390/nu15061396

Academic Editors: Birgit-Christiane Zyriax and Nataliya Makarova

Received: 20 February 2023
Revised: 8 March 2023
Accepted: 9 March 2023
Published: 14 March 2023

1. Introduction

With continual improvements in cancer outcomes, both patients and clinicians are shifting their focus from survival alone towards improving health-related quality of life (HRQoL) and patient-centred functional outcomes [1]. HRQoL is defined as the impact a disease and its treatment have on a patient's physical, functional, psychological, social, and financial well-being [2–4]. In cancer care, there is a growing recognition of the significance of HRQoL, as reduced HRQoL may result in lower treatment adherence [5] and an increased risk of mortality [6]. A more comprehensive definition of HRQoL could encompass dispositional optimism, which is a psychological attribute associated with health advantages [7]. Different aspects of HRQoL have been associated with chronic inflammation, i.e., decreased physical [8] and cognitive functioning [9], increased fatigue [10] and higher pain levels [11].

Pre-treatment inflammatory status may predict the development of common cancer treatment side effects [12], e.g., aromatase inhibitor-induced musculoskeletal syndrome in women with pre-existing musculoskeletal pain. Most importantly, elevated inflammatory markers have been associated with adverse cancer outcomes [13,14], potentially by promoting a microenvironment for tumour growth and metastasis [15]. The quantity, quality, and composition of foods have been shown to regulate inflammation [16–18]. This has prompted research into developing a literature-derived index to reflect the inflammatory potential of diets; the Dietary Inflammatory Index (DII) [19] scores an individual's diet on a continuum from anti- to pro-inflammatory. A pro-inflammatory diet has been linked to an increased cardiovascular risk and mortality [20], and it increases the likelihood of both metabolic syndrome (MetS) [21] and various types of cancer [22–25]. Recent studies indicate a negative association between a pro-inflammatory diet and HRQoL [26–28]. A diet associated with low DII scores is the Mediterranean diet (MD) [29]. MD is characterized by high consumption of fruits, vegetables, legumes, grains, and polyunsaturated fats from olive oil and nuts, moderate consumption of fish and dairy products, and low intake of red meat and processed foods [30]. MD has been shown to be associated with reduced cardiovascular risk [31], prevent MetS and lower cancer risk [32]. Furthermore, adherence to MD has been linked to improved HRQoL in healthy individuals [33,34], as well as cancer survivors [35].

Breast cancer (BC) is the most common type of cancer in women [36]. Particularly vulnerable are women with a germline (g)*BRCA1/2* mutation have a risk of 69–72% of developing breast cancer and a risk of 17–44% of developing ovarian cancer by the age of 80 years [37]. These women are exposed to cancer treatments and/or prophylactic surgeries with detrimental short- and long-term effects on their health [38–42] and HRQoL [6,43–45]. Recent studies suggest that beneficial dietary changes after completing primary cancer treatment, as opposed to during treatment, might be most effective in improving HRQoL [46]. Dietary factors to reduce chronic inflammation and improve metabolic profile may be an approach to improving HRQoL, functional capacity, and cancer outcomes in women with a g*BRCA1/2* mutation. A first step in addressing this issue is to determine the relationship of DII, MD, MetS, and different aspects of HRQoL in g*BRCA1/2* mutation carriers with and without a previous history of cancer.

2. Methods

2.1. Study Design and Participants

The present study is a cross-sectional secondary analysis of the baseline data from the randomized controlled LIBRE-2 trial (a lifestyle intervention study in women with hereditary breast and ovarian cancer) and the associated feasibility study LIBRE-1 [47,48]. The trials are registered at ClinicalTrial.gov (NCT numbers: NCT02087592–registered on 14 March 2014, NCT02516540–registered on 6 August 2015). The LIBRE-2 trial is an ongoing, two-armed randomized (1:1) controlled multicentre trial conducted in Germany aimed at determining the impact of a structured one-year lifestyle intervention program on adherence to MD, cardiorespiratory fitness, and body mass index (BMI) among g*BRCA1/2* mutation carriers. The study cohort includes both women with a previous diagnosis of early stage cancer in remission (diseased) and without a prior cancer diagnosis (non-diseased). Details on the study design have been published elsewhere [47,48]. A total of 312 participants were available for the current analysis.

2.2. Instruments

Blood samples, anthropometric measurements, and medical history. At baseline, participants completed a standardized questionnaire to collect information on their medical history, socio-demographic factors, as well as lifestyle factors. Furthermore, all participants underwent a physical examination to determine systolic and diastolic blood pressure, heart rate, and anthropometric measurements such as height (in m), body weight (in kg), and waist and hip circumferences (in cm). These were used to calculate BMI (kg/m^2) and the waist-to-hip ratio (waist circumference in cm/hip circumference in cm). Blood samples were taken after a 12-h fasting period, and analysed by the affiliated laboratories of the local institutions. MetS was defined according to the International Diabetes Federation criteria by the presence of a waist circumference $\geq$ 80 cm and at least two metabolic abnormalities, i.e., fasting glucose $\geq$ 100 mg/dL, systolic blood pressure $\geq$ 130 mmHg and/or diastolic blood pressure $\geq$ 85 mmHg, triglycerides $\geq$ 150 mg/dL, HDL-cholesterol < 50 mg/dL and/or treatment with lipid-lowering, glucose-lowering or antihypertensive drugs. Cardiopulmonary exercise testing was conducted to assess cardiorespiratory fitness via peak oxygen uptake (VO_{2peak}).

FFQ, MEDAS and Dietary Inflammatory Index. Dietary intake was determined by two validated questionnaires. The participants completed the German version of the PREDIMED questionnaire, the Mediterranean diet adherence screener (MEDAS), a 14-item questionnaire that captures adherence to MD [49–51]. We calculated the MEDAS score as the percentage of positively answered questions [52]. Additionally, the German version of the EPIC food frequency questionnaire (FFQ) was applied to collect information on the quantity and frequency of 148 food items consumed over the previous year [53,54]. Data from the FFQ were then used to calculate DII using the method reported by Shivappa et al. [19]. Briefly, the DII is based on 1943 scientific papers scoring 45 food parameters according to whether they increased (+1), decreased (−1), or had no effect (0) on six inflammatory biomarkers (IL-1β, IL-4, IL-6, IL-10, TNF-α, and CRPs). As reported in previous studies [22,55–57], not all required food items were assessed by the German FFQ. Hence, the DII was calculated using the corresponding 30 food parameters available from the FFQ used in our study. Those were carbohydrates, protein, saturated fat, polyunsaturated fatty acids (PUFA), monounsaturated fatty acids (MUFA), n-3-fatty-acids, n-6-fatty-acids, cholesterol, total fat, energy, fibre, alcohol, iron, magnesium, zinc, vitamin A, thiamin, vitamin B12, riboflavin, niacin, vitamin B6, folic acid, vitamin C, vitamin D, vitamin E, flavonones, anthocyanidins, flavan-3-ol, flavonols, and flavones.

Psychological questionnaires. All LIBRE trial participants completed several psychological questionnaires. To assess optimism and pessimism as a personality trait, the revised 10-item life orientation test (LOT-R) was applied [58]. The "optimism score" (LOTR-O) ranging from 0 (minimally optimistic) to 12 (maximally optimistic) was calculated as the sum of the three positively formulated items. The "pessimism score" (LOTR-P) was calculated accordingly. The EORTC QLQ-C30 (questionnaire for quality of life assessment in patients with cancer, Version 3.0) [3] was used to evaluate HRQoL. This questionnaire consists of 30 items and is designed for patients receiving cancer treatment regardless of cancer type and location. It measures five functional dimensions (physical, role, emotional, cognitive, and social), three symptom items (fatigue, nausea or vomiting, and pain), six single items (dyspnea, insomnia, appetite loss, constipation, diarrhea, and financial impact), and a global health status, which is the mean of two questions regarding overall health and overall quality of life. The BKAE ("Bewertung körperlicher Aktivität und Ernährung", in English "evaluation of physical activity and nutrition") is a questionnaire designed specifically for the LIBRE trials [47,48] to analyse attitudes and views on physical activity and dietary intake. It is based on the concept of "planned behaviour" by Fishbein & Ajzen (1975) [59] promoting that attitude, subjective norms and perceived behaviour control contribute to behavioural intention, which leads to actual behaviour. We only used the dietary information of the questionnaire for our analysis. The scores *BKAE-AT* (attitude towards healthy eating), *BKAE-SN* (subjective norms about healthy eating), *BKAE-PBC* (perceived behaviour control over healthy eating), *BKAE-IT* (intention to eat healthy in the future) and *BKAE-PB* (past behaviour with regard to healthy eating) range from 0 (minimum) to 100 (maximum). The physical activity part of the questionnaire has been evaluated previously; the strongest predictor for cardiopulmonary fitness was attitudes towards physical activity [60].

2.3. Statistical Analysis

SPSS Version 29.0.0 (IBM Corp., Armonk, NY, USA) was used to analyse data. Descriptive statistics are presented as mean ± standard deviation (SD) for continuous variables or as proportions for categorical variables. The distributions of continuous variables between diseased and non-diseased women were compared using Student's *t*-test. The distributions of categorical variables were compared using the Chi-square test. Linear regression models were created to detect associations between dietary intake and HRQoL. EORTC-QC30 scores were evaluated in diseased women only since the questionnaire was validated for cancer patients. Logistic regression models were performed to estimate odds ratios (ORs) and their associated 95% confidence intervals (95% CI) between MetS, dietary intake and different aspects of HRQoL. Multivariate analyses were carried out to control for potential confounding variables. These analyses were adjusted for body composition (BMI), physical fitness (VO_{2peak}), adherence to MD, and/or dietary inflammatory potential (DII). All p values were based on two-sided tests and were considered significant if $p \leq 0.05$.

2.4. Ethics

The study was approved by the ethics committees of both the host institutions Technical University of Munich (Reference No. 5685/13), the University Hospital Cologne (Reference No. 13-053), the University Hospital Schleswig-Holstein in Kiel (Reference No. B-235/13), and the participating study centres. Written consent from all study participants was obtained. All methods were carried out in accordance with relevant guidelines and regulations.

3. Results

A total of 312 women with a *gBRCA1* and/or *gBRCA2* mutation were included in the study. Table 1 summarizes the selected participants' characteristics by health status (diseased vs. non-diseased). The mean age of the entire study cohort was 43.5 years (SD ± 10.3 years). Of all the women, 59.6% had a previous diagnosis of cancer. Among these, breast cancer accounted for 88.7% and ovarian cancer for 7.0% of all cancer cases. Women with a history of breast cancer were older (46.5 years vs. 39.1 years, $p < 0.001$), more likely married (67% vs. 55%, $p = 0.026$), and less educated (high school diploma: 58% vs. 75%, $p = 0.002$). Diseased women had significantly lower hsCRP levels (1.7 vs. 3.3 mg/L, $p = 0.045$) and lower DII scores (−1.1 vs. −0.5, $p = 0.011$) compared to non-diseased women. Non-diseased mutation carriers had better physical fitness (17.3 vs. 16 mL/min/kg, $p = 0.029$), reported significantly higher quality of life (QL2 72.2 vs. 57.7, $p = 0.041$), role (RF 90.3 vs. 79.8, $p < 0.001$), cognitive (CF 82.5 vs. 72.9, $p < 0.001$) and social functioning (SF 85.2 vs. 72.0, $p < 0.001$), and experienced less pain (PA 15.7 vs. 25.6, $p = 0.001$), dyspnea (DY 9.6 vs. 16.1, $p = 0.015$), insomnia (SL 28.0 vs. 39.4, $p = 0.003$), and fewer financial difficulties (FI 4.3 vs. 18.5, $p < 0.001$). On the other hand, diseased mutation carriers reported stronger social norms about healthy eating (BKAE-SN 79.6 vs. 73.7, $p = 0.008$) and greater behavioural control over healthy eating (BKAE-PBC 86.9 vs. 84.2, $p = 0.010$). They also reported a more frequent consumption of healthy foods compared to women without a prior history of cancer (BKAE-PB 58.5 vs. 51.6; $p = 0.008$).

We then analysed associations between DII and various metabolic and lifestyle factors using linear regressions. The results are presented in Table 2. A lower DII score was significantly associated with higher adherence to MD ($p < 0.001$). Among diseased women, higher DII scores were associated with better role functioning (RF) ($p = 0.032$), cognitive functioning (CF) ($p = 0.003$), and social functioning (SF) ($p = 0.012$) as well as decreased fatigue (FA) ($p = 0.046$), dyspnea (DY) ($p = 0.029$) and appetite loss (AP) ($p = 0.007$).

Table 1. Baseline characteristics for participants with and without history of cancer.

Characteristic	Diseased	Non-Diseased	p-Value
n (%)	186 (59.6%)	126 (40.4%)	
	Socio-demographic Data		
Age, years, mean ± SD	46.5 ± 9.2	39.1 ± 10.4	**<0.001** *
Married, n (%)	(125) 67 %	70 (55%)	**0.026** *
Number of children, mean ± SD	1.3 ± 0.9	1.1 ± 1.1	0.110
Education, n (%)			
• General university entrance qualification	104 (58%)	95 (75%)	**0.002** *
• University degree	82 (44%)	68 (54%)	0.074
Net income, EUR, mean ± SD, n	4043.8 ± 1982.3; n = 126	3851.4 ± 2258.8; n = 84	0.515
	Anthropometric Data		
BMI, kg/m^2, mean ± SD	25.6 ± 4.9	25 ± 5.5	0.301
Waist-to-hip ratio, mean ± SD	0.8 ± 0.1	0.8 ± 0.1	0.256
	Metabolic Data		
Metabolic syndrome, n (%)	45 (24%)	26 (21%)	0.426
VO_{2peak}, mL/min/kg, mean ± SD	16.0 ± 5.0	17.3 ± 4.7	**0.029** *
	Nutritional Data		
DII, mean ± SD	−1.1 ± 1.8	−0.5 ± 1.2	**0.011** *
MEDAS, mean ± SD	47.9 ± 16.6	46.5 ± 15.3	0.478
	Psychological Data		
LOTR-O, mean ± SD	4.2 ± 2.9	4.1 ± 3.0	0.792
LOTR-P, mean ± SD	4.3 ± 2.1	4.0 ± 2.4	0.267
Quality of life (QL2), mean ± SD	67.7 ± 19.1	72.2 ± 8.6	**0.041** *
Physical Functioning (PF2), mean ± SD	88.8 ± 12.6	91.4 ± 11.0	0.054
Role Functioning (RF2), mean ± SD	79.8 ± 24.0	90.3 ± 18.5	**<0.001** *
Emotional Functioning (EF), mean ± SD	61.7 ± 27.3	62.6 ± 24.2	0.770
Cognitive Functioning (CF), mean ± SD	72.9 ± 25.9	82.5 ± 19.2	**<0.001** *
Social Functioning (SF), mean ± SD	72.0 ± 30.0	85.2 ± 23.0	**<0.001** *
Fatigue (FA), mean ± SD	33.6 ± 26.1	28.9 ± 19.9	0.085
Nausea and vomiting (NV), mean ± SD	3.9 ± 9.6	6.3 ± 14.9	0.096
Pain (PA), mean ± SD	25.6 + 28.3	15.7 ± 21.6	**0.001** *
Dyspnea (DY), mean ± SD	16.1 ± 24.6	9.6 ± 20.7	**0.015** *
Insomnia (SL), mean ± SD	39.4 ± 35.7	28.0 ± 29.8	**0.003** *
Appetite loss (AP), mean ± SD	6.1 ± 16.9	4.0 ± 10.9	0.223
Constipation (CO), mean ± SD	10.0 ± 22.9	8.5 ± 20.3	0.543
Diarrhea (DI), mean ± SD	6.6 ± 15.8	11.5 ± 21.6	**0.033** *
Financial difficulties (FI), mean ± SD	18.5 ± 29.0	4.3 ± 15.3	**<0.001** *
BKAE-AT, mean ± SD	79.8 ± 6.8	78.9 ± 7.2	0.308
BKAE-SN, mean ± SD	79.6 ± 15.8	73.7 ± 18.7	**0.008** *
BKAE-PBC, mean ± SD	86.9 ± 7.8	84.2 ± 10.7	**0.010** *
BKAE-IT, mean ± SD	78.2 ± 9.8	76.2 ± 11.3	0.108
BKAE-PB, mean ± SD	58.5 ± 20.0	51.6 ± 23.0	**0.008** *

* Results are statistically significant at a p-value of $\leq$0.05 (in bold).

Table 2. Associations between DII and metabolic, lifestyle and HRQoL factors using linear regression models.

Characteristic	Mean ± SD	Unadjusted Estimate [a] (95% CI)	Unadjusted p-Value [a]	Adjusted Estimate [b] (95% CI)	Adjusted p-Value [b]
MEDAS	47.3 ± 16.8	−2.340 (−3.579; −1.101)	**<0.001** *	−2.266 (−3.520; −1.011)	**<0.001** *
Role functioning (RF2) [1]	79.8 ± 24.0	0.012 (0.001; 0.023)	**0.032** *	0.014 (0.003; 0.025)	**0.010** *
Cognitive functioning (CF) [1]	72.9 ± 25.9	0.015 (0.005; 0.025)	**0.003** *	0.016 (0.006; 0.026)	**0.002** *
Social functioning (SF) [1]	72.0 ± 30.0	0.011 (0.002; 0.020)	**0.012** *	0.013 (0.004; 0.021)	**0.005** *
Fatigue (FA) [1]	33.6 ± 26.1	−0.010 (−0.020; 0.000)	**0.046** *	−0.012 (−0.022; −0.002)	**0.017** *
Pain (PA) [1]	25.6 ± 28.3	−0.009 (−0.018; 0.000)	0.057	−0.011 (−0.021; −0.002)	**0.017** *
Dyspnea (DY) [1]	16.1 ± 24.6	−0.012 (−0.022; −0.001)	**0.029** *	−0.016 (−0.027; −0.005)	**0.004** *
Appetite loss (AP) [1]	6.1 ± 16.9	−0.021 (−0.036; −0.006)	**0.007** *	−0.021 (−0.036; −0.006)	**0.008** *

* Results are statistically significant at a p-value of $\leq$0.05 (in bold); [1] only for 'Diseased'; [a] univariate linear regression unadjusted for BMI, VO_{2peak} and MEDAS; [b] multivariate linear regression adjusted for BMI, VO_{2peak} and MEDAS.

Associations between adherence to MD (MEDAS) and various factors were carried out using linear regressions (see Table 3). Adherence to MD was associated with higher VO_{2peak} (p = 0.024), as well as lower DII scores (p = <0.001). Furthermore, adherence to MD was associated with dispositional optimism (p = 0.001).

Table 3. Associations between MEDAS and metabolic, lifestyle, and HRQoL factors using linear regression models.

Characteristic	Mean ± SD	Unadjusted Estimate [a] (95% CI)	Unadjusted p-Value [a]	Adjusted Estimate [b] (95% CI)	Adjusted p-Value [b]
VO_{2peak}, mL/min/kg	16.7 ± 4.9	0.005 (0.001; 0.009)	**0.014** *	0.004 (0.000; 0.008)	0.053
DII	−0.9 ± 1.9	−0.018 (−0.028; −0.009)	**<0.001** *	−0.017 (−0.027; −0.008)	**<0.001** *
LOTR-O	4.2 ± 2.9	0.011 (0.004; 0.017)	**<0.001** *	0.010 (0.004; 0.016)	**0.002** *

* Results are statistically significant at a p-value of ≤0.05 (in bold); [a] univariate linear regression unadjusted for BMI, VO_{2peak} and DII; [b] multivariate linear regression adjusted for BMI, VO_{2peak} and DII.

We then carried out logistic regression models to estimate odds ratios (OR) and their associated 95% confidence intervals (95% CI) of having metabolic syndrome (MetS) by different dietary variables and different aspects of HRQoL (see Table 4). Higher adherence to MD (MEDAS ≥ 0.50) reduced odds for MetS (OR = 0.538, p = 0.024). Women with dispositional pessimism had increased odds for MetS (OR = 1.147, p = 0.023). Among diseased women, those who had poorer physical functioning (OR = 0.955, p < 0.001) or experienced more dyspnea (OR = 1.017, p = 0.012) had increased odds for MetS.

Table 4. Associations between MetS, lifestyle and HRQoL factors using logistic regression models.

Predictor	Unadjusted OR [a] (95% CI)	Unadjusted p-Value [a]	Adjusted OR [b] (95% CI)	Adjusted p-Value [b]
High adherence to MD (MEDAS ≥ 0.50) vs. low adherence to MD (<0.50)	0.538 (0.314; 0.922)	**0.024** *	0.602 (0.343; 1.058)	0.078
LOTR-P	1.147 (1.019; 1.292)	**0.023** *	1.150 (1.012; 1.307)	**0.032** *
Physical functioning (PF2) [1]	0.955 (0.930; 0.980)	**<0.001** *	0.963 (0.937; 0.990)	**0.007** *
Dyspnea (DY) [1]	1.017 (1.004; 1.030)	**0.012** *	1.013 (0.999; 1.026)	0.061

* Results are statistically significant at a p-value of ≤0.05 (in bold); [1] only for 'Diseased'; [a] univariate logistic regression unadjusted for VO_{2peak} and MEDAS; [b] multivariate logistic regression adjusted for VO_{2peak} and MEDAS.

4. Discussion

The aim of this analysis was to evaluate the relationship between anti-inflammatory diet, metabolic syndrome (MetS), and different aspects of health-related quality of life (HRQoL) in g*BRCA1/2* mutation carriers.

g*BRCA1/2* mutation carriers have a very high lifetime risk of developing breast and/or ovarian cancers. The average age of cancer diagnosis is substantially younger than in the general population (37). *BRCA*-associated cancers exhibit pathological features suggestive of an aggressive phenotype [61–63], and therefore, most patients undergo chemotherapy with detrimental side effects. When diagnosed with ER-positive breast cancer, patients might benefit from an extended adjuvant endocrine therapy [64,65], especially premenopausal women [66]. However, adjuvant endocrine therapy impacts HRQoL negatively [67]. Thus, identifying modifiable lifestyle factors to improve HRQoL is of particular relevance to g*BRCA1/2* mutation carriers, possibly resulting in better treatment adherence and (cancer-free) survival.

We observed that diseased women consumed a more anti-inflammatory diet compared to non-diseased women (DII −1.1 vs. −0.5, p = 0.011). Moreover, diseased participants perceived greater behavioural control over selecting healthier food options and were more likely to make healthier food choices than women without a history of cancer. This conforms to previous research that a breast cancer diagnosis can lead to beneficial dietary changes [68]. The German breast cancer guideline issued by the German Association of the Scientific Medical Societies (AWMF) and the German Agency for Quality in Medicine (AeZQ) acknowledges the importance of lifestyle factors, such as diet and physical activity, in the aftercare of breast cancer patients. However, it was not until 2017 that the guideline included this recommendation [69]. The guideline suggests adhering to the dietary guidelines set by the German Society for Nutrition (DGE), which emphasize the consumption of plant-based foods such as cereal, grains, fruits, and vegetables as the foundation of a healthy diet, with small portions of animal products such as dairy, eggs, meat, and fish [70]. Compared to the MD, the consumption of olive oil, fish, seafood, and red wine is less emphasized in the DGE guidelines. As four items of the MEDAS focus on these food groups, it is possible that the lack of emphasis on them in the DGE guidelines may explain why diseased women in our study did not report a higher adherence to the MD compared to non-diseased women.

Of interest is the significant inverse association between adherence to MD and DII, indicating that MD is an anti-inflammatory diet (p = <0.001). In a prospective study, Hodge et al. (2016) identified MD as an anti-inflammatory diet that significantly reduced the risk of lung cancer [29]. The PREDIMED trial was the first randomized controlled trial to support these findings in a group of postmenopausal females; adherence to MD reduced the risk of breast cancer by 68% (95% CI 0.13–0.79) [69].

Porciello et al. [35] reported that adherence to MD in breast cancer survivors was associated with better HRQoL, i.e., improved physical functioning, better sleep quality and lower pain. We were not able to show an association between adherence to MD and HRQoL among diseased g*BRCA1/2* mutation carriers in our univariate and multivariate analyses. In our analysis, the median time from cancer diagnosis to study enrolment was four years (range: 1–48 years). To be eligible for participation in the LIBRE study, women had to be physically fit and functional, and several criteria that could hinder participation in the intervention program had to be excluded at study entry. These criteria included ongoing chemotherapy and/or radiation therapy, metastatic tumor disease, Karnofsky index below 60%, and exercise capacity below 50 watts. Consequently, our study participants were likely much fitter and more functional than those in Porciello et al.'s study, where women had to be diagnosed with breast cancer within the previous twelve months and had a mean age that was ten years older than our study participants.

In our study, adherence to MD was positively associated with dispositional optimism ($p \leq 0.001$). This finding is consistent with the results of a study by Ait-Hadad et al. [70], which investigated the relationship between dietary intake and dispositional optimism in a sample of over 32,000 participants. The authors reported a positive association between optimism and overall diet quality; high intake of fruits, vegetables, legumes, whole grains, seafood, and fats was positively associated with optimism, while high intake of meat and dairy products was negatively associated with optimism. Dispositional optimism is characterized as a general expectation or belief in positive outcomes in the future [71]. It has been associated with improved cardiovascular health and reduced all-cause as well as cause-specific mortality in large epidemiological studies [72,73]. Among breast cancer patients, optimism has been linked to psychological well-being and improved quality of life [74,75]. Boehm et al. found that dispositional optimism was associated with higher serum levels of antioxidants [76]. This association was partially influenced by dietary intake. Scheier and Carver [77] suggest that there are two underlying mechanisms linking optimism to health. Firstly, dispositional optimism facilitates the engagement in health promoting behaviours, i.e., diet and physical activity. Secondly, optimistic individuals better cope with adverse life events better than pessimistic individuals, which results in reduced stress levels and increased physiological wellbeing. Although dispositional optimism is considered to be relatively stable across one's lifespan, some studies found that cognitive therapy can increase optimism levels [78,79]. Since dispositional optimism and MD are linked in g*BRCA1/2* mutation carriers, identifying further strategies to increase dispositional optimism might help to implement a healthy diet.

Moreover, adherence to MD was associated with reduced odds for MetS (OR = 0.54, $p = 0.024$). An Italian randomized controlled trial found that an MD-based dietary intervention in *gBRCA1/2* mutation carriers improved adherence to MD and reduced components of MetS [80]. Recent studies suggest that MetS is associated with impaired HRQoL [81–83]. Cohen et al. [84] reported a positive association between pessimism and the prevalence of MetS in patients with coronary heart disease. In our analysis, we found a positive association between MetS and dispositional pessimism (OR = 1.15, $p = 0.023$). In univariate analyses, MetS was associated with poorer physical functioning (OR = 0.96, $p = <0.001$) and higher levels of dyspnea (OR = 1.02; $p = 0.012$) among diseased women. However, these associations diminished following adjustment for physical fitness (VO_{2max}).

In women with a history of cancer, higher DII scores were associated with better role functioning (RF), cognitive functioning (CF), and social functioning (SF), as well as reduced fatigue (FA), reduced dyspnea (DY), and reduced appetite loss (AP). These findings were robust after adjustment for body composition (BMI), physical fitness (VO_{2peak}), and adherence to MD. To rule out that different types of cancer treatment or time since diagnosis influenced these associations, we calculated further multivariate regression models (see Appendix A). Our results were similar (see Tables A1–A3). This is surprising since it is contradictory to prior findings indicating that a more pro-inflammatory diet is associated with reduced HRQoL [28]. A possible explanation could be that diseased women with better HRQoL were not concerned about healthy eating. According to the theory of planned behaviour by Fishbein and Ajzen, behaviours are influenced by intentions, which are determined by three factors: attitudes, subjective norms, and perceived behavioural control [59]. To test our hypothesis, we adjusted our multivariate regression models for associations between DII and different dimensions of EORTC QLQ-C30 for the three core components of the Fishbein and Ajzen model, i.e., attitudes, subjective norms and perceived behavioural control. None of the three factors were associated with a pro-inflammatory diet nor did they influence the link between greater DII scores and reduced role, cognitive and social functioning (see Appendix A—Table A4). After inserting the variables attitudes, social norms and perceived behavioural control into the linear regression models for DII and fatigue, dyspnea and appetite loss, the models no longer reached significant levels ($p = 0.059$–0.143). Thus, in our analysis, the positive associations between pro-inflammatory diet patterns and HRQoL were likely not influenced by attitudes and beliefs towards healthy eating.

Strengths and Limitations

The strengths of the current study include the comprehensive evaluation of predictors that could be linked to HRQoL in *gBRCA1/2* mutation carriers. After adjusting for body composition, physical fitness and eating patterns, the adjusted and unadjusted results did not differ significantly. Therefore, any additional confounding was likely to be small. This supports our hypothesis that dietary intake is linked to different aspects of HRQoL. Although our results provide an interesting direction for HRQoL research, this study had several limitations. Firstly, the nature of a cross-sectional secondary analysis cannot establish a cause-and-effect relationship. The prospective nature of the LIBRE trials will allow for evaluating the impact of dietary changes on HRQoL. Secondly, as the number and type of food components to compute DII vary between studies, our results can hardly be compared to other populations of *gBRCA1/2* mutation carriers. Our study cohort was not representative for the average German population regarding education, net income, marital status, and parity [85–88]. Considering that our study cohort consisted of health-conscious females [89], the results obtained in this analysis likely underestimate the true associations between a pro-inflammatory diet and HRQoL outcomes. Finally, our cohort was not sufficiently powered to conduct analyses stratified by the *gBRCA* mutation type.

5. Conclusions

We were able to show that adherence to MD is linked to a more anti-inflammatory diet, dispositional optimism, and lower MetS prevalence among g*BRCA1/2* mutations carriers. Further research is needed to determine the long-term clinical implications of these findings.

Author Contributions: A.E. participated in the conceptualization of the study, was involved in the acquisition of data, carried out data analyses, and drafted the initial manuscript. J.L. designed and conceptualized this study, analysed and interpreted data, and critically revised the manuscript. L.N., S.G., S.C.B., M.H., M.S. and M.Y.-D. were involved in the acquisition and interpretation of patient data and contributed to the critical revision of the manuscript. M.K. conceived and supervised the study and acquired funding, and critically revised the manuscript. All authors have read and agreed to the published version of the manuscript.

Funding: This research was funded in part by the German Cancer Aid (Deutsche Krebshilfe: http://www.krebshilfe.de, accessed on 15 February 2023) within the Priority Program "Primary Prevention of Cancer" (Grant no. 110013).

Institutional Review Board Statement: The study was approved by the institutional ethics review boards of both the host institutions (Technical University of Munich: Reference No. 5686/13, University Hospital Cologne: Reference No. 13-053 and University Hospital Schleswig-Holstein in Kiel: Reference No. B-235/13) and the participating study centres. Trial registrations: NCT02087592; NCT02516540.

Informed Consent Statement: Written informed consent was obtained from all study subjects.

Data Availability Statement: Data are available upon reasonable request to the corresponding author.

Conflicts of Interest: The authors declare no conflict of interest.

Appendix A

Table A1. Characteristics of participants with a prior history of cancer (59.6%).

Characteristic	Mean ± SD/*n* (%)
Type of cancer	
• breast cancer (BC)	• 165 (88.7%)
• ovarian cancer (OC)	• 13 (7.0%)
• other	• 8 (4.3%)
Age at diagnosis, years	40.3 ± 9.0
Time since diagnosis, years	6.1 ± 6.9
Tumour biology of breast cancer	
• hormone receptor-positive	• 60 (36.3%)
• Her2-positive	• 6 (3.6%)
• triple-negative	• 85 (51.5%)
Breast cancer treatment	
• chemotherapy	• 138 (83.6%)
• chest radiation therapy	• 112 (67.9%)
• antihormonal treatment	• 65 (39.4%)
• HER2-targeted therapy	• 6 (3.6%)

Table A2. Associations between DII and EORTC-QC30 scores adjusted for different types of cancer treatment.

Characteristic		**Estimates (95% CI) [a]**	***p*-Value**
Quality of life (QL2) [1]	unadjusted	0.006 (−0.008; 0.021)	0.395
	adjusted for 'chemotherapy'	0.006 (−0.008; 0.021)	0.392
	adjusted for 'chest radiation therapy'	0.006 (−0.008; 0.021)	0.391
	adjusted for 'antihormonal treatment'	0.006 (−0.008; 0.021)	0.380
	adjusted for 'HER2-targeted therapy'	0.006 (−0.008; 0.020)	0.409
Physical Functioning (PF2) [1]	unadjusted	−0.004 (−0.027; 0.019)	0.752
	adjusted for 'chemotherapy'	−0.004 (−0.027; 0.020)	0.761
	adjusted for 'chest radiation therapy'	−0.004 (−0.027; 0.020)	0.753
	adjusted for 'antihormonal treatment'	−0.004 (−0.027; 0.019)	0.749
	adjusted for 'HER2-targeted therapy'	−0.004 (−0.027; 0.019)	0.869
Role Functioning (RF2) [1]	unadjusted	0.012 (0.001; 0.024)	**0.037** *
	adjusted for 'chemotherapy'	0.013 (0.001; 0.024)	**0.037** *
	adjusted for 'chest radiation therapy'	0.013 (0.001; 0.026)	**0.031** *
	adjusted for 'antihormonal treatment'	0.013 (0.001; 0.024)	**0.037** *
	adjusted for 'HER2-targeted therapy'	0.013 (0.001; 0.024)	**0.034** *
Emotional Functioning (EF) [1]	unadjusted	0.005 (−0.006; 0.015)	0.379
	adjusted for 'chemotherapy'	0.005 (−0.006; 0.015)	0.380
	adjusted for 'chest radiation therapy'	0.005 (−0.006; 0.015)	0.379
	adjusted for 'antihormonal treatment'	0.005 (−0.006; 0.015)	0.377
	adjusted for 'HER2-targeted therapy'	0.004 (−0.006; 0.015)	0.425
Cognitive Functioning (CF) [1]	unadjusted	0.015 (0.005; 0.026)	**0.005** *
	adjusted for 'chemotherapy'	0.016 (0.005; 0.026)	**0.005** *
	adjusted for 'chest radiation therapy'	0.016 (0.005; 0.026)	**0.005** *
	adjusted for 'antihormonal treatment'	0.016 (0.005; 0.026)	**0.004** *
	adjusted for 'HER2-targeted therapy'	0.015 (0.005; 0.026)	**0.004** *
Social Functioning (SF) [1]	unadjusted	0.011 (0.001; 0.020)	**0.024** *
	adjusted for 'chemotherapy'	0.011 (0.001; 0.020)	**0.024** *
	adjusted for 'chest radiation therapy'	0.011 (0.002; 0.021)	**0.021** *
	adjusted for 'antihormonal treatment'	0.011 (0.002; 0.021)	**0.023** *
	adjusted for 'HER2-targeted therapy'	0.011 (0.002; 0.021)	**0.020** *
Fatigue (FA) [1]	unadjusted	−0.010 (−0.021; 0.000)	0.057
	adjusted for 'chemotherapy'	−0.010 (−0.021; 0.000)	0.057
	adjusted for 'chest radiation therapy'	−0.010 (−0.021; 0.000)	0.058
	adjusted for 'antihormonal treatment'	−0.010 (−0.021; 0.000)	0.058
	adjusted for 'HER2-targeted therapy'	−0.010 (−0.020; 0.001)	0.070
Nausea and Vomiting (NV) [1]	unadjusted	−0.009 (−0.037; 0.020)	0.544
	adjusted for 'chemotherapy'	−0.009 (−0.038; 0.020)	0.539
	adjusted for 'chest radiation therapy'	−0.009 (−0.038; 0.020)	0.543
	adjusted for 'antihormonal treatment'	−0.009 (−0.038; 0.020)	0.540
	adjusted for 'HER2-targeted therapy'	−0.011 (−0.039; 0.018)	0.468
Pain (PA) [1]	unadjusted	−0.010 (−0.019; 0.000)	0.053
	adjusted for 'chemotherapy'	−0.010 (−0.019; 0.000)	0.054
	adjusted for 'chest radiation therapy'	−0.010 (−0.019; 0.000)	0.052
	adjusted for 'antihormonal treatment'	−0.009 (−0.019; 0.000)	0.055
	adjusted for 'HER2-targeted therapy'	−0.009 (−0.019; 0.000)	0.060
Dyspnea (DY) [1]	unadjusted	−0.010 (−0.021; −0.002)	0.089
	adjusted for 'chemotherapy'	−0.010 (−0.021; −0.002)	0.086
	adjusted for 'chest radiation therapy'	−0.010 (−0.021; −0.002)	0.089
	adjusted for 'antihormonal treatment'	−0.010 (−0.021; −0.002)	0.091
	adjusted for 'HER2-targeted therapy'	−0.010 (−0.021; −0.002)	0.087
Insomnia (SL) [1]	unadjusted	−0.007 (−0.015; 0.001)	0.088
	adjusted for 'chemotherapy'	−0.007 (−0.015; 0.001)	0.089
	adjusted for 'chest radiation therapy'	−0.007 (−0.015; 0.001)	0.088
	adjusted for 'antihormonal treatment'	−0.007 (−0.015; 0.001)	0.091
	adjusted for 'HER2-targeted therapy'	−0.007 (−0.014; 0.001)	0.094

Table A2. *Cont.*

Characteristic		Estimates (95% CI) [a]	*p*-Value
Appetite Loss (AP) [1]	unadjusted	−0.021 (−0.037; −0.006)	**0.010** *
	adjusted for 'chemotherapy'	−0.021 (−0.037; −0.005)	**0.010** *
	adjusted for 'chest radiation therapy'	−0.021 (−0.037; −0.005)	**0.010** *
	adjusted for 'antihormonal treatment'	−0.021 (−0.037; −0.005)	**0.010** *
	adjusted for 'HER2-targeted therapy'	−0.022 (−0.038; −0.006)	**0.08** *
Constipation (CO) [1]	unadjusted	0.001 (−0.012; 0.013)	0.884
	adjusted for 'chemotherapy'	0.001 (−0.012; 0.013)	0.884
	adjusted for 'chest radiation therapy'	0.001 (−0.012; 0.013)	0.886
	adjusted for 'antihormonal treatment'	0.001 (−0.012; 0.013)	0.890
	adjusted for 'HER2-targeted therapy'	0.002 (−0.012; 0.013)	0.795
Diarrhea (DI) [1]	unadjusted	−0.007 (−0.025; 0.010)	0.451
	adjusted for 'chemotherapy'	−0.007 (−0.025; 0.010)	0.413
	adjusted for 'chest radiation therapy'	−0.007 (−0.025; 0.010)	0.416
	adjusted for 'antihormonal treatment'	−0.007 (−0.026; 0.010)	0.429
	adjusted for 'HER2-targeted therapy'	−0.007 (−0.025; 0.011)	0.423
Financial Difficulties (FI) [1]	unadjusted	−0.002 (−0.012; 0.007)	0.637
	adjusted for 'chemotherapy'	−0.002 (−0.012; 0.007)	0.637
	adjusted for 'chest radiation therapy'	−0.002 (−0.012; 0.007)	0.636
	adjusted for 'antihormonal treatment'	−0.002 (−0.012; 0.007)	0.619
	adjusted for 'HER2-targeted therapy'	−0.002 (−0.012; 0.008)	0.682

* Results are statistically significant at a *p*-value of ≤0.05 (in bold); [1] only for women with a prior history of breast cancer, [a] linear regressions models.

Table A3. Associations between DII and EORTC-QC30 scores adjusted for 'time since cancer diagnosis'.

Characteristic	Unadjusted Estimate [a] [95% CI]	Unadjusted *p*-Value [a]	Adjusted Estimate [b] (96% CI)	Adjusted *p*-Value [b]
Quality of Life (QL2) [1]	0.008 [−0.005; 0.022]	0.234	0.008 (−0.048; 0.034)	0.730
Physical Functioning (PF2) [1]	0.000 [−0.021; 0.020]	0.966	−0.002 (−0.025; 0.022)	0.896
Role Functioning (RF2) [1]	0.012 [0.001; 0.023]	**0.032** *	0.014 (0.002; 0.026)	**0.026** *
Emotional Functioning (EF) [1]	0.006 [−0.003; 0.016]	0.203	0.006 (−0.004; 0.016)	0.224
Cognitive Functioning (CF) [1]	0.015 [0.005; 0.025]	**0.003** *	0.017 (0.006; 0.028)	**0.002** *
Social Functioning (SF) [1]	0.011 [0.002; 0.020]	**0.012** *	0.013 (0.004; 0.022)	**0.005** *
Fatigue (FA) [1]	−0.010 [−0.020; 0.000]	**0.046** *	−0.012 (−0.022; −0.001)	**0.031** *
Nausea and Vomiting (NV) [1]	−0.003 [−0.031; 0.024]	0.802	−0.009 (−0.038; 0.020)	0.542
Pain (PA) [1]	−0.009 [−0.018; 0.000]	0.057	−0.010 (−0.020; −0.001)	**0.034** *
Dyspnea (DY) [1]	−0.012 [−0.022; −0.001]	**0.029** *	−0.014 (−0.025; −0.002)	**0.018** *
Insomnia (SL) [1]	−0.006 [−0.014; 0.001]	0.095	−0.007 (−0.015; 0.001)	0.079
Appetite Loss (AP) [1]	−0.021 [−0.036; −0.006]	**0.007** *	−0.026 (−0.042; −0.010)	**0.002** *
Constipation (CO) [1]	0.000 [−0.011; 0.012]	0.999	−0.001 (−0.014; 0.012)	0.876
Diarrhea (DI) [1]	−0.008 [−0.025; 0.008]	0.331	−0.009 (−0.027; 0.009)	0.323
Financial Difficulties (FI) [1]	−0.006 [−0.015; 0.003]	0.217	−0.006 (−0.016; 0.004)	0.205

* Results are statistically significant at a *p*-value of ≤0.05 (in bold); [1] only for 'Diseased'; [a] univariate linear regression unadjusted for 'time since cancer diagnosis'; [b] multivariate linear regression adjusted for 'time since cancer diagnosis'.

Table A4. Associations between DII and EORTC-QC30 scores adjusted for attitudes towards, social norms about, and perceived behavioural control over healthy eating.

Characteristic	Mean ± SD	Unadjusted Estimate [a] (95% CI)	Unadjusted p-Value [a]	Adjusted Estimate [b] (96% CI)	Adjusted p-Value [b]
BKAE-AT	79.4 ± 7.0	−0.027 (−0.060; 0.005)	0.102	-	-
BKAE-SN	77.2 ± 17.3	−0.008 (−0.021; 0.006)	0.265	-	-
BKAE-PBC	85.8 ± 9.2	−0.010 (−0.034; 0.014)	0.394	-	-
Quality of life (QL2) [1]	67.7 ± 19.1	0.008 (−0.005; 0.022)	0.234	0.009 (−0.007; 0.025)	0.265
Physical Functioning (PF2) [1]	88.8 ± 12.6	0.000 (−0.021; 0.020)	0.966	−0.005 (−0.029; 0.019)	0.671
Role Functioning (RF2) [1]	79.8 ± 24.0	0.012 (0.001; 0.023)	**0.032** *	0.014 (0.001; 0.027)	**0.040** *
Emotional Functioning (EF) [1]	61.7 ± 27.3	0.006 (−0.003; 0.016)	0.203	0.008 (−0.003; 0.019)	0.148
Cognitive Functioning (CF) [1]	72.9 ± 25.9	0.015 (0.005; 0.025)	**0.003** *	0.011 (0.000; 0.023)	**0.047** *
Social Functioning (SF) [1]	72.0 ± 30.0	0.011 (0.002; 0.020)	**0.012** *	0.014 (0.004; 0.024)	**0.005** *
Fatigue (FA) [1]	33.6 ± 26.1	−0.010 (−0.020; 0.000)	**0.046** *	−0.009 (−0.020; 0.003)	0.143
Nausea and Vomiting (NV) [1]	3.9 ± 9.6	−0.003 (−0.031; 0.024)	0.802	0.001 (−0.030; 0.033)	0.928
Pain (PA) [1]	25.6 ± 28.3	−0.009 (−0.018; 0.000)	0.057	−0.010 (−0.020; 0.001)	0.086
Dyspnea (DY) [1]	16.1 ± 24.6	−0.012 (−0.022; −0.001)	**0.029** *	−0.010 (−0.022; 0.003)	0.131
Insomnia (SL) [1]	39.4 ± 35.7	−0.006 (−0.014; 0.001)	0.095	−0.004 (−0.012; 0.005)	0.411
Appetite Loss (AP) [1]	6.1 ± 16.9	−0.021 (−0.036; −0.006)	**0.007** *	−0.019 (−0.039; 0.001)	0.059
Constipation (CO) [1]	10.0 ± 22.9	0.000 (−0.011; 0.012)	0.999	0.000 (−0.014; 0.014)	0.971
Diarrhea (DI) [1]	6.6 ± 15.8	−0.008 (−0.025; 0.008)	0.331	−0.010 (−0.028; 0.008)	0.284
Financial Difficulties (FI) [1]	18.5 ± 29.0	−0.006 (−0.015; 0.003)	0.217	−0.008 (−0.018; 0.002)	0.131

* Results are statistically significant at a p-value of $\leq$0.05 (in bold); [1] only for 'Diseased'; [a] univariate linear regression unadjusted for BKAE-AT, BKAE-SN and BKAE-PBC; [b] multivariate linear regression adjusted for BKAE-AT, BKAE-SN and BKAE-PBC.

References

1. Ewertz, M.; Jensen, A.B. Late effects of breast cancer treatment and potentials for rehabilitation. *Acta Oncol.* **2011**, *50*, 187–193. [CrossRef] [PubMed]
2. Sitlinger, A.; Zafar, S.Y. Health-Related Quality of Life: The Impact on Morbidity and Mortality. *Surg. Oncol. Clin. N. Am.* **2018**, *27*, 675–684. [CrossRef] [PubMed]
3. Aaronson, N.K.; Ahmedzai, S.; Bergman, B.; Bullinger, M.; Cull, A.; Duez, N.J.; Filiberti, A.; Flechtner, H.; Fleishman, S.B.; de Haes, J.C.; et al. The European Organization for Research and Treatment of Cancer QLQ-C30: A quality-of-life instrument for use in international clinical trials in oncology. *J. Natl. Cancer Inst.* **1993**, *85*, 365–376. [CrossRef] [PubMed]
4. Ware, J.E., Jr.; Sherbourne, C.D. The MOS 36-item short-form health survey (SF-36). I. Conceptual framework and item selection. *Med. Care* **1992**, *30*, 473–483. [CrossRef] [PubMed]
5. Chantzaras, A.; Yfantopoulos, J. Association between medication adherence and health-related quality of life of patients with diabetes. *Hormones* **2022**, *21*, 691–705. [CrossRef]
6. Park, J.; Rodriguez, J.L.; O'Brien, K.M.; Nichols, H.B.; Hodgson, M.E.; Weinberg, C.R.; Sandler, D.P. Health-related quality of life outcomes among breast cancer survivors. *Cancer* **2021**, *127*, 1114–1125. [CrossRef]
7. Carver, C.S.; Scheier, M.F. Dispositional optimism. *Trends Cogn. Sci.* **2014**, *18*, 293–299. [CrossRef]
8. Imayama, I.; Alfano, C.M.; Neuhouser, M.L.; George, S.M.; Wilder Smith, A.; Baumgartner, R.N.; Baumgartner, K.B.; Bernstein, L.; Wang, C.Y.; Duggan, C.; et al. Weight, inflammation, cancer-related symptoms and health related quality of life among breast cancer survivors. *Breast Cancer Res. Treat.* **2013**, *140*, 159–176. [CrossRef]
9. Sartori, A.C.; Vance, D.E.; Slater, L.Z.; Crowe, M. The impact of inflammation on cognitive function in older adults: Implications for healthcare practice and research. *J. Neurosci. Nurs.* **2012**, *44*, 206–217. [CrossRef]
10. Alfano, C.M.; Imayama, I.; Neuhouser, M.L.; Kiecolt-Glaser, J.K.; Smith, A.W.; Meeske, K.; McTiernan, A.; Bernstein, L.; Baumgartner, K.B.; Ulrich, C.M.; et al. Fatigue, inflammation, and ω-3 and ω-6 fatty acid intake among breast cancer survivors. *J. Clin. Oncol.* **2012**, *30*, 1280–1287. [CrossRef] [PubMed]
11. Panis, C.; Pavanelli, W.R. Cytokines as Mediators of Pain-Related Process in Breast Cancer. *Mediat. Inflamm.* **2015**, *2015*, 129034. [CrossRef] [PubMed]
12. Henry, N.L.; Pchejetski, D.; A'Hern, R.; Nguyen, A.T.; Charles, P.; Waxman, J.; Li, L.; Storniolo, A.M.; Hayes, D.F.; Flockhart, D.A.; et al. Inflammatory cytokines and aromatase inhibitor-associated musculoskeletal syndrome: A case-control study. *Br. J. Cancer* **2010**, *103*, 291–296. [CrossRef]
13. Pierce, B.L.; Ballard-Barbash, R.; Bernstein, L.; Baumgartner, R.N.; Neuhouser, M.L.; Wener, M.H.; Baumgartner, K.B.; Gilliland, F.D.; Sorensen, B.E.; McTiernan, A.; et al. Elevated biomarkers of inflammation are associated with reduced survival among breast cancer patients. *J. Clin. Oncol.* **2009**, *27*, 3437–3444. [CrossRef] [PubMed]
14. Villaseñor, A.; Flatt, S.W.; Marinac, C.; Natarajan, L.; Pierce, J.P.; Patterson, R.E. Postdiagnosis C-reactive protein and breast cancer survivorship: Findings from the WHEL study. *Cancer Epidemiol. Biomark. Prev.* **2014**, *23*, 189–199. [CrossRef]
15. Hanahan, D.; Weinberg, R.A. Hallmarks of cancer: The next generation. *Cell* **2011**, *144*, 646–674. [CrossRef] [PubMed]

16. Giugliano, D.; Ceriello, A.; Esposito, K. The effects of diet on inflammation: Emphasis on the metabolic syndrome. *J. Am. Coll. Cardiol.* **2006**, *48*, 677–685. [CrossRef]
17. Cui, X.; Jin, Y.; Singh, U.P.; Chumanevich, A.A.; Harmon, B.; Cavicchia, P.; Hofseth, A.B.; Kotakadi, V.; Poudyal, D.; Stroud, B.; et al. Suppression of DNA damage in human peripheral blood lymphocytes by a juice concentrate: A randomized, double-blind, placebo-controlled trial. *Mol. Nutr. Food Res.* **2012**, *56*, 666–670. [CrossRef]
18. Roager, H.M.; Vogt, J.K.; Kristensen, M.; Hansen, L.B.S.; Ibrügger, S.; Mærkedahl, R.B.; Bahl, M.I.; Lind, M.V.; Nielsen, R.L.; Frøkiær, H.; et al. Whole grain-rich diet reduces body weight and systemic low-grade inflammation without inducing major changes of the gut microbiome: A randomised cross-over trial. *Gut* **2019**, *68*, 83–93. [CrossRef]
19. Shivappa, N.; Steck, S.E.; Hurley, T.G.; Hussey, J.R.; Hébert, J.R. Designing and developing a literature-derived, population-based dietary inflammatory index. *Public Health Nutr.* **2014**, *17*, 1689–1696. [CrossRef]
20. Shivappa, N.; Godos, J.; Hébert, J.R.; Wirth, M.D.; Piuri, G.; Speciani, A.F.; Grosso, G. Dietary Inflammatory Index and Cardiovascular Risk and Mortality—A Meta-Analysis. *Nutrients* **2018**, *10*, 200. [CrossRef]
21. Zhang, X.; Guo, Y.; Yao, N.; Wang, L.; Sun, M.; Xu, X.; Yang, H.; Sun, Y.; Li, B. Association between dietary inflammatory index and metabolic syndrome: Analysis of the NHANES 2005–2016. *Front. Nutr.* **2022**, *9*, 991907. [CrossRef]
22. Vahid, F.; Shivappa, N.; Hatami, M.; Sadeghi, M.; Ameri, F.; Jamshidi Naeini, Y.; Hebert, J.R.; Davoodi, S.H. Association between Dietary Inflammatory Index (DII) and Risk of Breast Cancer: A Case-Control Study. *Asian Pac. J. Cancer Prev.* **2018**, *19*, 1215–1221. [CrossRef] [PubMed]
23. Shivappa, N.; Hébert, J.R.; Paddock, L.E.; Rodriguez-Rodriguez, L.; Olson, S.H.; Bandera, E.V. Dietary inflammatory index and ovarian cancer risk in a New Jersey case-control study. *Nutrition* **2018**, *46*, 78–82. [CrossRef]
24. Vahid, F.; Shivappa, N.; Faghfoori, Z.; Khodabakhshi, A.; Zayeri, F.; Hebert, J.R.; Davoodi, S.H. Validation of a Dietary Inflammatory Index (DII) and Association with Risk of Gastric Cancer: A Case-Control Study. *Asian Pac. J. Cancer Prev.* **2018**, *19*, 1471–1477. [CrossRef] [PubMed]
25. Shivappa, N.; Godos, J.; Hébert, J.R.; Wirth, M.D.; Piuri, G.; Speciani, A.F.; Grosso, G. Dietary Inflammatory Index and Colorectal Cancer Risk—A Meta-Analysis. *Nutrients* **2017**, *9*, 1043. [CrossRef] [PubMed]
26. Golmohammadi, M.; Kheirouri, S.; Ebrahimzadeh Attari, V.; Moludi, J.; Sulistyowati, R.; Nachvak, S.M.; Mostafaei, R.; Mansordehghan, M. Is there any association between dietary inflammatory index and quality of life? A systematic review. *Front Nutr.* **2022**, *9*, 1067468. [CrossRef]
27. Toopchizadeh, V.; Dolatkhah, N.; Aghamohammadi, D.; Rasouli, M.; Hashemian, M. Dietary inflammatory index is associated with pain intensity and some components of quality of life in patients with knee osteoarthritis. *BMC Res. Notes* **2020**, *13*, 448. [CrossRef]
28. Yaseri, M.; Alipoor, E.; Hafizi, N.; Maghsoudi-Nasab, S.; Shivappa, N.; Hebert, J.R.; Hosseinzadeh-Attar, M.J. Dietary Inflammatory Index Is a Better Determinant of Quality of Life Compared to Obesity Status in Patients with Hemodialysis. *J. Ren. Nutr.* **2021**, *31*, 313–319. [CrossRef] [PubMed]
29. Hodge, A.M.; Bassett, J.K.; Shivappa, N.; Hébert, J.R.; English, D.R.; Giles, G.G.; Severi, G. Dietary inflammatory index, Mediterranean diet score, and lung cancer: A prospective study. *Cancer Causes Control* **2016**, *27*, 907–917. [CrossRef] [PubMed]
30. Willett, W.C.; Sacks, F.; Trichopoulou, A.; Drescher, G.; Ferro-Luzzi, A.; Helsing, E.; Trichopoulos, D. Mediterranean diet pyramid: A cultural model for healthy eating. *Am. J. Clin. Nutr.* **1995**, *61*, 1402s–1406s. [CrossRef]
31. Martínez-González, M.A.; Gea, A.; Ruiz-Canela, M. The Mediterranean Diet and Cardiovascular Health. *Circ. Res.* **2019**, *124*, 779–798. [CrossRef] [PubMed]
32. Di Daniele, N.; Noce, A.; Vidiri, M.F.; Moriconi, E.; Marrone, G.; Annicchiarico-Petruzzelli, M.; D'Urso, G.; Tesauro, M.; Rovella, V.; De Lorenzo, A. Impact of Mediterranean diet on metabolic syndrome, cancer and longevity. *Oncotarget* **2017**, *8*, 8947–8979. [CrossRef] [PubMed]
33. Bonaccio, M.; Di Castelnuovo, A.; Bonanni, A.; Costanzo, S.; De Lucia, F.; Pounis, G.; Zito, F.; Donati, M.B.; de Gaetano, G.; Iacoviello, L. Adherence to a Mediterranean diet is associated with a better health-related quality of life: A possible role of high dietary antioxidant content. *BMJ Open* **2013**, *3*, e003003. [CrossRef] [PubMed]
34. Galilea-Zabalza, I.; Buil-Cosiales, P.; Salas-Salvadó, J.; Toledo, E.; Ortega-Azorín, C.; Díez-Espino, J.; Vázquez-Ruiz, Z.; Zomeño, M.D.; Vioque, J.; Martínez, J.A.; et al. Mediterranean diet and quality of life: Baseline cross-sectional analysis of the PREDIMED-PLUS trial. *PLoS ONE* **2018**, *13*, e0198974. [CrossRef]
35. Porciello, G.; Montagnese, C.; Crispo, A.; Grimaldi, M.; Libra, M.; Vitale, S.; Palumbo, E.; Pica, R.; Calabrese, I.; Cubisino, S.; et al. Mediterranean diet and quality of life in women treated for breast cancer: A baseline analysis of DEDiCa multicentre trial. *PLoS ONE* **2020**, *15*, e0239803. [CrossRef] [PubMed]
36. WHO. *WHO Report on Cancer: Setting Priorities, Investing Wisely and Providing Care for All*; World Health Organization: Geneva, Switzerland, 2020.
37. Kuchenbaecker, K.B.; Hopper, J.L.; Barnes, D.R.; Phillips, K.A.; Mooij, T.M.; Roos-Blom, M.J.; Jervis, S.; van Leeuwen, F.E.; Milne, R.L.; Andrieu, N.; et al. Risks of Breast, Ovarian, and Contralateral Breast Cancer for BRCA1 and BRCA2 Mutation Carriers. *JAMA* **2017**, *317*, 2402–2416. [CrossRef]
38. Shapiro, C.L.; Recht, A. Side Effects of Adjuvant Treatment of Breast Cancer. *N. Engl. J. Med.* **2001**, *344*, 1997–2008. [CrossRef]
39. Bouillon, K.; Haddy, N.; Delaloge, S.; Garbay, J.R.; Garsi, J.P.; Brindel, P.; Mousannif, A.; Lê, M.G.; Labbe, M.; Arriagada, R.; et al. Long-term cardiovascular mortality after radiotherapy for breast cancer. *J. Am. Coll. Cardiol.* **2011**, *57*, 445–452. [CrossRef]

40. Tao, J.J.; Visvanathan, K.; Wolff, A.C. Long term side effects of adjuvant chemotherapy in patients with early breast cancer. *Breast* **2015**, *24* (Suppl. S2), S149–S153. [CrossRef]
41. Ramchand, S.K.; Cheung, Y.M.; Yeo, B.; Grossmann, M. The effects of adjuvant endocrine therapy on bone health in women with breast cancer. *J. Endocrinol.* **2019**, *241*, R111–R124. [CrossRef]
42. Barish, R.; Gates, E.; Barac, A. Trastuzumab-Induced Cardiomyopathy. *Cardiol. Clin.* **2019**, *37*, 407–418. [CrossRef] [PubMed]
43. Schmidt, M.E.; Scherer, S.; Wiskemann, J.; Steindorf, K. Return to work after breast cancer: The role of treatment-related side effects and potential impact on quality of life. *Eur. J. Cancer Care* **2019**, *28*, e13051. [CrossRef]
44. Hickey, I.; Jha, S.; Wyld, L. The psychosexual effects of risk-reducing bilateral salpingo-oophorectomy in female BRCA1/2 mutation carriers: A systematic review of qualitative studies. *Gynecol. Oncol.* **2021**, *160*, 763–770. [CrossRef]
45. D'Alonzo, M.; Piva, E.; Pecchio, S.; Liberale, V.; Modaffari, P.; Ponzone, R.; Biglia, N. Satisfaction and Impact on Quality of Life of Clinical and Instrumental Surveillance and Prophylactic Surgery in BRCA-mutation Carriers. *Clin. Breast Cancer* **2018**, *18*, e1361–e1366. [CrossRef] [PubMed]
46. Kassianos, A.P.; Raats, M.M.; Gage, H.; Peacock, M. Quality of life and dietary changes among cancer patients: A systematic review. *Qual. Life Res.* **2015**, *24*, 705–719. [CrossRef]
47. Kiechle, M.; Engel, C.; Berling, A.; Hebestreit, K.; Bischoff, S.; Dukatz, R.; Gerber, W.D.; Siniatchkin, M.; Pfeifer, K.; Grill, S.; et al. Lifestyle intervention in BRCA1/2 mutation carriers: Study protocol for a prospective, randomized, controlled clinical feasibility trial (LIBRE-1 study). *Pilot Feasibil. Stud.* **2016**, *2*, 74. [CrossRef]
48. Kiechle, M.; Engel, C.; Berling, A.; Hebestreit, K.; Bischoff, S.C.; Dukatz, R.; Siniatchkin, M.; Pfeifer, K.; Grill, S.; Yahiaoui-Doktor, M.; et al. Effects of lifestyle intervention in BRCA1/2 mutation carriers on nutrition, BMI, and physical fitness (LIBRE study): Study protocol for a randomized controlled trial. *Trials* **2016**, *17*, 368. [CrossRef]
49. Schröder, H.; Fitó, M.; Estruch, R.; Martínez-González, M.A.; Corella, D.; Salas-Salvadó, J.; Lamuela-Raventós, R.; Ros, E.; Salaverría, I.; Fiol, M.; et al. A short screener is valid for assessing Mediterranean diet adherence among older Spanish men and women. *J. Nutr.* **2011**, *141*, 1140–1145. [CrossRef] [PubMed]
50. Martínez-González, M.; Corella, D.; Salas-Salvadó, J.; Ros, E.; Covas, M.I.; Fiol, M.; Wärnberg, J.; Arós, F.; Ruíz-Gutiérrez, V.; Lamuela-Raventós, R.M.; et al. Cohort profile: Design and methods of the PREDIMED study. *Int. J. Epidemiol.* **2012**, *41*, 377–385. [CrossRef]
51. Hebestreit, K.; Yahiaoui-Doktor, M.; Engel, C.; Vetter, W.; Siniatchkin, M.; Erickson, N.; Halle, M.; Kiechle, M.; Bischoff, S.C. Validation of the German version of the Mediterranean Diet Adherence Screener (MEDAS) questionnaire. *BMC Cancer* **2017**, *17*, 341. [CrossRef]
52. Seethaler, B.; Basrai, M.; Vetter, W.; Lehnert, K.; Engel, C.; Siniatchkin, M.; Halle, M.; Kiechle, M.; Bischoff, S.C. Fatty acid profiles in erythrocyte membranes following the Mediterranean diet—Data from a multicenter lifestyle intervention study in women with hereditary breast cancer (LIBRE). *Clin. Nutr.* **2020**, *39*, 2389–2398. [CrossRef]
53. Boeing, H.; Bohlscheid-Thomas, S.; Voss, S.; Schneeweiss, S.; Wahrendorf, J. The relative validity of vitamin intakes derived from a food frequency questionnaire compared to 24-hour recalls and biological measurements: Results from the EPIC pilot study in Germany. European Prospective Investigation into Cancer and Nutrition. *Int. J. Epidemiol.* **1997**, *26* (Suppl. S1), S82–S90. [CrossRef]
54. Bohlscheid-Thomas, S.; Hoting, I.; Boeing, H.; Wahrendorf, J. Reproducibility and relative validity of energy and macronutrient intake of a food frequency questionnaire developed for the German part of the EPIC project. European Prospective Investigation into Cancer and Nutrition. *Int. J. Epidemiol.* **1997**, *26* (Suppl. S1), S71–S81. [CrossRef]
55. Gholamalizadeh, M.; Afsharfar, M.; Fathi, S.; Tajadod, S.; Mohseni, G.K.; Shekari, S.; Vahid, F.; Doaei, S.; Shafaei Kachaei, H.; Majidi, N.; et al. Relationship between breast cancer and dietary inflammatory index; a case-control study. *Clin. Nutr. ESPEN* **2022**, *51*, 353–358. [CrossRef]
56. Jalali, S.; Shivappa, N.; Hébert, J.R.; Heidari, Z.; Hekmatdoost, A.; Rashidkhani, B. Dietary Inflammatory Index and Odds of Breast Cancer in a Case-Control Study from Iran. *Nutr. Cancer* **2018**, *70*, 1034–1042. [CrossRef]
57. Huang, W.Q.; Mo, X.F.; Ye, Y.B.; Shivappa, N.; Lin, F.Y.; Huang, J.; Hébert, J.R.; Yan, B.; Zhang, C.X. A higher Dietary Inflammatory Index score is associated with a higher risk of breast cancer among Chinese women: A case-control study. *Br. J. Nutr.* **2017**, *117*, 1358–1367. [CrossRef] [PubMed]
58. Glaesmer, H.; Hoyer, J.; Klotsche, J.; Herzberg, P.Y. Die deutsche Version des Life-Orientation-Tests (LOT-R) zum dispositionellen Optimismus und Pessimismus. *Z. Gesundh.* **2008**, *16*, 26–31. [CrossRef]
59. Flanders, N.A.; Fishbein, M.; Ajzen, I. *Belief, Attitude, Intention, and Behavior: An Introduction to Theory and Research*; Addison-Wesley Publishing Company: Boston, MA, USA, 1975.
60. Berling-Ernst, A.; Yahiaoui-Doktor, M.; Kiechle, M.; Engel, C.; Lammert, J.; Grill, S.; Dukatz, R.; Rhiem, K.; Baumann, F.T.; Bischoff, S.C.; et al. Predictors of cardiopulmonary fitness in cancer-affected and -unaffected women with a pathogenic germline variant in the genes BRCA1/2 (LIBRE-1). *Sci. Rep.* **2022**, *12*, 2907. [CrossRef]
61. Lee, E.; McKean-Cowdin, R.; Ma, H.; Spicer, D.V.; Van Den Berg, D.; Bernstein, L.; Ursin, G. Characteristics of triple-negative breast cancer in patients with a BRCA1 mutation: Results from a population-based study of young women. *J. Clin. Oncol.* **2011**, *29*, 4373–4380. [CrossRef] [PubMed]

62. Mavaddat, N.; Barrowdale, D.; Andrulis, I.L.; Domchek, S.M.; Eccles, D.; Nevanlinna, H.; Ramus, S.J.; Spurdle, A.; Robson, M.; Sherman, M.; et al. Pathology of breast and ovarian cancers among BRCA1 and BRCA2 mutation carriers: Results from the Consortium of Investigators of Modifiers of BRCA1/2 (CIMBA). *Cancer Epidemiol. Biomark. Prev.* **2012**, *21*, 134–147. [CrossRef] [PubMed]
63. Guzmán-Arocho, Y.D.; Rosenberg, S.M.; Garber, J.E.; Vardeh, H.; Poorvu, P.D.; Ruddy, K.J.; Kirkner, G.; Snow, C.; Tamimi, R.M.; Peppercorn, J.; et al. Clinicopathological features and BRCA1 and BRCA2 mutation status in a prospective cohort of young women with breast cancer. *Br. J. Cancer* **2022**, *126*, 302–309. [CrossRef] [PubMed]
64. Davies, C.; Pan, H.; Godwin, J.; Gray, R.; Arriagada, R.; Raina, V.; Abraham, M.; Medeiros Alencar, V.H.; Badran, A.; Bonfill, X.; et al. Long-term effects of continuing adjuvant tamoxifen to 10 years versus stopping at 5 years after diagnosis of oestrogen receptor-positive breast cancer: ATLAS, a randomised trial. *Lancet* **2013**, *381*, 805–816. [CrossRef] [PubMed]
65. Gray, R.G.; Rea, D.; Handley, K.; Bowden, S.J.; Perry, P.; Earl, H.M.; Poole, C.J.; Bates, T.; Chetiyawardana, S.; Dewar, J.A. *aTTom: Long-Term Effects of Continuing Adjuvant Tamoxifen to 10 Years versus Stopping at 5 Years in 6953 Women with Early Breast Cancer*; American Society of Clinical Oncology: Alexandria, VA, USA, 2013.
66. Burstein, H.; Griggs, J.; Prestrud, A.; Temin, S. American society of clinical oncology clinical practice guideline update on adjuvant endocrine therapy for women with hormone receptor-positive breast cancer. *J. Oncol. Pract.* **2010**, *5*, 243–246. [CrossRef] [PubMed]
67. Condorelli, R.; Vaz-Luis, I. Managing side effects in adjuvant endocrine therapy for breast cancer. *Expert Rev. Anticancer Ther.* **2018**, *18*, 1101–1112. [CrossRef] [PubMed]
68. Maunsell, E.; Drolet, M.; Brisson, J.; Robert, J.; Deschênes, L. Dietary change after breast cancer: Extent, predictors, and relation with psychological distress. *J. Clin. Oncol.* **2002**, *20*, 1017–1025. [CrossRef]
69. Leitlinienprogramm Onkologie (Deutsche Krebsgesellschaft, Deutsche Krebshilfe, AWMF): S3-Leitlinie Früherkennung, Diagnose, Therapie und Nachsorge des Mammakarzinoms, Version 4.0, 2017 AWMF Registernummer: 032-045OL. Available online: http://www.leitlinienprogramm-onkologie.de/leitlinien/mammakarzinom/ (accessed on 5 March 2023).
70. Deutsche Gesellschaft für Ernährung e. V., DGE-Ernährungskreis. 2019. Available online: https://www.dge-ernaehrungskreis.de/orientierungswerte/ (accessed on 5 March 2023).
71. Scheier, M.F.; Carver, C.S. Optimism, coping, and health: Assessment and implications of generalized outcome expectancies. *Health Psychol.* **1985**, *4*, 219–247. [CrossRef]
72. Tindle, H.A.; Chang, Y.F.; Kuller, L.H.; Manson, J.E.; Robinson, J.G.; Rosal, M.C.; Siegle, G.J.; Matthews, K.A. Optimism, cynical hostility, and incident coronary heart disease and mortality in the Women's Health Initiative. *Circulation* **2009**, *120*, 656–662. [CrossRef]
73. Kim, E.S.; Hagan, K.A.; Grodstein, F.; DeMeo, D.L.; De Vivo, I.; Kubzansky, L.D. Optimism and Cause-Specific Mortality: A Prospective Cohort Study. *Am. J. Epidemiol.* **2017**, *185*, 21–29. [CrossRef]
74. Friedman, L.C.; Kalidas, M.; Elledge, R.; Chang, J.; Romero, C.; Husain, I.; Dulay, M.F.; Liscum, K.R. Optimism, social support and psychosocial functioning among women with breast cancer. *Psychooncology* **2006**, *15*, 595–603. [CrossRef]
75. Thieme, M.; Einenkel, J.; Zenger, M.; Hinz, A. Optimism, pessimism and self-efficacy in female cancer patients. *Jpn. J. Clin. Oncol.* **2017**, *47*, 849–855. [CrossRef]
76. Boehm, J.K.; Williams, D.R.; Rimm, E.B.; Ryff, C.; Kubzansky, L.D. Association between optimism and serum antioxidants in the midlife in the United States study. *Psychosom. Med.* **2013**, *75*, 2–10. [CrossRef]
77. Scheier, M.F.; Carver, C.S. Dispositional optimism and physical health: A long look back, a quick look forward. *Am. Psychol.* **2018**, *73*, 1082–1094. [CrossRef] [PubMed]
78. Peters, M.L.; Flink, I.K.; Boersma, K.; Linton, S.J. Manipulating optimism: Can imagining a best possible self be used to increase positive future expectancies? *J. Posit. Psychol.* **2010**, *5*, 204–211. [CrossRef]
79. Meevissen, Y.M.C.; Peters, M.L.; Alberts, H.J.E.M. Become more optimistic by imagining a best possible self: Effects of a two week intervention. *J. Behav. Ther. Exp. Psychiatry* **2011**, *42*, 371–378. [CrossRef] [PubMed]
80. Bruno, E.; Manoukian, S.; Venturelli, E.; Oliverio, A.; Rovera, F.; Iula, G.; Morelli, D.; Peissel, B.; Azzolini, J.; Roveda, E.; et al. Adherence to Mediterranean Diet and Metabolic Syndrome in BRCA Mutation Carriers. *Integr. Cancer Ther.* **2018**, *17*, 153–160. [CrossRef] [PubMed]
81. Ford, E.S.; Li, C. Metabolic syndrome and health-related quality of life among U.S. adults. *Ann. Epidemiol.* **2008**, *18*, 165–171. [CrossRef]
82. Tziallas, D.; Kastanioti, C.; Kostapanos, M.S.; Skapinakis, P.; Elisaf, M.S.; Mavreas, V. The impact of the metabolic syndrome on health-related quality of life: A cross-sectional study in Greece. *Eur. J. Cardiovasc. Nurs.* **2012**, *11*, 297–303. [CrossRef]
83. Lee, H.; Kim, J.S. Factors Affecting the Health-Related Quality of Life of Cancer Survivors According to Metabolic Syndrome. *Cancer Nurs.* 2022; *adhead of print*. [CrossRef]
84. Cohen, B.E.; Panguluri, P.; Na, B.; Whooley, M.A. Psychological risk factors and the metabolic syndrome in patients with coronary heart disease: Findings from the Heart and Soul Study. *Psychiatry Res.* **2010**, *175*, 133–137. [CrossRef]
85. Bundesamt, S. Bevölkerung Nach FamilienStand 2011 bis 2021 Deutschland. 2022. Available online: https://www.destatis.de/DE/Themen/Gesellschaft-Umwelt/Bevoelkerung/Bevoelkerungsstand/Tabellen/familienstand-jahre-5.html (accessed on 15 February 2023).
86. Bundesamt, S. Zusammengefasste Geburtenziffer nach Kalenderjahren. Available online: https://www.destatis.de/DE/Themen/Gesellschaft-Umwelt/Bevoelkerung/Geburten/Tabellen/geburtenziffer.html (accessed on 15 February 2023).

87. Bundesamt, S. Bevölkerung im Alter von 15 Jahren und Mehr Nach Allgemeinen und Beruflichen Bildungsabschlüssen nach Jahren. Available online: https://www.destatis.de/DE/Themen/Gesellschaft-Umwelt/Bildung-Forschung-Kultur/Bildungsstand/Tabellen/bildungsabschluss.html (accessed on 15 February 2023).
88. Bundesamt, S. Statistik zu Einkommen und Lebensbedingungen (Mikrozensus-Unterstichprobe zu Einkommen und Lebensbedingungen), Fachserie 15, Reihe 3, EU-SILC 2021. Statistisches Bundesamt: Wiesbaden, Germany, 2022; p. 23. Available online: https://www.destatis.de/DE/Themen/Gesellschaft-Umwelt/Einkommen-Konsum-Lebensbedingungen/Lebensbedingungen-Armutsgefaehrdung/Publikationen/Downloads-Lebensbedingungen/einkommen-lebensbedingungen-2150300217004.html (accessed on 15 February 2023).
89. Grill, S.; Yahiaoui-Doktor, M.; Dukatz, R.; Lammert, J.; Ullrich, M.; Engel, C.; Pfeifer, K.; Basrai, M.; Siniatchkin, M.; Schmidt, T.; et al. Smoking and physical inactivity increase cancer prevalence in BRCA-1 and BRCA-2 mutation carriers: Results from a retrospective observational analysis. *Arch. Gynecol. Obs.* **2017**, *296*, 1135–1144. [CrossRef]

 nutrients

Article

Human Milk-Derived Levels of let-7g-5p May Serve as a Diagnostic and Prognostic Marker of Low Milk Supply in Breastfeeding Women

Steven D. Hicks [1], Desirae Chandran [1], Alexandra Confair [1], Anna Ward [2] and Shannon L. Kelleher [2,*]

[1] Department of Pediatrics, Penn State College of Medicine, Hershey, PA 17033, USA
[2] Department of Biomedical and Nutritional Sciences, University of Massachusetts Lowell, Lowell, MA 01854, USA
* Correspondence: shannon_kelleher@uml.edu

Abstract: Low milk supply (LMS) is associated with early breastfeeding cessation; however, the biological underpinnings in the mammary gland are not understood. MicroRNAs (miRNAs) are small non-coding RNAs that post-transcriptionally downregulate gene expression, and we hypothesized the profile of miRNAs secreted into milk reflects lactation performance. Longitudinal changes in milk miRNAs were measured using RNAseq in women with LMS (n = 47) and adequate milk supply (AMS; n = 123). Relationships between milk miRNAs, milk supply, breastfeeding outcomes, and infant weight gain were assessed, and interactions between milk miRNAs, maternal diet, smoking status, and BMI were determined. Women with LMS had lower milk volume (p = 0.003), were more likely to have ceased breast feeding by 24 wks (p = 0.0003) and had infants with a lower mean weight-for-length z-score (p = 0.013). Milk production was significantly associated with milk levels of miR-16-5p (R = −0.14, adj p = 0.044), miR-22-3p (R = 0.13, adj p = 0.044), and let-7g-5p (R = 0.12, adj p = 0.046). Early milk levels of let-7g-5p were significantly higher in mothers with LMS (adj p = 0.0025), displayed an interaction between lactation stage and milk supply (p < 0.001), and were negatively related to fruit intake (p = 0.015). Putative targets of let-7g-5p include genes important to hormone signaling, RNA regulation, ion transport, and the extracellular matrix, and down-regulation of two targets (PRLR and IGF2BP1/IMP1) was confirmed in mammary cells overexpressing let-7g-5p in vitro. Our data provide evidence that milk-derived miRNAs reflect lactation performance in women and warrant further investigation to assess their utility for predicting LMS risk and early breastfeeding cessation.

Keywords: lactation; human milk; breastfeeding; miRNAs; low milk supply

Citation: Hicks, S.D.; Chandran, D.; Confair, A.; Ward, A.; Kelleher, S.L. Human Milk-Derived Levels of let-7g-5p May Serve as a Diagnostic and Prognostic Marker of Low Milk Supply in Breastfeeding Women. *Nutrients* **2023**, *15*, 567. https://doi.org/10.3390/nu15030567

Academic Editors: Birgit-Christiane Zyriax and Nataliya Makarova

Received: 1 January 2023
Revised: 13 January 2023
Accepted: 18 January 2023
Published: 21 January 2023

1. Introduction

Low milk supply is associated with early breastfeeding cessation. Nearly 40% of breastfeeding women cite concerns over low milk supply as a primary reason for not meeting their breastfeeding goals [1]. The etiology of suboptimal lactation is clearly multi-faceted. Abnormal breast conditions and previous breast surgeries [2] are structural factors that contribute to suboptimal lactation. In addition, numerous social, psychological, and behavioral factors are associated with early breastfeeding cessation [3–6]. Several studies reported associations between maternal metabolic conditions such as malnutrition [7], excessive maternal fat mass [8], and gestational diabetes [9] and milk supply. Moreover, we and others determined that genetic variations in genes critical for mammary gland function (i.e., prolactin (*PRL*) [10], prolactin receptor (*PRLR*) [11], ZnT2 (*SLC30A2*) [12], and milk fat globule-epidermal factor 8 protein (*MFGE8*) [13]) are associated with the ability to produce milk, providing evidence that biological factors are also responsible for low milk supply.

MicroRNAs (miRNAs) are small, non-coding nucleic acid sequences (~22 nucleotides) that post-transcriptionally regulate gene expression by binding to specific mRNA targets and either inhibiting translation or promoting degradation, thus affecting corresponding protein expression. Strong evidence indicates miRNAs control normal physiology and miRNAs have been associated with various pathological states [14]. MiRNAs are found in all bodily fluids, including saliva, plasma, and urine [15,16], although milk is one of the richest sources of miRNAs in humans, containing ~1400 mature miRNAs [17]. Many of these miRNAs are released by secreting mammary epithelial cells (MECs) in exosomes. Mammary gland miRNA profiles differentiate discrete stages of mammary gland development in rodents and dairy animals, and parallel gene expression required for ion transport, G protein signaling, translation, and intracellular protein transport, and oxidative phosphorylation [18–20]. Several previous studies suggested the profile of human milk miRNAs may reflect breast function [17,21]. Consistent with this hypothesis, we and others [22,23] previously showed that several maternal factors associated with milk production and composition (i.e., diet, genotype, preterm birth, and stage of lactation) are associated with the profile of milk-derived miRNAs, implicating milk miRNAs as bioreporters of lactation performance in humans [24]. Here we hypothesized that specific milk-derived miRNAs are associated with low milk supply, and using a genome-wide computational approach, identified milk-derived miRNAs that may serve as novel genetic drivers or reporters of low milk supply and confirmed regulation of two targets in MECs in vitro.

2. Materials and Methods

2.1. Participants

This longitudinal cohort study involved a convenience sample of 221 women, ages 19–42 years. Mothers of full-term, singleton infants (37–42 weeks gestation) who planned to breastfeed at least six months were eligible. Exclusion criteria and the study design were previously reported [13]. Participants were dichotomized into two groups: low milk supply (LMS) or adequate milk supply (AMS) as we have previously reported. Ultimately, 47 women with LMS were compared against mothers with AMS (n = 123) who either maintained exclusive breast feeding throughout the duration of the study or reported formula introduction for reasons other than "decreased or low breastmilk production" (Figure 1).

2.2. Participant Characteristics

Participant characteristics were collected via electronic surveys administered by research staff at enrollment. Survey responses were confirmed through review of the medical record where possible. The following medical and demographic characteristics were collected: maternal age, maternal race, parity, pre-pregnancy body mass index, tobacco use, maternal educational level, marital status, duration of previous breastfeeding, infant gestational age, infant sex, and infant birth weight as previously reported [13]. No women reported the use of galactagogue supplements.

Maternal nutrition was assessed alongside each milk collection (at 1, 4, 16 wks) through electronic administration of the Dietary Screener Questionnaire (DSQ), developed as part of the National Health and Nutrition Examination Survey (NHANES) [25]. Published guidelines were used to compute consumption of fruit, vegetables, dairy, added sugars, and calcium. Infant feeding characteristics were collected through electronic administration of the modified Infant Feeding Practices survey (IFP), and as we have previously reported, milk production was approximated based on maternal report of pumping volumes and infant feeding practices [13]. Mothers who were unable to estimate milk production volumes (n = 57/340 data-points; 16%) were excluded from analyses of milk production. For mothers who reported infant weaning (or failed to provide a milk sample due to low milk supply; n = 25), a milk production volume of 0 oz/day was assigned. Infant weight and length were abstracted from the medical record at birth and four weeks post-delivery. For each infant, the change in weight-for-length Z-score was calculated at each time point using standardized curves from the World Health Organization.

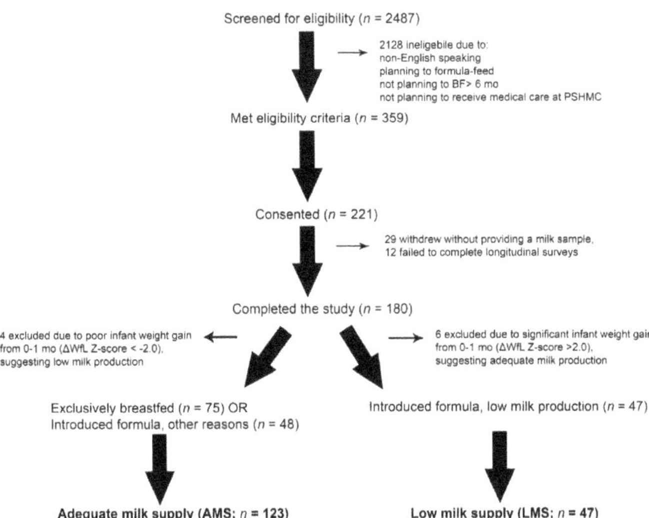

Figure 1. CONSORT Diagram. Research staff screened 2487 mother–infant dyads, approached 359 eligible dyads, and obtained consent from 221 dyads. There were 180 mothers that provided at least one milk sample and completed sufficient longitudinal surveys to determine whether they experienced low milk supply (LMS) or adequate milk supply (AMS). There were 4 or 6 mothers excluded from each group for excessive infant weight gain or weight loss (defined as a change in weight-for-length (WfL) Z-score > 2.0), which suggested milk production may have been over- or under-estimated. This left 47 mothers with LMS, and 123 mothers with AMS. Of the 123 mothers with AMS, 48 introduced formula into their infant's diet prior to 12 months for reasons other than concerns about milk supply.

2.3. Milk Collection

One milk sample was collected from each mother at 1, 4, and 16 wks post-delivery (or until breastfeeding ceased). The 170 mothers provided 453 milk samples: 170 samples in the first wk post-delivery, 158 samples 4 wks post-delivery, and 125 samples 16 wks post-delivery. Maternal milk (1–5 mL) was manually expressed from a sterilized nipple surface into RNAse-free tubes prior to feeding (i.e., fore-milk). Foremilk samples were exclusively collected from the same breast to minimize confounding impacts of fore- and hind-milk differences [17]. Samples were immediately transferred to −20 °C, underwent one freeze–thaw cycle for aliquoting, and stored at −80 °C.

2.4. RNA Processing

Milk was skimmed by centrifugation for 20 min at 4 °C at 800 rpm and the lipid fraction was used for RNA extraction [21,26]. RNA was purified, sequenced, and analyzed as previously described [26,27]. The 30 miRNA features with the most robust expression (present in raw counts >10 in all 453 samples) were quantile normalized and mean-center scaled.

2.5. Cell Culture and let-7g-5p Transfection

Mouse mammary epithelial (HC11) cells were gifted by Dr. Jeffrey Rosen (Department of Molecular and Cellular Biology, Baylor College of Medicine, Houston, TX, USA) and used with permission of Dr. Bernd Groner (Institute for Biomedical Research, Frankfurt, Germany). Cells were maintained in RPMI 1640 growth medium supplemented with 10% fetal bovine serum, 5 µg/mL insulin (Sigma-Aldrich, Burlington, MA, USA), 10 ng/mL EGF (EMD Millipore, Burlington, MA, USA) and 1% Penicillin-Streptomycin (Sigma-Aldrich, Burlington, MA, USA). Cells were plated in antibiotic-free growth medium in 6-well plates (2.7×10^5 cells/well) and cultured for 2 d. Cells were transfected for 5 h at 50%

confluency with 10 nM or 30 nM of hsa-let-7g-mirVana miRNA mimic in antibiotic-free Opti-MEM reduced serum medium (ThermoFisher, Waltham, MA, USA) using Lipofectamine RNAiMAX (ThermoFisher, Waltham, MA, USA). Opti-MEM was replaced with antibiotic-free growth medium and total protein was extracted from transfected cells 25 h later for immunoblotting.

2.6. Immunoblotting

Cells were washed with ice-cold PBS, scraped into a RIPA lysis buffer containing protease/phosphatase inhibitors and vortexed every 5 min for 30 min until well-solubilized. The cell extract was centrifuged for 20 min at 12,000× g (4 °C) to pellet cellular debris and the supernatant was collected. Protein concentration was determined using the Pierce BCA protein assay kit (ThermoFisher, Waltham, MA, USA). Protein (20 µg) was solubilized in Laemmli sample buffer containing dithiothreitol (DTT; 100 mM) at 95 °C for 5 min. Proteins were separated by SDS-PAGE for 1 h (200 V) and transferred to nitrocellulose for 1 h (100 V). Immunoblotting was performed with prolactin receptor (PRLR) polyclonal antibody (1:5000; ThermoFisher, Waltham, MA, USA) and anti-rabbit horseradish peroxidase (HRP)-conjugated IgG antibody (1:10,000) or anti-IGF2BP1/IMP1 antibody (1:1000; generous gift from Dr. S. Andres, Oregon Health and Science University, Portland, OR, USA) and donkey anti-goat HRP-conjugated antibody (1:5000). The membrane was stripped with Restore Plus Western Blot Stripping Buffer (ThermoFisher, Waltham, MA, USA) and re-probed using monoclonal anti-beta actin antibody (1:8000) and anti-mouse HRP-conjugated IgG antibody (1:20,000). Proteins were visualized by chemiluminescence using SuperSignal West Femto Maximum Sensitivity Substrate (ThermoFisher, Waltham, MA, USA). Relative band intensity was quantified using ImageJ software.

2.7. Statistical Methods

Medical and demographic traits were compared between LMS and AMS groups using a student's t-test or chi-square test, as appropriate. Levels of miRNAs in the initial milk sample (at 1 wk) were compared between LMS and AMS groups using a Wilcoxon Rank Sum test. p-values were adjusted for multiple testing using the false detection rate (FDR) method, and values less than 0.05 were considered significant. Candidate miRNAs identified on Wilcoxon Rank Sum testing underwent the following secondary analyses: (1) Relationships between milk miRNA levels and milk production (oz/day) were assessed with Spearman Rank Correlation testing; (2) longitudinal changes in milk levels of miRNA candidates were assessed across 1, 4, and 16 wks post-delivery using a linear mixed model fit by restricted maximum likelihood (with each miRNA as the dependent variable, participant ID as the clustering variable, and maternal characteristics as covariates). Interactions between LMS/AMS group and time (wks post-delivery) were assessed with fixed effects omnibus tests; (3) the effects of modifiable maternal characteristics (nutrition, BMI, tobacco use) on milk levels of candidate miRNAs were assessed with a mixed model (with miRNA level was the dependent variable, participant ID as the clustering variable, and maternal characteristics as covariates); (4) the ability of candidate miRNA levels to differentiate participants at risk for LMS was assessed relative to maternal factors with a hierarchical logistic regression. LMS/AMS group served as the dependent variable, and candidate miRNAs without collinearity were used in a feed-forward model building approach. Sensitivity, specificity, and area under a receiver operator characteristic curve were reported; (5) physiologic functions of the candidate miRNAs were explored in DIANA miRPath v3 [28], through identification of putative mRNA targets (Tarbase algorithm). Enrichment of Kyoto Encyclopedia Genes and Genomes (KEGG) pathways was compared to that expected by chance using a Fisher's exact test with FDR correction.

For cell experiments, data represent mean ± SD. Statistical analysis was carried out using GraphPad Prism software (Version 9.0; GraphPad Prism Software, Inc., San Diego, CA, USA). Differences between means were determined by students t-test and were considered statistically significant at p-value < 0.05.

3. Results

3.1. Participants

Participating mothers were, on average 30 (±4) years of age, with a mean BMI of 27 (±6) kg/m^2 (Table 1). The majority were white (135/170; 79%), married (134/170; 78%), multi-parous (120/170; 70%), had never used tobacco (144/170; 84%), and had obtained a college or post-graduate degree (121/170; 71%). Few experienced gestational diabetes (20/170, 11%). Approximately half had previously breastfed more than four months (90/170; 52%). Compared to mothers with AMS, mothers with LMS were older (d = 0.39, p = 0.023), were more likely to have a post-graduate degree (X^2 = 12.1, p = 0.017), and were less likely to have previously breastfed (X^2 = 11.9, p = 0.018). There were no differences in other medical or demographic factors.

Table 1. Participant characteristics.

	All (n = 170)	AMS (n = 123)	LMS (n = 47)
Age in years, mean (SD)	30 (4.6)	29.5 (4.5)	31.3 (4.7) *
Asian, n (%)	8 (4.7)	6 (4.9)	2 (4.3)
Bi-racial, n (%)	6 (3.5)	4 (3.3)	2 (4.3)
African American, n (%)	12 (7.1)	8 (6.5)	4 (8.5)
Other—not specified, n (%)	3 (1.8)	3 (2.4)	0 (0.0)
Other—specified, n (%)	6 (3.5)	4 (3.3)	2 (4.3)
White, n (%)	135 (79.4)	98 (79.7)	37 (78.7)
Parity, # (SD)	2 (1)	2 (1)	2 (1)
Pre-pregnancy BMI, kg/m^2, mean (SD)	27.8 (6.7)	27.2 (6.1)	29.2 (8.0)
Gestational diabetes, n (%)	20 (11.8)	13 (10.6)	7 (14.9)
Prior tobacco use, n (%)	26 (15.3)	17 (13.8)	9 (19.1)
Some HS, n (%)	2 (1.2)	2 (1.6)	0 (0.0)
Completed HS, n (%)	22 (12.9)	19 (15.4)	3 (6.4) *
Some college, n (%)	25 (14.7)	16 (13.0)	9 (19.1)
Completed college, n (%)	69 (40.6)	56 (45.5)	13 (27.7)
Post-graduate degree, n (%)	52 (30.6)	30 (24.4)	22 (46.8) *
Single, n (%)	11 (6.5)	9 (7.3)	2 (4.3)
Divorced, n (%)	2 (1.2)	1 (0.8)	1 (2.1)
Co-habitating, n (%)	23 (13.5)	16 (13.0)	7 (14.9)
Married, n (%)	134 (78.8)	97 (78.9)	37 (78.7)
Never, n (%)	67 (39.4)	42 (34.1)	25 (53.2) *
<1 month, n (%)	4 (2.4)	1 (0.8)	3 (6.4)
1–2 months, n (%)	4 (2.4)	4 (3.3)	0 (0.0)
2–4 months, n (%)	5 (2.9)	4 (3.3)	1 (2.1)
>4 months, n (%)	90 (52.9)	72 (58.5)	18 (38.3) *
Gestational age, weeks (SD)	38.9 (1.0)	38.9 (1.0)	39.0 (1.0)
Infant Sex, female, n (%)	99 (58.2)	74 (60.1)	25 (53.1)
Birth weight, grams (SD)	3364 (440)	3377 (440)	3329 (443)

* Denotes $p < 0.05$ on Student's t-test or chi-square test. Abbreviations: Body mass index (BMI); High school (HS).

3.2. Milk Production and Infant Weight Trajectory

Infants of mothers with LMS had a similar mean weight-for-length z-score at birth (−0.76 ± 1.0) as infants of mothers with AMS (−0.71 ± 1.1; d = 0.03, p = 0.41; Table 2). On average, mothers with LMS first reported difficulties with milk supply 3 (±3) months after delivery. However, 4 wks after delivery, infants of mothers with LMS displayed a lower mean weight-for-length z-score (0.05 ± 1.2) than infants of mothers with AMS (0.50 ± 1.1; d = 0.38, p = 0.013). At 4 wks, mothers with LMS reported lower volumes of daily milk production (20.6 ± 11.8 oz/day) than mothers with AMS (28.1 ± 15.6 oz/day; d = 0.50, p = 0.003), and they were more likely to have introduced formula (28/47, 59%) than mothers with AMS (27/127, 21%; X^2 = 27.8, $p = 2.1 \times 10^{-7}$). Mothers with LMS continued to report lower volumes of daily milk production at 16 weeks (20.1 ± 14.1 oz/day) than mothers with AMS (33.3 ± 24.0 oz/day; d = 0.60, p = 0.0007). Mothers with LMS were also more likely to have ceased breast feeding completely at 24 wks (18/47, 38.2%), compared to mothers with AMS (22/127, 17%; $X^2 = 11.4$, $p = 0.0003$).

Table 2. Infant weight trajectory and feeding characteristics.

	All (n = 170)	AMS (n = 123)	LMS (n = 47)
Formula intro by 4 weeks, n (%)	55 (32.3)	27 (21.9)	28 (59.5) *
Breastfeeding at 6 months, n (%)	134 (78.8)	105 (85.4)	29 (61.7) *
Milk production at 4 weeks, oz/day (SD)	26.0 (15.0)	28.1 (15.6)	20.6 (11.8) *
Milk production at 16 weeks, oz/day (SD)	29.6 (22.5)	33.3 (24.0)	20.1 (14.1) *
Infant WfL Z-score at birth	−0.72 (1.1)	−0.71 (1.1)	−0.76 (1.0)
Infant WfL Z-score at 4 weeks	0.38 (1.2)	0.50 (1.1)	0.05 (1.2) *
Δ WfL Z-score from birth to 4 weeks	1.1 (1.4)	1.2 (1.4)	0.8 (1.3) *

* $p < 0.05$ on Student's t-test or chi-square test. Abbreviations: Introduction (intro), Weight for length (WfL).

3.3. Milk miRNA Profiles

There was no discernible difference between the overall milk miRNA profiles the first week after delivery (Figure 2A). However, 5 of the 30 most robustly expressed milk miRNAs (i.e., >10 counts in all 453 samples), displayed nominal differences (raw $p \leq 0.01$) between groups (Figure 2B–F). Maternal estimates of daily milk production were associated with levels of three miRNAs: miR-16-5p (R = −0.14, $p = 0.0088$, adj $p = 0.044$), miR-22-3p (R = 0.13, $p = 0.011$, adj $p = 0.044$), and let-7g-5p (R = 0.12, $p = 0.023$, adj $p = 0.046$). Only levels of let-7g-5p survived multiple testing correction (V = 1765, adj $p = 0.0025$), indicating higher levels in the milk of mothers with LMS.

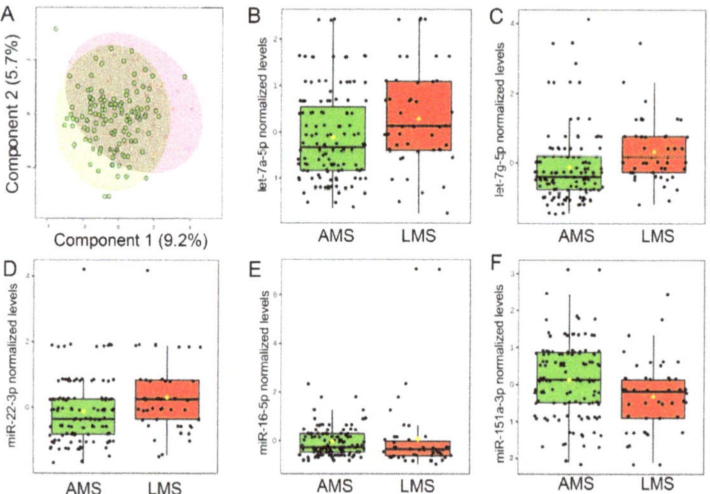

Figure 2. Levels of miRNAs in breast milk differ among women with low milk supply. (**A**) Results of a two-dimensional partial least squares discriminant analysis (PLSDA) employing levels of 30 miRNAs with the most robust concentrations in maternal breast milk in the first week after delivery. Note that total milk miRNA profiles do not differ between mothers with adequate milk supply (AMS; green) and mothers with low milk supply (LMS; red). However, milk levels of five miRNAs did differ in the milk of mothers with LMS. In the first week after delivery, levels of (**B**) let-7a-5p (V = 2194, $p = 0.015$, adj $p = 0.090$), (**C**) let-7g-5p (V = 1765, $p = 8.5 \times 10^{-5}$, adj $p = 0.0025$), and (**D**) miR-22-3p (V = 2135, $p = 0.0080$, adj $p = 0.080$) were higher in the milk of mothers with LMS, whereas levels of (**E**) miR-16-5p (V = 0.3595, $p = 0.013$ adj $p = 0.090$) and (**F**) miR-151a-3p (V = 3670, $p = 0.0063$, adj $p = 0.080$) were lower.

3.4. Longitudinal Changes in LMS-Related miRNAs

Linear mixed models were used to assess candidate miRNAs for changes in abundance over the course of lactation, while controlling for maternal age, education, and breastfeeding experience. Milk levels of let-7g-5p increased over the course of lactation (F = 21.4, $p < 0.001$), and there was a significant interaction between lactation period and group (F = 7.5, $p < 0.001$; Figure 3A). Levels of let-7g-5p generally increased in the AMS group over time but remained stable in mothers with LMS. Levels of let-7a-5p (F = 15.7, $p < 0.001$), miR-22-3p (F = 71.2, $p < 0.001$), and miR-16-5p (F = 62.7, $p < 0.001$) also changed over the course of lactation but did not display an interaction between lactation stage and milk supply.

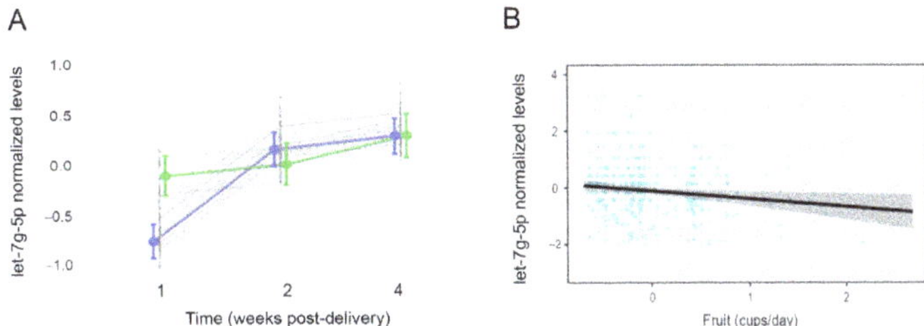

Figure 3. Milk levels of let-7g-5p differ over time among mothers with low milk supply and are related to maternal fruit intake. The effects plot displays normalized concentrations of let-7g-5p in 453 samples from 170 lactating mothers across three time points: 1 wk, 4 wks, and 16 wks after delivery (**A**). Mean concentrations are shown for mothers with low milk supply (LMS, green), and mothers with adequate milk supply (AMS, blue). There was a significant interaction effect ($p < 0.001$) between LMS/AMS group and time (F = 7.5), with let-7g-5p levels increasing between 1 wk and 4 wks post-delivery in the AMS group only. (**B**) A second mixed effects model also revealed a significant effect of maternal fruit consumption on milk levels of let-7g-5p (F = 5.99, $p = 0.015$). Higher levels of let-7g-5p were associated with lower fruit consumption, as reported longitudinally on the Dietary Screener Questionnaire.

3.5. Effect of Modifiable Maternal Characteristics on LMS-Related miRNAs

Mixed models were used to assess the effect of nutrition, BMI, and tobacco use on candidate miRNAs. Nutrition displayed an association with milk miRNA levels over the course of lactation. Lower milk levels of let-7g-5p were associated with higher maternal fruit consumption (F = 5.99, $p = 0.015$, Figure 3B). There was no interaction between fruit consumption and milk supply (F = 0.22, $p = 0.63$). Higher milk levels of miR-22-3p were associated with lower consumption of calcium (F = 7.31, $p = 0.007$) and dairy (F = 6.48, $p = 0.011$).

3.6. Predicting Low Milk Supply Status

Hierarchical logistic regression was used to assess the ability of milk miRNAs to predict low milk supply, relative to medical and demographic traits. The three maternal characteristics that differed between mothers with LMS and mothers with AMS (i.e., age, education, breastfeeding experience) accounted for 15.5% of the variance between groups, ($X^2 = 31.0$, $p < 0.001$), and accurately identified 35/47 mothers with AMS (74% sensitivity) and 80/123 mothers with AMS (65% specificity; AUC = 0.763). Addition of three miRNAs that lacked co-linearity (i.e., miR-22-3p, let-7a-5p, and let-7g-5p) accounted for an additional 8.1% of variance between groups ($X^2 = 47.3$, $p < 0.001$), and significantly improved the model ($X^2 = 16.2$, $p = 0.001$). The combined model displayed an AUC of 0.816.

3.7. Pathway Analysis

The messenger RNA targets of the five miRNAs of interest (miR-22-3p, miR-16-5p, let-7a-5p, let-7g-5p, miR-151a-3p) were interrogated in DIANA miRPATH using the Tarbase algorithm. The five miRNAs targeted 13 physiologic pathways with greater frequency than would be expected by chance alone, including cell cycle, fatty acid biosynthesis, adherens junctions, Hippo signaling, TGFβ signaling, and ECM-receptor interaction (Table 3).

Table 3. Physiologic pathways targeted by the five candidate miRNAs.

KEGG Pathway	*p*-Value	Genes (#)	miRNAs (#)
Prion diseases	6.88×10^{-17}	13	2
Cell cycle	1.11×10^{-16}	63	3
Hepatitis B	3.77×10^{-15}	72	3
Proteoglycans in cancer	4.77×10^{-14}	85	2
Fatty acid biosynthesis	1.63×10^{-13}	4	2
Adherens junction	1.29×10^{-11}	46	2
Hippo signaling pathway	1.42×10^{-10}	57	3
TGF-beta signaling pathway	1.32×10^{-09}	31	1
Lysine degradation	1.77×10^{-09}	15	1
Oocyte meiosis	6.19×10^{-09}	35	1
Viral carcinogenesis	9.29×10^{-9}	102	4
ECM-receptor interaction	8.02×10^{-08}	16	2
Transcriptional misregulation in cancer	2.10×10^{-05}	38	1

p-values represent multiple correction adjusted *p*-values on Fisher's exact *t*-tests. The number of genes targeted by each of the five miRNAs was determined using the Tarbase algorithm in DIANA miRPath software. The top 13 pathways (adj. $p < 1.0 \times 10^{-5}$) are shown.

Hierarchical clustering showed that let-7a-5p and let-7g-5p had the most closely related physiologic targets, and miR-22-3p and miR-16-5p displayed the second highest physiologic relatedness (Figure 4).

Due to the interaction between let-7g-5p, lactation stage, and milk supply, we further queried molecular pathways affected by this miRNA. Key KEGG pathways predicted to be downregulated by let-7g-5p (https://genome.jp, accessed on 2 November, 2022) include Metabolic, PI3K-Akt signaling, MAPK signaling, and JAK-STAT signaling pathways (Table 4), and include numerous genes associated with lactation traits such as those involved in hormone signaling (*PRLR, INSR*), extracellular matrix (*COL1A1 2, COL3A1, COL4A1-3* and *A6, COL5A2, COL14A1, COL27A1*), zinc transport (*SLC30A4*), H^+ transport (*ATP6V1G1, ATP6V1C1*), and calcium transport (*ATP2A2, ATP2B4*). Moreover, the top mRNAs identified included several novel lactogenic targets involved in mRNA binding (*IGF2BP1-3*), chromatin binding (*HMGA2*), and actin assembly (*STARD13*) (Supplementary Materials).

3.8. Confirmation of PRLR and IGF2BP1 Regulation in Mammary Cells

To directly confirm the effect of elevated levels of let-7g-5p in MECs, cells were transiently transfected with a let-7g-5p mimic. Protein expression of two predicted mRNA targets, PRLR and IGF2BP1/IMP1, was measured 30 h later by immunoblot (Table S1). PRLR was selected for confirmation of its regulation by let-7g-5p in MECs because of its well-known importance in lactation, and IGF2BP1/IMP1 was selected to assess effects on a novel molecular target. Both the long and short forms of IGF2BP1/IMP1 were identified in HC11 cells [29], and both isoforms were significantly lower ($p < 0.01$) in cells transfected with 30 nM hsa-let-7g mimic compared to non-transfected cells (Figure 5A–C). In addition, PRLR protein expression was significantly lower in cells transfected with 30 nM hsa-let-7g mimic ($p < 0.05$) compared to non-transfected cells (Figure 5D,E). These results confirm that expression of PRLR and IGF2BP1/IMP1 are negatively regulated by let-7g-5p in MECs and provide evidence of two discrete molecular mechanisms that may underlie effects of let-7g-5p on lactation and milk supply.

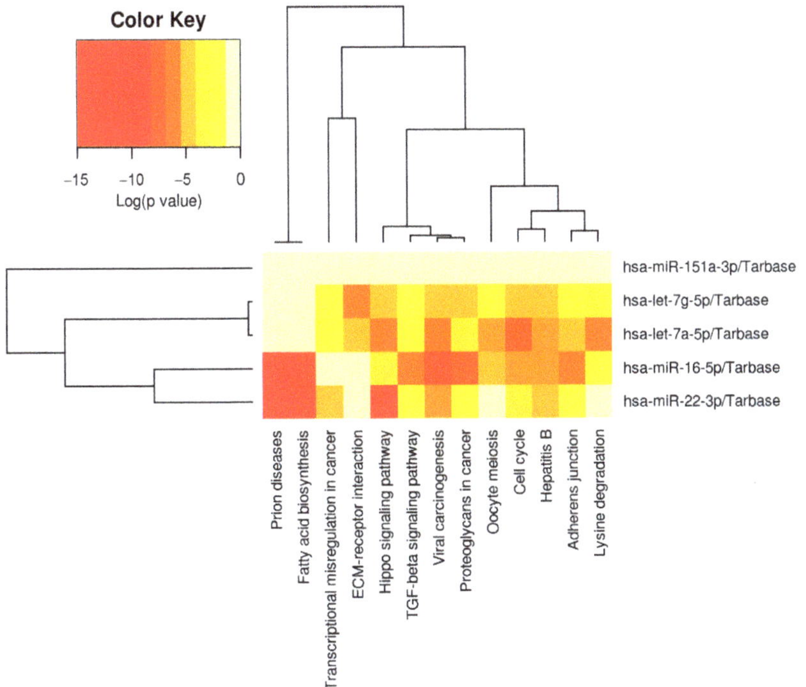

Figure 4. Physiologic pathways targeted by the five milk miRNAs implicated in low milk supply. The heatmap displays the relative level of messenger RNA targets for 13 Kyoto Encyclopedia of Genes and Genomes (KEGG) pathways, where red represents the highest number of targets. These 13 pathways all displayed a significantly greater number of targets than would be expected by chance alone on Fisher Exact Testing. Note that pathways and miRNAs have been clustered using the Ward method, and the dendrograms display relatedness of each pathway or miRNA.

Table 4. Top twenty physiologic pathways targeted by let-7g-5p.

KEGG Pathway	Genes (#)
Metabolic pathways	72
Pathways in cancer	39
PI3K-Akt signaling pathway	33
Human Papillomavirus infection	30
Herpes simplex virus infection	29
Pathways of neurodegeneration	28
Proteoglycans in cancer	23
MicroRNAs in cancer	23
Focal adhesion	21
Axon guidance	21
Human T-cell leukemia virus 1 infection	20
Cytokine-cytokine receptor interaction	20
Transcriptional misregulation in cancer	19
FOXO signaling pathway	19
Ras signaling pathway	19
Calcium signaling pathway	18
mTOR signaling pathway	18
JAK-STAT signaling pathway	18
Lipid and atherosclerosis	18
AGE-RAGE signaling pathway	18

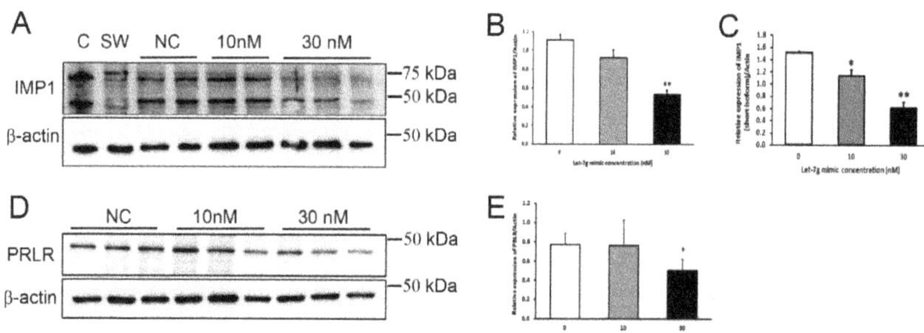

Figure 5. Let-7g-5p reduced IGF2BP1/IMP1 and prolactin receptor (PRLR) expression in mammary epithelial (HC11) cells. (**A**) Representative immunoblot of IGF2BP1/IMP1 in protein lysates (20 μg protein/well) from HC11 cells transfected with 10 nM or 30 nM of let-7g-5p mimic compared with non-transfected (NC) cells. Caco-2 (**C**) and SW480 (SW) protein lysates were used as positive controls. Membranes were striped and re-probed for β-actin. (**B**) Relative expression of the long form of IGF2BP1/IMP1 (~70 kDa) after normalization to β-actin. Results represent the mean ratio ± SD (n = 2–3 samples/group). ** $p < 0.01$ (**C**) Relative expression of the short form of IGF2BP1/IMP1 (~42 kDa) after normalization to β-actin. Results represent the mean ratio ± SD (n = 2–3 samples/group). * $p < 0.05$, ** $p < 0.01$ (**D**) Representative immunoblot of PRLR in protein lysates (20 μg protein/well) from HC11 cells transfected with 10 nM or 30 nM of let-7g-5p mimic compared with non-transfected (NC€cells. (**E**) Relative PRLR protein expression of after normalization to β-actin. Results represent the mean ratio ± SD (n = 3 samples/group) from two separate experiments. * $p < 0.05$.

4. Discussion

Here, we present a novel approach for identifying biological factors that underpin low milk supply in breastfeeding women. We established that the profile of milk-derived miRNAs reflects lactation performance in humans, and for the first time identified miRNAs associated with low milk supply. Importantly, we determined that let-7g-5p may serve as a potential regulon of breast function and predicted key genes involved in metabolism, hormone signaling, ion transport, mRNA binding, and tissue remodeling in the lactating mammary gland, supporting previous studies that have posited a role for let-7g-5p in breast function [30]. Importantly, bioinformatic prediction of PRLR and IGF2BP1/IMP1 as let-7g-5p targets was confirmed in MECS. We detected a significant interaction between let-7g-5p, milk volume, and lactation stage, suggesting that measurement of let-7g-5p levels in milk during early lactation may be useful in predicting risk for low milk supply. Interestingly, our data suggest the let-7g-5p regulon may be modifiable as there was a negative relationship between let-7g-5p levels and fruit intake in our population of breastfeeding women.

Milk is one of the richest sources of miRNAs in humans and contains ~1400 mature miRNAs [17] and the profile of milk miRNAs has been proposed to reflect mammary gland function [17–21]. We found maternal estimates of daily milk production were indeed associated with levels of miR-16-5p, miR-22-3p, and let-7g-5p; however, only levels of let-7g-5p survived multiple testing correction. The *let-7* family of miRNAs is one of the earliest discovered miRNA clusters and is conserved across species [31]. It is comprised of ten miRNAs (*let-7a, let-7b, let-7c, let-7d, let-7e, let-7f, let-7g, let-7i, miR-98,* and miR-202) and plays key roles in suppressing proliferation and differentiation [32] and is a key regulator of glucose metabolism [33]. Let-7g-5p levels were enriched in the milk of women with low milk supply, consistent with a putative role for let-7g-5p suppression in motivating proliferation, suppressing mammary gland differentiation [30] and affecting mammary gland metabolism. Importantly, let-7g-5p is conserved between human and mouse [31], suggesting the opportunity to use preclinical mouse models to understand mechanistic implications of let-7g-5p on MEC proliferation, differentiation, and milk production and secretion.

A critical gap in knowledge is how let-7g-5p is regulated in mammary epitheial cells. Let-7g is found on chromosome 3 in humans [31] and is post-transcriptionally repressed through binding of the RNA binding protein Lin28 to its terminal loop, thereby either inhibiting binding of Dicer and Drosha or re-routing pre-let-7g for degradation [34]. Thus, factors that affect Lin28 regulation would have major impacts on mammary gland function. One potential factor that regulates the Lin28/let-7 axis is inflammation [35]. Interestingly, inflammation is known to compromise lactation and milk supply [36,37], thus therapeutic strategies to reduce inflammation may be key to maximizing milk supply. Additionally, let-7g expression is repressed by DNA methylation [38], suggesting intake of methyl donors such as folate [39] and B_{12} may play a role. Several studies suggest estrogen may regulate let-7g-5p; however, results are inconsistent [40,41]. Positive regulation by estrogen would be consistent with suppression of let-7g-5p levels upon estrogen withdrawal at partition and the subsequent activation of lactogenesis II. Given the importance of robust activation lactogenesis II in the maintenance of copious milk production going forward, identification of factors that regulate let-7g-5p is critical to our understanding of this regulon and its intriguing role in lactation [42].

Numerous mRNA targets of let-7g-5p associated with lactation traits include hormone signaling (*PRLR, INSR*) [43–45], extracellular matrix (*COL1A1-2, COL3A1, COL4A1-3* and *A6, COL5A2, COL14A1, COL27A1*) [46], zinc transport (*SLC30A4*) [47], H^+ transport (*ATP6V1G1, ATP6V1C1*) [48], and calcium transport (*ATP2A2, ATP2B4*) [49], further implicating it as a regulatory hub for maintaining milk supply. Importantly, our study confirmed that PRLR expression was indeed regulated by let-7g-5p in MECs, providing direct evidence that let-7g-5p is a critical regulator of lactation. Three examples of novel targets of let-7g-5p that may have potential lactogenic implications include steriodogenic acute regulatory protein-related lipid transfer domain-containing protein 13 (STARD13), high mobility group AT-hook 2 (HMGA2), and insulin-like growth factor 2 messenger RNA-binding protein (IGF2BP1/IMP1). STARD13 serves as a Rho-GTPase activating protein that selectively regulates RhoA and cdc42 to inhibit actin assembly. STARD13 attenuation leaves RhoA constitutively active which inhibits Rac and thus inhibits motility [50]. To our knowledge, a role for STARD13 in mammary gland development or lactation has yet to be explored; however, this finding may have important implications for breast cancer detection as STARD13 is a tumor suppressor and Rac1 plays a major role in cancer cell motility [50]. HMGA2 is a group of small chromatin-associated proteins that act as an architectural transcription factor that directly binds to DNA and modulates the transcription of target genes; however, a putative role for HMGA2 in the breast requires exploration. IGF2BP1/IMP1 is a highly conserved RNA-binding protein that regulates RNA processing at several levels, including localization, translation, and stability. Key targets of IGF2BP1/IMP1 with potential consequences on lactation include β-actin, STAT3, c-myc, glutathione peroxidases, and several mitochondrial proteins [51]. Two IGF2BP1/IMP1 isoforms have been identified in MECs [29], a long isoform (~70 kDa) and a short isoform (~40 kDa) resulting from an *N*-terminal truncation with currently unknown function. Herein, we confirmed that let-7g-5p downregulated both isoforms, suggesting key molecular functions such as morphogenesis, oxidative stress, and ATP production are targets of the let-7g-5p regulon, which implicates IGF2BP1/IMPs as a novel molecular target for low milk supply.

While our genome-wide approach identified critical milk-derived miRNAs and predicted novel genes and molecular pathways for further exploration, an exciting finding from this study was that high milk levels of let-7g-5p during early lactation may be useful as a bioreporter of low milk supply and risk for early breastfeeding cessation. This observation warrants further study as >40% of women cite concerns over low milk supply as a primary reason for not meeting their breastfeeding goals [1] and early identification could inform interventions to improve breastfeeding success. One such intervention might include dietary modification, as intriguing findings suggest levels of let-7g-5p may be modifiable. For example, ursolic acid (UA) is a natural triterpene found in various fruits and vegetables that suppresses let-7g-5p [52] and there is a growing interest in UA because of its beneficial effects, which include pro- and anti-inflammatory, antioxidant, anti-apoptotic, and anti-carcinogenic effects [53]. Additionally, quercetin is a polyphenol ubiquitously present in certain fruits (e.g., apples and grapes) and vegetables (e.g., onions, kale, broccoli, lettuce, and tomatoes) that also has pro- and anti-inflammatory and antioxidant capacity, and let-7g-5p levels are positively associated with a quercetin-rich diet [54]. Further studies are required to reproduce our findings and determine how dietary polyphenols and secondary metabolites affect the let-7g-5p regulon in the mammary gland, which may eventually offer a therapeutic option with an evidence-based rationale for women with low milk supply.

There are several strengths of the present study: (1) a large sample size with longitudinal collections; (2) uniformity insample collection and processing; (3) high throughput sequencing; and (4) the use of mixed effects models to assess relative impacts of maternal characteristics. However, there are several limitations. The scatter in our data suggest that trends may be driven more heavily by samples at the fringe ends of the dataset, and additional studies are required to confirm our findings. The current cohort was mostly white and included only mothers delivering at term, which may limit generalizability of the findings. In addition, given our finding that let-7g-5p may be a potential regulator, further studies should include potential modifiers such as caloric intake, physical activity, quality of life, and comprehensive dietary analysis to better understand the role and regulation of let-7g-5p during lactation. Finally, our results (which include both exosomal and non-exosomal miRNA) may differ from studies focused solely on exosomal miRNAs [55] and while prior studies have demonstrated minimal miRNA differences across milk fractions [56,57], results from miRNAs in skim versus cellular fractions may differ.

5. Conclusions

In conclusion, the results of this study advance our collective understanding of the biological contributors to low milk supply by identifying miRNAs, novel molecular pathways, and new gene targets associated with poor milk production. Predicted transcripts highlight both classical and novel lactogenic targets, therefore future studies should interrogate consequences of these miRNAs on mammary gland function. Importantly, this study is the first to identify an interaction between milk levels of miRNAs, low milk supply, and early cessation of breastfeeding, and suggests that measuring milk levels of let-7g-5p during the first weeks after birth may be a useful tool in predicting risk for low milk supply and identifying women who need targeted lactation support.

Supplementary Materials: The following supporting information can be downloaded at: https://www.mdpi.com/article/10.3390/nu15030567/s1, Table S1: predicted mRNA targets.

Author Contributions: Conceptualization, S.L.K. and S.D.H.; methodology, S.L.K. and S.D.H.; formal analysis, D.C., A.C. and A.W.; investigation, D.C., A.C. and A.W.; resources, S.D.H. and S.L.K.; data curation, D.C., A.C. and A.W.; writing—original draft preparation, S.L.K. and S.D.H.; writing—review and editing, A.W.; supervision, S.D.H. and S.L.K.; project administration, S.D.H.; funding acquisition, S.D.H. and S.L.K. All authors have read and agreed to the published version of the manuscript.

Funding: This study was funded by grants from the Gerber Foundation (grant number 5295) and Center for Research in Women and Newborns (CROWN, grant number 2020) Foundation to S.D.H. and intramural funds to S.L.K.

Institutional Review Board Statement: The study was conducted in accordance with the Declaration of Helsinki and approved by the Institutional Review Board of Penn State University (STUDY00008657 approved 18 January 2018).

Informed Consent Statement: Informed consent was obtained from all subjects involved in the study.

Data Availability Statement: The RNA sequencing data presented in the study are deposited in the Gene Expression Omnibus repository, accession number GSE192543. GEO repository link: https://www.ncbi.nlm.nih.gov/geo/query/acc.cgi?acc=GSE192543; 1 January, 2022.

Acknowledgments: The authors wish to thank the mothers and infants that participated in this study. We thank Jessica Beiler for assistance with participant recruitment and data collection and Kaitlyn Warren for assistance with RNA extraction. We gratefully acknowledge Sarah Andres for the generous gift of IMP1 antibody.

Conflicts of Interest: S.D.H. is a consultant and scientific advisory board member for Spectrum Solutions and Quadrant Biosciences. S.L.K. is the owner/founder of LactoGenix, LLC, and is a technical advisory board member for Biomilq. The funders had no role in the design of the study; in the collection, analyses, or interpretation of data; in the writing of the manuscript; or in the decision to publish the results.

References

1. Ahluwalia, I.B.; Morrow, B.; Hsia, J. Why do women stop breastfeeding? Findings from the pregnancy risk assessment and monitoring system. *Pediatrics* **2005**, *116*, 1408–1412. [CrossRef] [PubMed]
2. Neifert, M.; DeMarzo, S.; Seacat, J.; Young, D.; Leff, M.; Orleans, M. The influence of breast surgery, breast appearance, and pregnancy-induced breast changes on lactation sufficiency as measured by infant weight gain. *Birth* **1990**, *17*, 31–38. [CrossRef] [PubMed]
3. Figueiredo, B.; Dias, C.C.; Brandão, S.; Canário, C.; Nunes-Costa, R. Breastfeeding and postpartum depression: State of the art review. *J. Pediatr.* **2013**, *89*, 332–338. [CrossRef] [PubMed]
4. Gilmour, C.; Hall, H.; McIntyre, M.; Gillies, L.; Harrison, B. Factors associated with early breastfeeding cessation in frankston, victoria: A descriptive study. *Breastfeed. Rev.* **2009**, *17*, 13–19.
5. Meier, P.; Patel, A.L.; Wright, K.; Engstrom, J.L. Management of breastfeeding during and after the maternity hospitalization for late preterm infants. *Clin. Perinatol.* **2013**, *40*, 689–705. [CrossRef]
6. Szabo, A.L. Review article: Intrapartum neuraxial analgesia and breastfeeding outcomes: Limitations of current knowledge. *Anesth. Analg.* **2013**, *116*, 399–405. [CrossRef]
7. Ettyang, G.A.; van Marken Lichtenbelt, W.D.; Esamai, F.; Saris, W.H.; Westerterp, K.R. Assessment of body composition and breast milk volume in lactating mothers in pastoral communities in pokot, kenya, using deuterium oxide. *Ann. Nutr. Metab.* **2005**, *49*, 110–117. [CrossRef]
8. Diana, A.; Haszard, J.J.; Houghton, L.A.; Gibson, R.S. Breastmilk intake among exclusively breastfed indonesian infants is negatively associated with maternal fat mass. *Eur. J. Clin. Nutr.* **2019**, *73*, 1206–1208. [CrossRef]
9. Suwaydi, M.A.; Wlodek, M.E.; Lai, C.T.; Prosser, S.A.; Geddes, D.T.; Perrella, S.L. Delayed secretory activation and low milk production in women with gestational diabetes: A case series. *BMC Pregnancy Childbirth* **2022**, *22*, 350. [CrossRef]
10. Moriwaki, M.; Welt, C. Mon-434 familial alactogenesis associated with a prolactin mutation. *J. Endocr. Soc.* **2019**, *3* (Suppl. 1), MON-434. [CrossRef]
11. Kobayashi, T.; Usui, H.; Tanaka, H.; Shozu, M. Variant prolactin receptor in agalactia and hyperprolactinemia. *N. Engl. J. Med.* **2018**, *379*, 2230–2236. [CrossRef] [PubMed]
12. Rivera, O.C.; Geddes, D.T.; Barber-Zucker, S.; Zarivach, R.; Gagnon, A.; Soybel, D.I.; Kelleher, S.L. A common genetic variant in zinc transporter znt2 (thr288ser) is present in women with low milk volume and alters lysosome function and cell energetics. *Am. J. Physiol. Cell Physiol.* **2020**, *318*, C1166–C1177. [CrossRef] [PubMed]
13. Chandran, D.; Confair, A.; Warren, K.; Kawasawa, Y.I.; Hicks, S.D. Maternal variants in the mfge8 gene are associated with perceived breast milk supply. *Breastfeed. Med.* **2022**, *17*, 331–340. [CrossRef] [PubMed]
14. Liu, B.; Li, J.; Cairns, M.J. Identifying mirnas, targets and functions. *Brief Bioinform.* **2014**, *15*, 1–19. [CrossRef] [PubMed]
15. Arantes, L.; De Carvalho, A.C.; Melendez, M.E.; Lopes Carvalho, A. Serum, plasma and saliva biomarkers for head and neck cancer. *Expert Rev. Mol. Diagn.* **2018**, *18*, 85–112. [CrossRef]
16. Tölle, A.; Blobel, C.C.; Jung, K. Circulating mirnas in blood and urine as diagnostic and prognostic biomarkers for bladder cancer: An update in 2017. *Biomark. Med.* **2018**, *12*, 667–676. [CrossRef]
17. Alsaweed, M.; Lai, C.T.; Hartmann, P.E.; Geddes, D.T.; Kakulas, F. Human milk cells contain numerous mirnas that may change with milk removal and regulate multiple physiological processes. *Int. J. Mol. Sci.* **2016**, *17*, 956. [CrossRef]
18. Avril-Sassen, S.; Goldstein, L.D.; Stingl, J.; Blenkiron, C.; Le Quesne, J.; Spiteri, I.; Karagavriilidou, K.; Watson, C.J.; Tavaré, S.; Miska, E.A.; et al. Characterisation of microrna expression in post-natal mouse mammary gland development. *BMC Genom.* **2009**, *10*, 548. [CrossRef]
19. Li, Z.; Liu, H.; Jin, X.; Lo, L.; Liu, J. Expression profiles of micrornas from lactating and non-lactating bovine mammary glands and identification of mirna related to lactation. *BMC Genom.* **2012**, *13*, 731. [CrossRef]
20. Izumi, H.; Kosaka, N.; Shimizu, T.; Sekine, K.; Ochiya, T.; Takase, M. Time-dependent expression profiles of micrornas and mrnas in rat milk whey. *PLoS ONE* **2014**, *9*, e88843. [CrossRef]
21. Alsaweed, M.; Lai, C.T.; Hartmann, P.E.; Geddes, D.T.; Kakulas, F. Human milk cells and lipids conserve numerous known and novel mirnas, some of which are differentially expressed during lactation. *PLoS ONE* **2016**, *11*, e0152610. [CrossRef]

MDPI

Article

Modifiable Risk Factors for Cardiovascular Disease among Women with and without a History of Hypertensive Disorders of Pregnancy

Kaylee Slater [1,2], Tracy L. Schumacher [2,3], Ker Nee Ding [1], Rachael M. Taylor [1,2], Vanessa A. Shrewsbury [1,2] and Melinda J. Hutchesson [1,2,*]

1 School of Health Sciences, College of Health, Medicine and Wellbeing, University of Newcastle, Callaghan, NSW 2308, Australia
2 Food and Nutrition Research Program, Hunter Medial Research Institute, Lot 1, Kookaburra Circuit, New Lambton Heights, NSW 2305, Australia
3 Department of Rural Health, College of Health, Medicine and Wellbeing, University of Newcastle, Tamworth, NSW 2340, Australia
* Correspondence: melinda.hutchesson@newcastle.edu.au

Abstract: Cardiovascular disease (CVD) is the leading cause of morbidity and mortality in women. Hypertensive disorders of pregnancy (HDP) affect 5–10% of pregnancies worldwide, and are an independent risk factor for CVD. A greater understanding of the rates of modifiable CVD risk factors in women with a history of HDP can inform CVD prevention priorities in this group. The aim of this study was to understand the rates of individual and multiple modifiable risk factors for CVD (body mass index, fruit and vegetable intake, physical activity, sitting time, smoking, alcohol consumption and depressive symptoms) among women with a history of HDP, and assess whether they differ to women without a history of HDP. This study is a cross-sectional analysis of self-reported data collected for the Australian Longitudinal Study of Women's Health (ALSWH). The sample included 5820 women aged 32–37 years old, who completed survey 7 of the ALSWH in 2015. Women with a history of HDP had a higher multiple CVD modifiable risk factor score compared to those without HDP (mean (SD): 2.3 (1.4) vs. 2.0 (1.3); $p < 0.01$). HDP history was significantly associated with a higher body mass index ($p < 0.01$), high-risk alcohol consumption ($p = 0.04$) and more depressive symptoms ($p < 0.01$). Understanding that women with a history of HDP have higher CVD risk factors, specifically body mass index, alcohol consumption and depressive symptoms, allows clinicians to provide appropriate and tailored CVD interventions for this group of women.

Keywords: hypertensive disorders of pregnancy; postpartum management; hypertension; cardiovascular disease; multiple CVD modifiable risk factors; CVD prevention; women's health

Citation: Slater, K.; Schumacher, T.L.; Ding, K.N.; Taylor, R.M.; Shrewsbury, V.A.; Hutchesson, M.J. Modifiable Risk Factors for Cardiovascular Disease among Women with and without a History of Hypertensive Disorders of Pregnancy. *Nutrients* **2023**, *15*, 410. https://doi.org/10.3390/nu15020410

Academic Editors: Birgit-Christiane Zyriax and Nataliya Makarova

Received: 24 November 2022
Revised: 8 January 2023
Accepted: 10 January 2023
Published: 13 January 2023

1. Introduction

Cardiovascular disease (CVD) is the leading cause of morbidity and mortality in women, accounting for approximately 33% of total mortality [1]. In 2017–2018, approximately 5% of Australian women aged 18 and over had experienced at least one cardiac event [2]. CVD, however, is preventable through risk factor modification.

Managing modifiable CVD risk factors (excess body weight, unhealthy diet, physical inactivity, sedentary behaviour, smoking, excessive alcohol intake, and poor mental health) reduces the incidence and recurrence of CVD events [2–4]. Tsai et al.'s meta-analysis of prospective studies ($n = 20$) demonstrated that adults who had a lower number of modifiable CVD risk factors (weight management, dietary intake, physical activity, smoking and alcohol consumption) had an overall lower risk of CVD, demonstrating the cumulative effect of modifiable CVD risk factors on overall CVD risk [5]. Tsai also found that improvements in those modifiable risk factors had more of a protective effect in younger adults (aged 37.1–49.9 years) than older adults (aged 60.0–72.9 years), suggesting a need to intervene early [5].

In addition to modifiable risk factors for CVD, women also experience unique sex-specific risk factors, such as hypertensive disorders of pregnancy (HDP). HDP includes chronic hypertension, gestational hypertension, preeclampsia, and eclampsia, and affects 5–10% of pregnancies worldwide [6]. A 2020 systematic review and meta-analysis of 73 studies, with over 13-million participants with HDP, revealed the relative risks of various CVD events after HDP, including hypertension [Relative Risk (RR): 3.2, 95% Confidence Interval (CI): 2.7–3.6], heart failure (RR: 2.9, 95% CI: 2.1–3.9), stroke (RR:1.7, 95% CI:1.5–1.8) and coronary heart disease (RR: 1.7, 95% CI: 1.5–1.8) [7]. Wu et al. also noted that the increased risk for all CVD events was greatest within 10 years postpartum [7]. Therefore, the first 10 years after HDP is a crucial period for engagement in assessment and management of modifiable CVD risk factors in women with a history of HDP.

A recent systematic review of observational studies (n = 11) examined whether modifiable risk factors for CVD post-pregnancy influenced the association between HDP and CVD outcomes. The review found consistent evidence that a higher post pregnancy body mass index (BMI) further amplified the risk of hypertension following HDP [8]. However, the influence of other modifiable risk factors (dietary intake, physical activity, smoking, alcohol, and mental health status) on CVD outcomes post-HDP could not be determined due to a lack of studies [8]. To our knowledge, there is a lack of research exploring the prevalence of a combination of modifiable CVD risk factors in women with a history of HDP and determining whether this differs to women without a history of HDP. This knowledge is crucial for informing CVD prevention strategies that are targeted to women with a history of HDP. Therefore, the aim of this study was to describe individual and modifiable risk factors for CVD (BMI, fruit and vegetable intake, physical activity, sitting time, smoking, alcohol consumption and depressive symptoms) among women with a history of HDP, and assess whether they differ to women without a history of HDP.

2. Materials and Methods

2.1. Study Design & Setting

This is a cross-sectional analysis of data from the Australian Longitudinal Study on Women's Health (ALSWH). The ALSWH is a prospective longitudinal population-based survey conducted every three years. The survey takes a comprehensive view of different aspects of women's health, including physical health and psychological well-being across the lifespan by assessing demographic, social, biological, behavioral, psychological, and lifestyle factors, as well as the use of and satisfaction with healthcare services [9–11]. Briefly, it includes follow-up of women (n = 58,000) across four generations in Australia. Three cohorts of women who were born between 1921–1926, 1946–1951, and 1973–1978 responded to the baseline survey sent out in 1996 [10]. A new cohort of women born between 1989–1995 were also recruited in 2012–2013. Data from the seventh ALSWH survey of the young cohort (born 1973–1978) when women (n = 7186) were aged 37–42 years were used in the current study due to their age at completion of the survey and the likelihood that more women in this sample would be $\geq$10 years postpartum compared to other samples from the ALSWH. The ALSWH survey program has ethical approval from the Human Research Ethics Committees (HRECs) of the Universities of Newcastle and Queensland (approval numbers H076-0795 and 2004000224, respectively, for the 1973–1978, 1946–1951 and 1921–1926 cohorts: and H-2012-0256 and 2012000950, for the 1989–1995 cohort). All participants consented to joining the study and were free to withdraw or suspend their participation at any time with no need to provide a reason.

2.2. Study Participants

Women with and without a history of HDP from the ALSWH were included. Pregnancy and birth data from Survey 7 of the 1973–1978 cohort, collected in 2015, were used.

Women who had been pregnant in the past were eligible for the analysis and ineligible if they had never been pregnant.

2.3. Exposure: Hypertensive Disorders of Pregnancy

Women who responded "yes" to the following question "Were you diagnosed with or treated for hypertension (high blood pressure) during pregnancy?" were classified as women with a history of HDP. Those who answered "Never experienced this" were classified as women with no prior HDP. Maternal recall of HDP has been reported to be relatively high in accuracy, and is not affected by time since pregnancy [12]. Women were excluded from the primary analysis if data for this question were missing.

2.4. Outcomes

2.4.1. Body Mass Index

BMI was calculated from self-reported weight divided by the square of self-reported height (kg/m^2) and categorised as underweight (BMI of <18.5 kg/m^2), healthy weight (BMI of 18.5–24.9 kg/m^2), overweight (BMI of 25.0–29.9 kg/m^2), and obese (BMI of $\geq$30.0 kg/m^2) based on the World Health Organization classification [13]. An overweight and obese BMI (BMI $\geq$ 25.0 kg/m^2) is associated with CVD risk factors, such as diabetes, dyslipidemia, hypertension and metabolic syndrome, as well as abdominal obesity [14].

2.4.2. Fruit and Vegetable Intake

Fruit and vegetable intake were assessed by short diet questions asking participants to self-report the number of fruit and vegetable serves consumed per day. Responses were compared with the Australian Guide to Healthy Eating (AGHE) [15] and for the purpose of this study, fruit and vegetable intake were combined to form three groups, <3 serves/day, 3–5 serves/day and $\geq$5 serves/day of fruit and vegetables. There is a dose-response relationship between the intake of fruit and vegetables and risk of CVD, where a daily consumption of <3 servings of fruit and vegetables pose the highest risk of stroke compared to >5 servings [16].

2.4.3. Physical Activity

Physical activity was assessed using the Active Australia survey [17] which included self-reported activity in hours (lasting >10 min) in the past week. The survey classified activity in four categories: light intensity (i.e., walking briskly), moderate leisure activity (e.g., swimming, exercise classes, dancing), vigorous leisure (i.e., aerobics, competitive sport, vigorous swimming, running, cycling), vigorous household or garden chores activities (i.e., gardening, household chores). Total physical activity was calculated using Metabolic Equivalent Task (MET) and computed by multiplying the sum of weekly physical activity (in minutes) by the assigned MET value (i.e., light intensity = 3.3 METs, moderate intensity = 4 METs, and vigorous intensity = 7.5). The MET values were adopted from the compendium of physical activity [18]. Four ordered categories of physical activity levels: level 1 (sedentary) = 0 $\leq$ 40 MET min/week, level 2 (insufficiently active) = 40 $\leq$ 600 MET min/week, level 3 (sufficiently active) = 600 $\leq$ 1200 MET min/week, and level 4 (very active) $\geq$ 1200 such MET min/week. Regular physical activity of at least 500–1000 MET minutes/week can reduce CVD risk factors, such as obesity, low-density lipoproteins (LDL-C) and triglycerides, increasing high-density lipoproteins (HDL-C) and insulin sensitivity and lower blood pressure [19,20].

2.4.4. Sitting Time

The time (hours and minutes) spent sitting each day on a usual weekday/weekend day, across eight domains (at home, work, transport, or for leisure such as visiting friends, reading, driving, watching television, and working on computer) was reported. This question is based on a validated questionnaire developed by Marshall et al. [21]. Higher levels of sedentary behaviour and sitting time of >10 h per day, versus ≤5 h per day are associated with elevated risk of CVD, independent of physical activity [19,22].

2.4.5. Smoking

Participants self-reported smoking status, where those indicating they smoke "daily", "weekly", or "less than weekly" classified as smokers, those who were past smokers as ex-smokers, and all others as non-smokers. Heavy smoking of ≥25 cigarettes a day increases CVD mortality by 5-fold, and light smoking (4–5 cigarettes per day) almost double's a person's risk of CVD mortality [23]. However, CVD risks are elevated for all levels of smoking, increasing with smoking intensity [23].

2.4.6. Alcohol Consumption

Alcohol consumption was assessed as frequency and quantity of alcohol consumption, e.g., the number of standard drinks usually consumed on a drinking occasion. The responses were compared to the Australian National Health and Medical Research Council (NHMRC) Alcohol Guidelines which recommends ≤2 standard drinks on any day [24]. Based on the NHMRC recommendations, participants were categorised as "non-drinker", "rarely drinks" (less than 1–2 drinks per day), "low risk" (up to 14 drinks per week/up to 2 drinks per day), "risky drinker" (15 to 28 drinks per week/3 to 4 drinks per day), as well as "high risk drinker" (5 or more drinks per day). Alcohol consumption has a dose-response relationship with CVD, where ≤2 standard drinks per day is associated with reduced risk of CVD, whereas amounts >2 increases the risk of high total cholesterol and triglycerides [19,24,25].

2.4.7. Depressive Symptoms

Depressive symptoms were measured using the Centre for Epidemiological Studies Depression Scale 10-item (CESD-10) [26]. A three-point scale was used for the CESD-10 tool to measure depression. A higher score indicates greater severity for depressive symptoms, with a score >10 being clinically significant and indicating major depression [27]. Clinical depression appears to worsen the prognosis of CVD, independent of traditional CVD risk factors. The risk of CVD, specifically coronary heart disease increases one to two-fold with minor depression and three to four-fold with major depression [28].

2.4.8. Multiple CVD Modifiable Risk Factor Score

All seven CVD modifiable risk factors were included in defining a multiple CVD modifiable risk factor score (Table 1). Participants were awarded one point if they did not meet population-based recommendations for CVD prevention, with points summed to an overall score. The score ranged from zero (no occurrence of any risk factors) to seven (occurrence of all risk factors). Participants with missing values for any of the modifiable risk factors were excluded from this analysis.

Table 1. Modifiable risk factor score.

Variable	Measurement Scale	Operationalize	Scoring
Body weight	Body Mass Index (BMI)	BMI (kg/m^2) < 18.5	1
		BMI = 18.5–24.9	0
		BMI = 25–29.9	1
		BMI ≥ 30	1
Physical Activity	Metabolic equivalent of task (MET) value	Level 1 (sedentary) = 0 <40 MET min/week	1
		Level 2 (insufficiently active) = 40 ≤ 600 MET min/week	1
		Level 3 (sufficiently active) = 600 ≤ 1200 MET min/week	0
		Level 4 (very active) = >1200 MET min/week	0

Table 1. *Cont.*

Variable	Measurement Scale	Operationalize	Scoring
Fruit & Vegetable Intake	Number of serves per day	≥5 serves of fruit and/or vegetables/day	0
		3–5 serves of fruit and/or vegetables/day	0
		<3 serves fruit and/or vegetables/day	1
Smoking	Yes/No	Current smoker/Ex-smoker	1
		Non-smoker	0
Alcohol	Standard Drink	≤1–2 standard drink/day	0
		>2 standard drinks per day	1
Sitting Time	Time spent sitting	≥8 h sitting time/day	1
		<8 h sitting time/day	0
Mental Health	Centre for Epidemiological Studies Depression Scale 10-item (CESD-10)	>10 points CESD-10	1
		≤10 points CESD-10	0

2.5. Sociodemographic Characteristics and Health Status

Sociodemographic variables, including age at the time of completing the survey, current residential area, marital status, employment status, highest education level, ability to manage income, number of children, and difficulty sleeping were included.

2.6. Statistical Methods

All statistical analyses were performed using the Stata 16.1 (StataCorp LLC, College Station, TX, USA) [29].To describe individual modifiable risk factors and the Multiple Modifiable Risk Factor Score among women with a history of HDP, results are presented as median with interquartile range for continuous data and percentages for categorical data.

To explore the difference in individual modifiable risk factors between women with and without a history of HDP, a univariate analysis using *t*-tests for continuous variables and χ^2 tests for categorical variables was undertaken, with *p*-values < 0.05 considered statistically significant. Individual risk factors demonstrating significant differences between women with and without a history of HDP in the univariate analysis were tested in multinominal logistic regression models to estimate risk of individual modifiable risk factors between women with and without a history of HDP. The models were adjusted for socio-demographic covariates found to differ between women with and without a history of HDP (weekly hours worked, area of residence, ability to manage on income and highest level of education) [30]. Results are presented as Odds Ratio (OR) with 95% confidence intervals.

To explore the difference in the Multiple Modifiable Risk Factor Score between women with and without a history of HDP, an χ^2 test was undertaken, with *p*-values < 0.05 considered statistically significant. A generalised linear model with a Poisson distribution for count data and a log link was used to estimate the association between history of HDP and Multiple Modifiable Risk Factor Score. Socio-demographic covariates found to differ between women with and without a history of HDP were included in the adjusted model (ability to manage on income, area of residence, hours worked per week, and education level) [30]. Data were presented using an Incidence Rate Ratio (IRR), with 95% confidence intervals.

3. Results

3.1. Selection of Participants

Overall, 7186 women responded to Survey 7, with 5820 participants who identified as parous eligible for inclusion in this analysis. Seven hundred and fifty-five women reported a history of HDP, and 4549 did not (Figure 1). Five hundred and sixteen women were excluded from the analysis as their history of HDP was unknown.

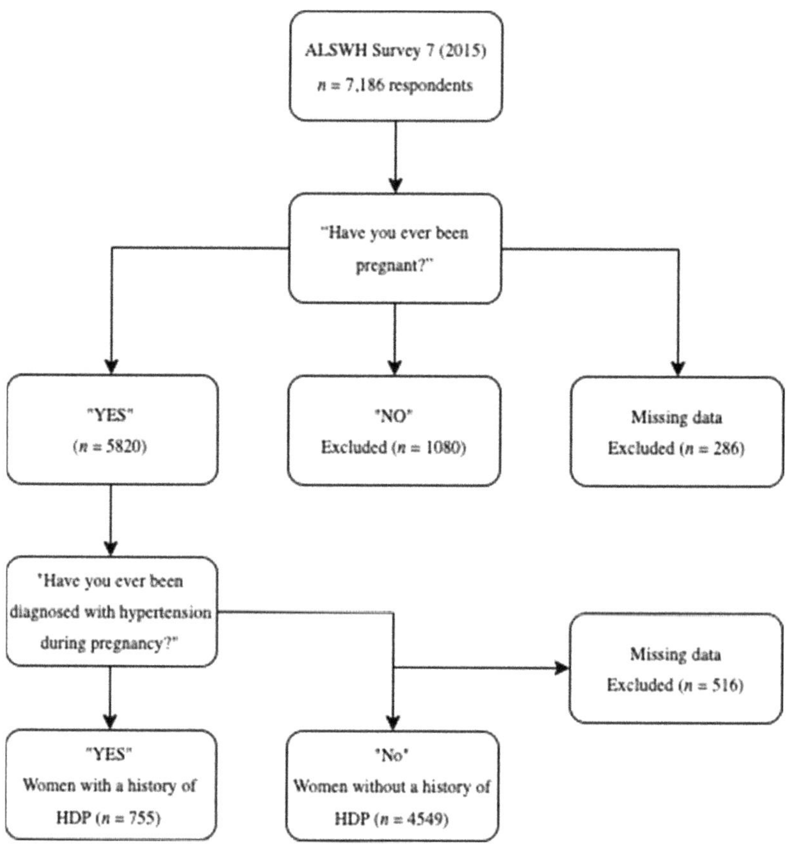

Figure 1. Summary of selection of the participants included in the analysis.

3.2. Participants' Sociodemographic Characteristics and Health Status

The sociodemographic characteristics of the women are presented in Table 2. Women who did and did not have a history of HDP were comparable in age and marital status. A significantly higher proportion of women with a history of HDP reported difficulties with managing income, residing outside of major cities, and had lower education qualifications ($p < 0.05$).

Table 2. Socio-demographic characteristics and health status of women categorised by their history of HDP.

Characteristic	Women without a History of HDP (n = 4549)	Women with a History of HDP (n = 755)	p-Value *
Age (years), mean (SD)	39.69 (1.5)	39.77 (1.5)	0.19
Current residential area (ARIA [(a)]+ Grouped into categories), n (%)			
Major cities	2544 (55.9)	386 (51.1)	0.02 *
Inner regional	1215 (26.7)	213 (28.2)	
Outer regional	557 (12.2)	121 (16.0)	

Table 2. *Cont.*

Characteristic	Women without a History of HDP (n = 4549)	Women with a History of HDP (n = 755)	p-Value *
Remote	85 (1.9)	20 (2.7)	
Very remote	30 (0.7)	6 (0.8)	
Missing	118 (2.6)	9 (1.2)	
Current Marital Status, n (%)			
Married/de facto—with partner	3967 (87.2)	642 (85.0)	0.07
Not married [(b)]	521 (11.5)	104 (13.8)	
Missing	61 (1.3)	9 (1.2)	
Weekly hours worked, n (%)			
Part-time	2169 (47.7)	308 (40.8)	<0.01 *
Full-time	1693 (37.2)	313 (41.5)	
Not in Labour Force	685 (15.1)	134 (17.8)	
Missing	2 (0.0)	0 (0.0)	
Ability to manage on income, n (%)			
Impossible/Difficult	650 (14.3)	159 (21.1)	<0.01 *
Difficult sometimes	1313 (28.9)	230 (30.5)	
Not too bad/Easy	2524 (55.5)	356 (47.2)	
Missing	62 (1.4)	10 (1.3)	
Highest qualification, n (%)			
No formal/Year 10 or equivalent	247 (5.4)	56 (7.4)	<0.01 *
Year 12 or equivalent	443 (9.7)	106 (14.0)	
Trade/apprenticeship/certificate/diploma	1208 (26.6)	241 (31.9)	
University/Higher university degree	2575 (56.6)	342 (45.3)	
Missing	76 (1.7)	10 (1.3)	
Number of Children, n (%)			
0	2 (0.0)	0 (0.0)	0.05
1	789 (17.3)	116 (15.4)	
2	2304 (50.7)	354 (46.9)	
3	1077 (23.7)	206 (27.3)	
4	295 (6.5)	59 (7.8)	
5	82 (1.8)	20 (2.7)	

* p-value of <0.05 was considered statistically significant. (a): ARIA: Accessibility/Remoteness Index of Australia; (b): Consists of separated/divorced/widowed/single.

3.3. Rates of Individual and Multiple Modifiable Risk Factors among Women with a History of HDP

The proportion of women reporting individual risk factors is summarized in Table 3. Many (69.1%) women with a history of HDP had a BMI $\geq$ 25 kg/m^2, 44.1% had physical activity levels <600 MET min/week, 38.8% were either ex-smokers or smoked at least monthly, 19.5% consumed <3 serves of fruit and vegetables per day, 7.6% consumed risky levels of alcohol, 23.4% sat for $\geq$8 h per day and 27.4% had CESD-10 scores above 10.

Table 3. Rates of individual CVD risk factors in both women with and without a history of HDP.

Variable	Women without a History of HDP (n = 5055)	Women with a History of HDP (n = 755)	p-Value *
BMI, n (%)			
BMI (kg/m^2) < 18.5	91 (2.0)	5 (0.7)	<0.01 *
BMI = 18.5–24.9 $^\beta$	2242 (49.3)	218 (28.9)	
BMI = 25–29.9	1199 (26.4)	209 (27.7)	
BMI ≥ 30	941 (20.7)	313 (41.5)	
Missing	76 (1.7)	10 (1.3)	
Fruit and vegetable intake, n (%)			
<3 servings a day	803 (17.7)	147 (19.5)	<0.20
3–5 servings a day $^\beta$	2737 (60.2)	461 (61.1)	
≥5 servings a day $^\beta$	944 (20.8)	138 (18.3)	
Missing	65 (1.4)	9 (1.2)	
Physical Activity [(a)], n (%)			
≤40 MET min/week	605 (13.3)	116 (15.4)	0.33
40–600 MET min/week	1283 (28.2)	217 (28.7)	
600–1200 MET min/week $^\beta$	950 (20.9)	145 (19.2)	
>1200 MET min/week $^\beta$	1332 (29.3)	210 (27.8)	
Missing	379 (8.3)	67 (8.9)	
Sitting Time, n (%)			
≥8 h	993 (21.8)	177 (23.4)	0.22
<8 h $^\beta$	2260 (73.9)	534 (70.7)	
Missing	196 (4.3)	44 (5.8)	
Smoking [(b)], n (%)			
Non-smoker $^\beta$	2788 (61.3)	458 (60.7)	0.14
Ex-smoker	1338 (29.4)	208 (27.6)	
Smoke, daily, weekly, or monthly	416 (9.1)	85 (11.3)	
Missing	7 (0.2)	4 (0.5)	
Alcohol consumption [(c)], n (%)			
Non-drinker $^\beta$	460 (10.1)	93 (12.3)	<0.01 *
Rarely drinker $^\beta$	1084 (23.8)	204 (27.0)	
Low risk drinker $^\beta$	2723 (59.7)	399 (52.9)	
Risky drinker	227 (5.0)	42 (5.6)	
High risk drinker	52 (1.1)	15 (2.0)	
Missing	3 (0.1)	2 (0.3)	
Depressive symptoms [(d)], n (%)			
Having a score of >10 CESD-10	805 (17.7)	184 (27.4)	<0.01 *
Having a score of ≤10 CESD-10 $^\beta$	3725 (81.9)	566 (75.0)	
Missing	19 (0.4)	5 (0.7)	

HDP: Hypertensive Disorders of Pregnancy. BMI: Body mass index. * Chi-Squared (χ^2) test for categorical variables and differences between HDP and non-HDP, p-value of <0.05 was considered statistically significant. $^\beta$ Indicates that the respective recommendation was met. Data are presented as mean (standard deviation) for continuous variables and frequency and percentage (%) for categorical variables. (a): Physical activity measured in metabolic minutes (MET min); (b): Smoking cessation of more than 15 years would be classified as non-smoker in the multiple CVD modifiable risk factor score (see Table 1), missing data on year quitting smoking assumed to be current smoker; (c): NHMRC: National Health and Medical Research Council alcohol classification; (d): CESD-10: 10-item Centre for Epidemiological Studies Depression Scales was used to identify depressive symptoms.

Figure 2 and Table 4 present the multiple CVD modifiable risk factor score among women with and without a history of HDP, which is the summative of participants not meeting population-based recommendations for individual CVD modifiable risk factors. Over one third (36.0%) of women with a history of HDP had ≥3 risk factors and 5.7% with ≥5 risk factors. Majority of women with a history of HDP had ≤2 risk factors (48.5%) where 7.3% had no CVD risk factors.

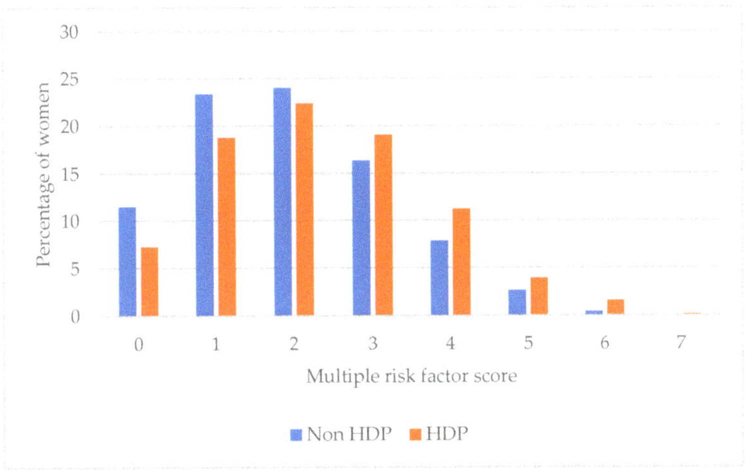

Figure 2. Multiple modifiable risk factor scores among women with and without HDP.

Table 4. Multiple modifiable risk factor scores among women with and without HDP.

Multiple CVD Modifiable Risk Factor Score, *n* (%) [(a)]	Women without a History of HDP (*n* = 3929)	Women with a History of HDP (*n* = 638)	*p*-Value *
0 points	524 (11.5)	55 (7.3)	<0.01 *
1 point	1065 (23.4)	142 (18.8)	
2 points	1093 (24.0)	169 (22.4)	
3 points	744 (16.4)	144 (19.1)	
4 points	359 (7.9)	85 (11.3)	
5 points	121 (2.7)	30 (4.0)	
6 points	21 (0.5)	12 (1.6)	
7 points	2 (0.0)	1 (0.1)	
Multiple Modifiable Risk Factor Score, mean (SD)	1.95 (1.3)	2.32 (1.4)	<0.01 *

(a): Multiple CVD Modifiable Risk Factor Score: Summative of participants not meeting population-based recommendations for individual CVD modifiable risk factors. * *p*-value of <0.05 was considered statistically significant. Participants with missing value for any of the above modifiable risk factors were excluded from this analysis.

3.4. Differences in Individual and Multiple Modifiable Risk Factors between Women with and without a History of HDP

In the univariate analysis (Table 3) when compared with women without a history of HDP, women with a history of HDP had significantly higher rates of overweight and obesity, high-risk alcohol intake, and higher rates of depressive symptoms.

In adjusted multinomial logistic regression models (Table 5) for women with a history of HDP, compared with those without, the relative risk of an overweight BMI would increase by 1.7, and obese BMI 3.1, relative to those with a healthy weight. For women with a history of HDP compared with those without, the relative risk of being a 'high risk drinker' would increase by 1.9, relative to low-risk drinkers. For women with a history of HDP compared with those without, the relative risk of having a CESD-10 score >10 would increase by 1.3, relative to those with a score ≤10.

Table 5. Individual modifiable risk factors comparing individual risk factors between women with and without a history of HDP.

	Relative Risk (95% CI)	***p*-Value ***
Body mass index		
BMI (kg/m^2) < 18.5	0.5 (0.2, 1.3)	0.16
BMI = 18.5–24.9 $^{\beta}$	Reference	
BMI = 25–29.9	1.7 (1.4, 2.1)	<0.01 *
BMI ≥ 30	3.1 (2.6, 3.8)	<0.01 *
NHMRC alcohol classification		
Non-drinker $^{\beta}$	1.2 (1.0, 1.6)	0.11
Rarely drinker $^{\beta}$	1.2 (1.0, 1.4)	0.15
Low risk drinker $^{\beta}$	Reference	
Risky drinker	1.2 (0.8, 1.)	0.45
High risk drinker	1.9 (1.0, 3.4)	0.04 *
CESD-10 score		
Having a score of >10 CESD-10	1.34 (1.1, 1.6)	<0.01 *
Having a score of ≤10 CESD-10 $^{\beta}$	Reference	

BMI: body mass index. NHMRC: National Health and Medical Research Council. CESD-10: Centre for Epidemiological Studies Depression Scales. * *p*-value of <0.05 was considered statistically significant. $^{\beta}$ Indicates that the respective recommendation was met.

In univariate analysis (Table 4), when compared to women without a history of HDP, women with a history of HDP had a significantly higher multiple modifiable risk factor score. This was confirmed in the adjusted model (IRR 1.1, 95% CI 1.1, 1.2, $p < 0.01$), where those with HDP had a 12% increase in risk factor score, although this value may vary up too 19% in similar samples.

4. Discussion

The aim of this study was to understand the rates of individual and multiple modifiable risk factors for CVD among women with a history of HDP and determine whether they differ to those without a history of HDP. Women with a history of HDP had a higher multiple CVD risk factor score, likely driven by a higher BMI ($p < 0.01$), higher levels of high-risk alcohol consumption ($p = 0.04$) and higher levels of depressive symptoms ($p < 0.01$). The results suggest that multiple CVD modifiable risk factors, but especially weight management, alcohol intake and improvements in mental health are important targets for CVD prevention among women with previous HDP.

This is the first observational study to explore multiple modifiable risk factors among women following HDP. In this representative sample of Australian women, those with a history of HDP had a higher mean multiple risk factor score and were more likely to have four or more modifiable risk factors than women without a history of HDP. Other studies also reported a higher prevalence of CVD modifiable risk factors in women with a history of HDP, including smoking [31], BMI [32,33], and hypertension [32,33], however the majority of these studies investigated CVD risk factors independently. This reinforces the notion that targeting modifiable risk factors could be a useful strategy for CVD prevention in women with a history of HDP.

In the current analyses, having a history of HDP was associated with a higher BMI. This was also shown in the Nord-Trøndelag Health Study, where higher BMI was associated with a 41% increased CVD risk in women following HDP (HR: 1.2, 95% CI: 1.1, 1.3) [32]. Similarly, findings from the Nurses' Health Study II found a BMI of 30–34.9 kg/m^2 in women with previous HDP contributed to 25% of the excess risk of chronic hypertension [33]. Our findings, supported by previous research, suggest that a healthy weight during the postpartum period is of importance to women with previous HDP, and interventions should target weight management for the reduction of future CVD risk.

In the current study having a history of HDP was also associated with greater depressive symptoms. A recent prospective cohort study explored the prevalence of mental health disorders including depression, posttraumatic stress disorder and anxiety in women with and without a history of HDP at 6-months postpartum [34]. This study similarly discovered that more women with a history of preeclampsia scored above the threshold for depression (7% vs. 2%, $p = 0.04$) on the Edinburg Postnatal Depression Scale and reported a traumatic birth experience (7% vs. 1%, $p = 0.01$) [34]. Despite an inconclusive pool of research, a 2013 systematic review of six cohort studies also suggested that there is an association between preeclampsia and depression [35]. Therefore, CVD prevention interventions for women with a history of HDP should also include an aspect of mental health support.

A history of HDP was also associated with high-risk alcohol intake, but not risky alcohol intake. Though, a previous cross-sectional study of self-reported alcohol intake determined that alcohol consumption was not statistically associated with preeclampsia risk [36]. Contrarily, a 2016 cohort study noted that binge-drinking (≥5 drinks on one occasion) increased the risk for pre-term preeclampsia specifically [37]. Although there are inconsistencies within the literature, alcohol intake above recommendations may still increase the risk of CVD, specifically through increases in blood pressure, and should be included as a target risk factor within CVD prevention after HDP [25].

Women with and without a history of HDP did not have statistically significant differences in the other modifiable risk factors (fruit and vegetable intake, physical activity, sitting time or smoking status in the current study). Despite non-significant results, women with a history of HDP still had high rates of modifiable CVD risk factors. Specifically, 19.7% of women with a history of HDP consumed <3 servings of fruit and vegetables combined per day. Additionally, 44.1% of women with a history of HDP had inadequate levels of physical activity (≤600 MET min/week) and 23.4% sat for >8 h per day. Encouragingly, a 2022 systematic review of observational and randomised controlled trials suggested that physical activity interventions for women after HDP were associated with positive improvements in CVD risk [38]. Our findings indicate that women with a history of HDP have high levels of modifiable CVD risk factors and therefore support the notion that the postpartum period is an important window of opportunity for participation in CVD interventions targeting these risk factors.

The key strengths of this study include the large sample size of women in the ALSWH, which contains a representative sample of Australian women. This study, however, did not differentiate between the subgroups of HDP (chronic hypertension, gestational hypertension, and preeclampsia). It is evident that regardless of HDP subtype, women will have an increased CVD risk within 10 years postpartum [7], however future research should examine differences in CVD modifiable risk factors by HDP sub-groups. Such research would be strengthened through inclusion of a more objective measure of HDP history (e.g., obtained from medical records). The ALSWH is a self-reported survey, which could inherently introduce reporting bias. A limitation of the current study is the creation of cut-points and categories used for the Multiple CVD Modifiable Risk Factor Score, which were structured based on the evidence of CVD risk. In addition, the scoring approach used for the Multiple CVD Modifiable Risk Factor Score (e.g., combining all seven risk factors together), was developed for the purpose of this study and has not been previously validated. Furthermore, this cross-sectional analysis does not consider time post-HDP, nor does it consider occurrence of CVD, modifiable CVD risk factors or other CVD risk factors (e.g., family history) present prior to a pregnancy complicated by HDP. Rather, this study delivers a snapshot of CVD risk factors present in women with a history of HDP and insight into what risk factors may need to be the focus of future CVD preventative strategies for women with a history of HDP.

5. Conclusions

The findings from this study suggest that a history of HDP is independently associated with CVD modifiable risk factors. Higher BMI, high-risk alcohol consumption and depressive symptoms appear to be important risk factors, highlighting the significance of weight management, responsible alcohol consumption and mental health services as CVD prevention strategies for women post-HDP. Understanding that following HDP women have higher rates of CVD risk factors allows clinicians to provide management and intervention for this group of women. Therefore, the postpartum period is a suitable opportunity to engage women in timely CVD prevention interventions and address these risk factors.

Author Contributions: Conceptualization, M.J.H., R.M.T., T.L.S. and V.A.S.; methodology, M.J.H. and T.L.S.; software, M.J.H. and T.L.S.; validation, M.J.H., R.M.T., T.L.S. and V.A.S.; formal analysis, K.S., M.J.H. and T.L.S.; investigation, M.J.H., R.M.T., T.L.S. and V.A.S.; resources, K.S., K.N.D. and M.J.H.; data curation, M.J.H. and K.S.; writing—original draft preparation, K.S. and K.N.D.; writing—review and editing, K.S., K.N.D., M.J.H., R.M.T., T.L.S. and V.A.S.; visualization, K.S.; supervision, M.J.H., R.M.T., T.L.S. and V.A.S.; project administration, M.J.H. All authors have read and agreed to the published version of the manuscript.

Funding: This research received no external funding. V.A.S. is supported by a Hunter Medical Research Institute grant (HMRI Grant: 1664).

Institutional Review Board Statement: The study was conducted in accordance with the Declaration of Helsinki, and approved by Human Research Ethics Committees of the Universities of Newcastle and Queensland (approval numbers H076-0795 and 2004000224, respectively, for the 1973-78, 1946-51 and 1921-26 cohorts: and H-2012-0256 and 2012000950, for the 1989-95 cohort) on the 31 January 2012.

Informed Consent Statement: Informed consent was obtained from all subjects involved in the study.

Data Availability Statement: The original contributions presented in the study are included in the article, further inquiries can be directed to the corresponding author.

Acknowledgments: The research on which this paper is based was conducted as part of the Australian Longitudinal Study on Women's Health by the University of Queensland and the University of Newcastle. We are grateful to the Australian Government Department of Health and Aged Care for funding and to the women who provided the survey data.

Conflicts of Interest: The authors declare no conflict of interest.

References

1. Institute for Health Metrics and Evaluation. Gbd Compare Data Visualization. Available online: https://vizhub.healthdata.org/gbd-compare/ (accessed on 11 May 2022).
2. Australian Institute of Health and Welfare (AIHW). *Cardiovascular Disease in Women*; ACT: Canberra, Australia, 2019.
3. Word Health Organization. Cardiovascular Disease Risk Factors. Available online: https://apps.who.int/iris/bitstream/handle/10665/37644/WHO_TRS_841.pdf;jsessionid=5E12914F766D796EB23B78CBBCD5A077?sequence=1 (accessed on 18 May 2022).
4. Garcia, M.; Mulvagh, S.L.; Bairey Merz, C.N.; Buring, J.E.; Manson, J.E. Cardiovascular Disease in Women. *Circ. Res.* **2016**, *118*, 1273–1293. [CrossRef] [PubMed]
5. Tsai, M.C.; Lee, C.C.; Liu, S.C.; Tseng, P.J.; Chien, K.L. Combined healthy lifestyle factors are more beneficial in reducing cardiovascular disease in younger adults: A meta-analysis of prospective cohort studies. *Sci. Rep.* **2020**, *10*, 18165. [CrossRef] [PubMed]
6. Ikem, E.; Halldorsson, T.I.; Birgisdóttir, B.E.; Rasmussen, M.A.; Olsen, S.F.; Maslova, E. Dietary patterns and the risk of pregnancy-associated hypertension in the Danish National Birth Cohort: A prospective longitudinal study. *BJOG* **2019**, *126*, 663–673. [CrossRef] [PubMed]
7. Wu, R.; Wang, T.; Gu, R.; Xing, D.; Ye, C.; Chen, Y.; Liu, X.; Chen, L. Hypertensive Disorders of Pregnancy and Risk of Cardiovascular Disease-Related Morbidity and Mortality: A Systematic Review and Meta-Analysis. *Cardiology* **2020**, *145*, 633–647. [CrossRef] [PubMed]
8. Hutchesson, M.; Campbell, L.; Leonard, A.; Vincze, L.; Shrewsbury, V.; Collins, C.; Taylor, R. Do modifiable risk factors for cardiovascular disease post-pregnancy influence the association between hypertensive disorders of pregnancy and cardiovascular health outcomes? A systematic review of observational studies. *Pregnancy Hypertens* **2022**, *27*, 138–147. [CrossRef]
9. Lee, C. Cohort Profile: The Australian Longitudinal Study on Women's Health. *Int. J. Epidemiol.* **2005**, *34*, 987–991. [CrossRef] [PubMed]

10. Brown, W.J.; Bryson, L.; Byles, J.E.; Dobson, A.J.; Lee, C.; Mishra, G.; Schofield, M. Women's Health Australia: Recruitment for a National Longitudinal Cohort Study. *J. Womens Health* **1999**, *28*, 23–40. [CrossRef]
11. Brown, W.; Bryson, L.; Byles, J.; Dobson, A.; Manderson, L.; Schofield, M.; WIlliams, G. Women's Health Australia: Establishment of The Australian Longitudinal Study on Women's Health. *J. Womens Health* **1996**, *5*, 467–472. [CrossRef]
12. Stuart, J.J.; Bairey Merz, C.N.; Berga, S.L.; Miller, V.M.; Ouyang, P.; Shufelt, C.L.; Steiner, M.; Wenger, N.K.; Rich-Edwards, J.W. Maternal recall of hypertensive disorders in pregnancy: A systematic review. *J. Womens Health* **2013**, *22*, 37–47. [CrossRef]
13. World Health Organisation. Body Mass Index—BMI. Available online: http://www.euro.who.int/en/health-topics/disease-prevention/nutrition/a-healthy-lifestyle/body-mass-index-bmi (accessed on 1 June 2022).
14. Rippe, J.M. Lifestyle Strategies for Risk Factor Reduction, Prevention, and Treatment of Cardiovascular Disease. *Am. J. Lifestyle Med.* **2018**, *13*, 204–212. [CrossRef]
15. Australian Government National Health and Medical Research Council (NHMRC). *Australian Dietary Guidelines*; National Health and Medical Research Council: Canberra, Australia, 2013.
16. He, F.J.; Nowson, C.A.; MacGregor, G.A. Fruit and vegetable consumption and stroke: Meta-analysis of cohort studies. *Lancet* **2006**, *367*, 320–326. [CrossRef]
17. Brown, W.J.; Burton, N.W.; Marshall, A.L.; Miller, Y.D. Reliability and validity of a modified self-administered version of the Active Australia physical activity survey in a sample of mid-age women. *Aust. N. Z. J. Public Health* **2008**, *32*, 535–541. [CrossRef] [PubMed]
18. Ainsworth, B.E.; Haskell, W.L.; Herrmann, S.D.; Meckes, N.; Bassett, D.R.; Tudor-Locke, C.; Greer, J.L.; Vezina, J.; Whitt-Glover, M.C.; Leon, A.S. 2011 Compendium of Physical Activities. *Med. Sci. Sports Exerc.* **2011**, *43*, 1575–1581. [CrossRef] [PubMed]
19. National Vascular Disease Prevention Alliance. *Guidelines for the Management of Absolute Cardiovascular Disease Risk*; The Heart Foundation: Subiaco, Australia, 2012.
20. Eckel, R.H.; Jakicic, J.M.; Ard, J.D.; de Jesus, J.M.; Houston Miller, N.; Hubbard, V.S.; Lee, I.M.; Lichtenstein, A.H.; Loria, C.M.; Millen, B.E.; et al. 2013 AHA/ACC guideline on lifestyle management to reduce cardiovascular risk: A report of the American College of Cardiology/American Heart Association Task Force on Practice Guidelines. *J. Am. Coll. Cardiol.* **2014**, *63*, 2960–2984. [CrossRef] [PubMed]
21. Marshall, A.; Miller, Y.; Burton, N.; Brown, W. Measuring Total and Domain-Specific Sitting. *Med. Sci. Sports Exerc.* **2009**, *42*, 1094–1102. [CrossRef]
22. Young, D.R.; Hivert, M.-F.; Alhassan, S.; Camhi, S.M.; Ferguson, J.F.; Katzmarzyk, P.T.; Lewis, C.E.; Owen, N.; Perry, C.K.; Siddique, J.; et al. Sedentary Behavior and Cardiovascular Morbidity and Mortality: A Science Advisory From the American Heart Association. *Circulation* **2016**, *134*, e262–e279. [CrossRef] [PubMed]
23. Banks, E.; Joshy, G.; Korda, R.J.; Stavreski, B.; Soga, K.; Egger, S.; Day, C.; Clarke, N.E.; Lewington, S.; Lopez, A.D. Tobacco smoking and risk of 36 cardiovascular disease subtypes: Fatal and non-fatal outcomes in a large prospective Australian study. *BMC Med.* **2019**, *17*, 128. [CrossRef] [PubMed]
24. National Health and Medical Research Council. Australian Guidelines to Reduce Health Risks from Drinking Alcohol. Available online: https://www.nhmrc.gov.au/about-us/publications/australian-guidelines-reduce-health-risks-drinking-alcohol#download (accessed on 26 May 2020).
25. Larsson, S.C.; Burgess, S.; Mason, A.M.; Michaëlsson, K. Alcohol Consumption and Cardiovascular Disease. *Circ. Genom. Precis. Med.* **2020**, *13*, e002814. [CrossRef] [PubMed]
26. Holden, L.; Dobson, A.; Byles, J.E.; Loxton, D.J.; Dolja-Gore, X.; Hockey, R.; Lee, C.; Chojenta, C.; Reilly, N.M.; Mishra, G.D.; et al. *Mental Health: Findings from the Australian Longitudinal Study on Women's Health*; Women's Health: Greenslopes, QLD, Australia, 2013.
27. Andresen, E.M.; Malmgren, J.A.; Carter, W.B.; Patrick, D.L. Screening for depression in well older adults: Evaluation of a short form of the CES-D. *Am. J. Prev. Med.* **1994**, *10*, 77–84. [CrossRef]
28. Bunker, S.J.; Colquhoun, D.M.; Esler, M.D.; Hickie, I.B.; Hunt, D.; Jelinek, V.M.; Oldenburg, B.F.; Peach, H.G.; Ruth, D.; Tennant, C.C.; et al. "Stress" and coronary heart disease: Psychosocial risk factors. *Med. J. Aust.* **2003**, *178*, 272–276. [CrossRef]
29. StataCorp. *Stata Statistical Software: Release 16*; StataCorp LLC.: College Station, TX, USA, 2019.
30. Schultz, W.M.; Kelli, H.M.; Lisko, J.C.; Varghese, T.; Shen, J.; Sandesara, P.; Quyyumi, A.A.; Taylor, H.A.; Gulati, M.; Harold, J.G.; et al. Socioeconomic Status and Cardiovascular Outcomes. *Circulation* **2018**, *137*, 2166–2178. [CrossRef] [PubMed]
31. Arnott, C.; Nelson, M.; Alfaro Ramirez, M.; Hyett, J.; Gale, M.; Henry, A.; Celermajer, D.S.; Taylor, L.; Woodward, M. Maternal cardiovascular risk after hypertensive disorder of pregnancy. *Heart* **2020**, *106*, 1927. [CrossRef] [PubMed]
32. Haug, E.B.; Horn, J.; Markovitz, A.R.; Fraser, A.; Klykken, B.; Dalen, H.; Vatten, L.J.; Romundstad, P.R.; Rich-Edwards, J.W.; Åsvold, B.O. Association of Conventional Cardiovascular Risk Factors With Cardiovascular Disease After Hypertensive Disorders of Pregnancy. *JAMA Cardiol.* **2019**, *4*, 628. [CrossRef] [PubMed]
33. Timpka, S.; Stuart, J.J.; Tanz, L.J.; Rimm, E.B.; Franks, P.W.; Rich-Edwards, J.W. Lifestyle in progression from hypertensive disorders of pregnancy to chronic hypertension in Nurses' Health Study II: Observational cohort study. *BMJ* **2017**, *358*, j3024. [CrossRef] [PubMed]
34. Roberts, L.; Henry, A.; Harvey, S.B.; Homer, C.S.E.; Davis, G.K. Depression, anxiety and posttraumatic stress disorder six months following preeclampsia and normotensive pregnancy: A P4 study. *BMC Pregnancy Childbirth* **2022**, *22*, 108. [CrossRef] [PubMed]

35. Delahaije, D.H.; Dirksen, C.D.; Peeters, L.L.; Smits, L.J. Anxiety and depression following preeclampsia or hemolysis, elevated liver enzymes, and low platelets syndrome. A systematic review. *Acta Obs. Gynecol. Scand.* **2013**, *92*, 746–761. [CrossRef]
36. Klonoff-Cohen, H.S.; Edelstein, S.L. Alcohol Consumption During Pregnancy and Preeclampsia. *J. Womens Health* **1996**, *5*, 225–230. [CrossRef]
37. Egeland, G.M.; Klungsøyr, K.; Øyen, N.; Tell, G.S.; Næss, Ø.; Skjærven, R. Preconception Cardiovascular Risk Factor Differences Between Gestational Hypertension and Preeclampsia: Cohort Norway Study. *Hypertension* **2016**, *67*, 1173–1180. [CrossRef]
38. Byrnes, M.; Buchholz, S.W. Physical Activity and Cardiovascular Risk Factor Outcomes in Women with a History of Hypertensive Disorders of Pregnancy: Integrative Review. *Worldviews Evid. Based Nurs.* **2022**, *19*, 47–55. [CrossRef]

Article

The Effect of Dextrose or Protein Ingestion on Circulating Growth Differentiation Factor 15 and Appetite in Older Compared to Younger Women

Catrin Herpich [1,2,3,*], Stephanie Lehmann [1], Bastian Kochlik [3], Maximilian Kleinert [4,5,6], Susanne Klaus [1,7], Ursula Müller-Werdan [2,8] and Kristina Norman [1,2,3,9]

1 Institute of Nutritional Science, University of Potsdam, 14558 Nuthetal, Germany
2 Department of Geriatrics and Medical Gerontology, Charité—Universitätsmedizin Berlin, Corporate Member of Freie Universität Berlin and Humboldt-Universität zu Berlin, 13347 Berlin, Germany
3 Department of Nutrition and Gerontology, German Institute of Human Nutrition Potsdam-Rehbrücke, 14558 Nuthetal, Germany
4 Muscle Physiology and Metabolism Group, German Institute of Human Nutrition, Potsdam-Rehbrücke, 14558 Nuthetal, Germany
5 Department of Nutrition, Faculty of Science, Section of Molecular Physiology, Exercise and Sports, University of Copenhagen, 1165 Copenhagen, Denmark
6 German Center for Diabetes Research (DZD), DIfE, Potsdam-Rehbrücke, 14558 Nuthetal, Germany
7 Department of Physiology of Energy Metabolism, German Institute of Human Nutrition Potsdam-Rehbrücke, 14558 Nuthetal, Germany
8 Evangelisches Geriatriezentrum Berlin gGmbH, 13347 Berlin, Germany
9 German Center for Cardiovascular Research (DZHK), Partner Site Berlin,14558 Berlin, Germany
* Correspondence: catrin.herpich@dife.de

Citation: Herpich, C.; Lehmann, S.; Kochlik, B.; Kleinert, M.; Klaus, S.; Müller-Werdan, U.; Norman, K. The Effect of Dextrose or Protein Ingestion on Circulating Growth Differentiation Factor 15 and Appetite in Older Compared to Younger Women. *Nutrients* **2022**, *14*, 4066. https://doi.org/10.3390/nu14194066

Academic Editors: Birgit-Christiane Zyriax and Nataliya Makarova

Received: 9 September 2022
Accepted: 26 September 2022
Published: 30 September 2022

Abstract: Growth differentiation factor 15 (GDF15) is a stress signal that can be induced by protein restriction and is associated with reduced food intake. Anorexia of aging, insufficient protein intake as well as high GDF15 concentrations often occur in older age, but it is unknown whether GDF15 concentrations change acutely after meal ingestion and affect appetite in older individuals. After an overnight fast, appetite was assessed in older (n = 20; 73.7 ± 6.30 years) and younger (n = 20; 25.7 ± 4.39 years) women with visual analogue scales, and concentrations of circulating GDF15 and glucagon-like peptide-1 (GLP-1) were quantified before and at 1, 2 and 4 h after ingestion of either dextrose (182 kcal) or a mixed protein-rich meal (450 kcal). In response to dextrose ingestion, appetite increased in both older and younger women, whereas GDF15 concentrations increased only in the older group. In older women, appetite response was negatively correlated with the GDF15 response (rho = −0.802, p = 0.005). Following high-protein ingestion, appetite increased in younger women, but remained low in the old, while GDF15 concentrations did not change significantly in either age group. GLP-1 concentrations did not differ between age groups or test meals. In summary, acute GDF15 response differed between older and younger women. Associations of postprandial appetite and GDF15 following dextrose ingestion in older women suggest a reduced appetite response when the GDF15 response is high, thus supporting the proposed anorectic effects of high GDF15 concentrations.

Keywords: aging; anorexia of aging; GDF15; GLP-1; postprandial

1. Introduction

Growth differentiation factor 15 (GDF15) is a cellular stress-induced cytokine, and higher circulating concentrations are found in various chronic and acute diseases [1] as well as in older age [2]. The role and effects of higher GDF15 concentrations, during aging in particular, are unclear.

While most cells and tissues are able to secrete GDF15 [3], the expression of the GDF15 receptor, glial cell-derived neurotrophic factor family receptor alpha-like (GFRAL), has been detected only in the brainstem [4]. Recently, GDF15 has been shown to regulate appetite [5] via GDF15-GFRAL signaling [1]. The activation of this signaling axis is associated with conditioned taste aversion [3,6] and the modulation of the vagal sympathetic nervous system, which controls, e.g., gastric emptying [7]. GDF15 expression is sensitive to various nutritional stimuli such as chronic high-fat overfeeding or lysine-deficient diets, which is mediated by the integrated stress response (ISR) [3].

Aging is frequently accompanied by anorexia of aging, which is characterized by reduced appetite as well as lower food intake, and is associated with an increased risk for malnutrition, since the lower energy intake often coincides with insufficient macro- and micronutrient supply, most importantly of protein [8,9]. The etiology is not clear, but mostly likely involves the interplay of several age-related sensory and metabolic changes [8]. Sensory decline, e.g., the loss of taste and sense of smell, can lead to reduced food palatability. Furthermore, various metabolic alterations contribute to overall increased satiety, and therefore lower appetite and food intake, in the old [8]. For example, postprandial hunger, satiety as well as gastric emptying differ in older compared to younger adults after ingestion of a mixed meal [10]. In this context it is interesting that bariatric surgery, which is known to alter satiety and gastric emptying, increases circulating GDF15 in both men and women [11].

To date, it is not known if the high GDF15 concentrations found in older ages affect appetite in humans, and whether there is an age-dependent difference in the GDF15 response to different meals. In addition, it remains to be elucidated if the GDF15 response to meal ingestion affects postprandial appetite. Overall, there are only a few studies addressing the postprandial response of GDF15 in humans [12–14]. Therefore, we investigated postprandial circulating GDF15 and its association with postprandial appetite in young and older women.

2. Materials and Methods

This is a sub-analysis of a larger study described elsewhere [15]. In brief, community-dwelling older and younger adults were recruited. In order to obtain a significant age gap between the groups, we pre-specified age ranges of 65 to 85 years for the older group and 18 to 35 years for the younger group. In the older group, we did not recruit adults aged above 85 out of ethical considerations, and in the younger group we did not recruit adults over 35 to preclude any early perimenopausal changes. The study was approved by the ethics committee of the University of Potsdam and registered at drks.de as DRKS00017090. All participants signed a written informed consent.

As the number of men was low and sex differences are known [16], men were excluded from this analysis. One younger woman had to be excluded from postprandial analysis since she did not complete the meal challenge. The postprandial GDF15 response was assessed after dextrose (50 g dextrose, total energy content: 182 kcal; $n = 10$ per age group) or high-protein ingestion (77 energy percent protein, total energy content: 450 kcal; 250 g curd cheese, 50 g protein powder, 100 g raspberries, 100 mL milk 1.5% fat; $n = 10$ older group, $n = 9$ younger group). Blood samples were drawn after an overnight fast and repeated blood samples were taken 30, 60, 120 and 240 min after meal ingestion. EDTA plasma was obtained and stored at -80 °C until analysis. Subjective appetite was assessed using a visual analogue scale. Participants were instructed to mark their current feeling of appetite (ranging from 0 = "no appetite" to 10 = "great appetite") every time a blood sample was drawn. Appetite sensation was displayed in cm on the VAS. Fat mass was estimated using bioelectrical impedance analysis and age-appropriate equations [17]. Fat mass index (FMI) was calculated by dividing the fat mass (kg) by height squared (m^2).

Plasma GDF15 (intra-assay CV: 6.3–7.2%; inter-assay CV: 2.9–5.6%; BioVendor, Brno, Czech Republic) concentrations were quantified using commercial ELISA assays. As an objective marker for appetite/satiety, we also measured glucagon-like peptide 1 (GLP-1) (intra-assay CV: 4.69–10.7%; inter-assay CV: 9.63–17.6%; Yanaihara Institute Inc, Shizuoka, Japan). As markers of the glucose metabolism, serum insulin (commercial ELISA, intra-assay CV: 4.8–6.0%; inter-assay CV: 8.1–9.0%; BioVendor, Brno, Czech Republic) as well as serum glucose concentrations (colorimetric method, ABX Pentra 400, Horiba, Ltd., Kyoto, Japan) were quantified. Homeostasis model assessment was used to estimate insulin resistance (HOMA-IR).

Statistical analyses were performed with SPSS (IBM version 27, SPSS Incorporated, Chicago, IL, USA) and GraphPad Prism (version 7.00 for Windows, GraphPad Software, La Jolla, CA, USA). Data are presented as mean ± standard deviation (SD). Group differences were calculated using as appropriate Student's *t*-test or Mann–Whitney U test and correlations with Pearson's correlation coefficient or Spearman's rho. GLP-1 concentrations were logarithmized for normalization. Changes over time and time × meal interactions were examined with repeated measures ANOVA. GDF15, GLP-1, glucose and insulin response, and increase in appetite after meal ingestion were evaluated using positive incremental area under the curve (iAUC). An acceptable level of statistical significance was established a priori at $p < 0.05$.

3. Results

A description of the participants is shown in Table 1. The study participants were overall healthy, with self-reported high blood pressure being the most frequent pre-existing condition in older women (50%). Despite having higher fasting glucose concentrations, older women exhibited similar insulin and HOMA-IR values to younger women. Baseline GDF15 concentrations were significantly higher in the older compared to the younger women (802 ± 227 versus 364 ± 125 pg/mL, $p < 0.001$).

Table 1. Characteristics of study participants at baseline.

	Older Women $n = 20$	Younger Women $n = 20$	*p*-Value
	Mean ± SD	**Mean ± SD**	
Age (years)	73.7 ± 6.30	25.7 ± 4.39	
BMI (kg/m^2)	23.9 ± 3.93	22.1 ± 2.33	0.090
FMI (kg/m^2)	8.65 ± 2.78	7.04 ± 1.73	0.036
Glucose (mmol/L)	4.98 ± 0.44	4.54 ± 0.42	0.003
Insulin (µUI/mL)	12.2 ± 11.5	10.3 ± 2.49	0.488
HOMA-IR	2.70 ± 2.48	2.09 ± 0.58	0.309
GDF15 (pg/mL)	802 ± 227	364 ± 125	<0.001
GLP-1 (ng/mL)	2.60 ± 1.37	3.62 ± 2.29	0.108 [a]

BMI: body mass index; FMI: fat mass index, GDF15: growth differentiation factor 15; GLP-1: glucagon-like peptide 1; HOMA-IR: Homeostasis Model Assessment—Insulin Resistance; SD: standard deviation, differences between groups calculated using Student's *t*-test; [a] differences between groups calculated using Mann–Whitney U test.

BMI was similar between both groups, but older women exhibited a higher fat mass (8.65 versus 7.0 kg/m^2, $p = 0.036$). Fat mass index was also positively correlated with fasting GDF15 concentrations (r = 0.346, $p = 0.029$), but not with BMI (Supplementary Material Figure S1). In older women, baseline GDF15 concentrations were negatively correlated with baseline appetite (r = −0.488, $p = 0.029$), whereas fasting GLP-1 concentrations were positively associated with baseline appetite (r = 0.461, $p = 0.041$) (Figure 1). Glycemic parameters were not correlated with GDF15 or GLP-1.

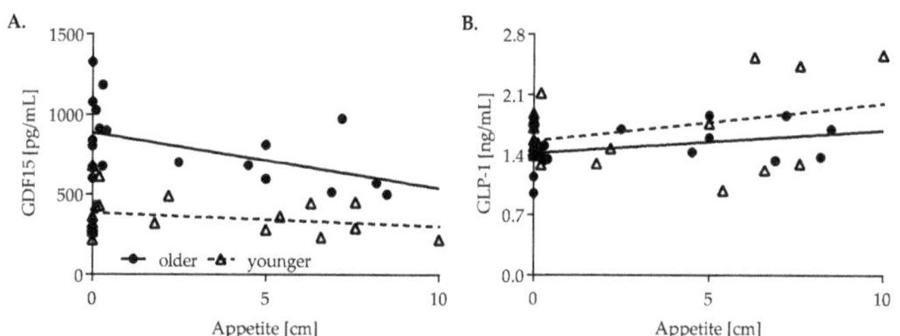

Figure 1. Correlation of baseline appetite and baseline (**A**) GDF15 (older: r = −0.488, p = 0.029; younger: r = −0.228, p = 0.347) and (**B**) GLP-1 concentrations (older: older, r = 0.461, p = 0.041; younger: r = 0.227, p = 0.350). GLP-1 was logarithmized for normalization. Correlations of GLP-1 were calculated using log-transformed values but are shown as untransformed values for better visualization. Closed circles represent older women, open triangles younger women.

3.1. Appetite

Overall, in both older and younger women, appetite changed over time (p = 0.015 versus p < 0.001; Figure 2A,D), but only in older women did postprandial appetite differ between the two test meals (p = 0.015 versus p = 0.383 in younger women). Following dextrose ingestion, appetite significantly increased in both age groups during the meal challenge. After protein ingestion, only younger women exhibited increasing appetite from 120 to 240 min, whereas appetite did not change over time in the older women. At the end of the meal challenge, appetite was similar for both test meals in younger women, but in older women appetite was higher after dextrose compared to protein ingestion (mean difference: 4 cm, p = 0.021).

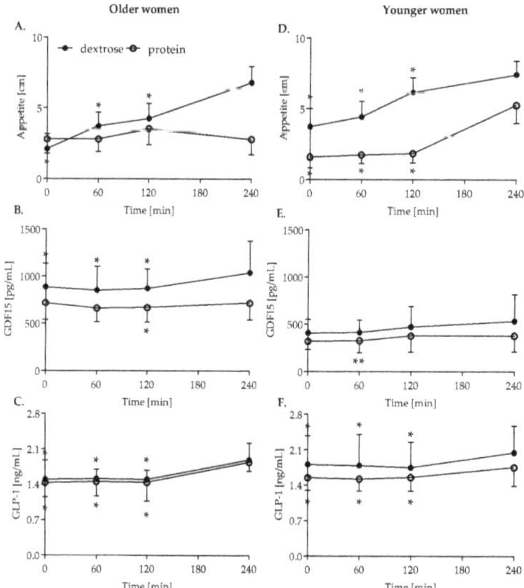

Figure 2. Postprandial appetite (**A**), GDF15 (**B**) and GLP-1 (**C**) concentrations in older and younger women ((**D**,**E**,**F**), respectively) following dextrose (closed circles) or protein (open circles) ingestion. GLP-1 concentrations were logarithmized for normalization. Repeated measures ANOVA, data are shown as mean ± SD. * indicates significant difference to 240 min, ** to 120 min, separately for both test meals. Postprandial changes of GLP-1 concentrations were calculated using log-transformed values but are shown as untransformed values for better visualization. n = 10 per group; n = 9 in younger high-protein group. Closed circles represent older women, open circles younger women.

3.2. GDF15

GDF15 concentrations significantly changed over time in both older ($p < 0.001$) and younger women ($p = 0.019$; Figure 2B,E). Only in older women did the test meal have an effect on postprandial GDF15 concentrations ($p = 0.026$). Following dextrose ingestion, GDF15 concentrations significantly increased in older women from 60 to 240 min, but not in younger women. After protein ingestion, GDF15 concentrations slightly increased in older women from 120 to 240 min and from 60 to 120 min in younger women. However, in older women, GDF15 concentrations were higher in response to dextrose compared to protein at 120 min (mean difference: 202 pg/mL, $p = 0.023$) and 240 min (mean difference: 320 pg/mL, $p = 0.017$) after meal ingestion. GDF15 concentrations during the meal challenge were not different between dextrose or high-protein ingestion in younger women.

3.3. GLP-1

In both age groups, GLP-1 concentrations increased after meal ingestion (older: $p < 0.001$, younger $p < 0.001$; Figure 2C,F), but were not different between the meals.

3.4. Glucose Metabolism

Postprandial glucose and insulin concentrations after meal ingestion are depicted in Figure 3. Glucose concentrations significantly changed over time in both older ($p < 0.001$, Figure 3A) as well as younger women ($p = 0.012$, Figure 3C) and were different between test meals (old: $p < 0.001$, young: $p = 0.001$). Postprandial insulin concentrations also significantly changed over time in both older and younger women ($p < 0.001$ in both age groups), but only in older women, the insulin concentration changes over time were different between the test meals ($p = 0.027$; Figure 3B,D).

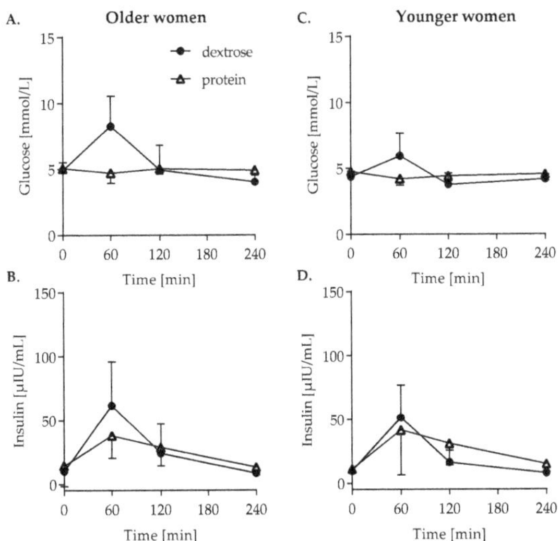

Figure 3. Postprandial glucose (**A**) and insulin (**B**) concentrations in older and younger women ((**C**,**D**) respectively) following dextrose (closed circles) and high protein (open triangles ingestion. Repeated measures ANOVA, data are shown as mean ± SD. $n = 10$ per group; $n = 9$ in the younger high-protein group. Closed circles represent older women, open triangles younger women.

3.5. Association of Appetite, GDF15, GLP-1, Glucose and Insulin Response

To evaluate associations among appetite, GDF15 and GLP-1 response, iAUCs were calculated. There was no age difference regarding appetite, GDF15, GLP-1, glucose and insulin iAUCs (Supplementary Material Figure S2). However, following dextrose ingestion in older women only, appetite iAUC was negatively correlated with GDF15 iAUC (rho = −0.802, $p = 0.005$), which suggests that the increase in appetite was less prominent when the postprandial increase in GDF15 was high (Figure 4). In younger women after protein ingestion, GDF15 iAUC was positively correlated with GLP-1 iAUC (rho = 0.729, $p = 0.026$) and insulin (rho = 0.949, $p < 0.001$), but this was not seen in older women. This might indicate that GDF15 behaves similarly to GLP-1 in the young but not in older women.

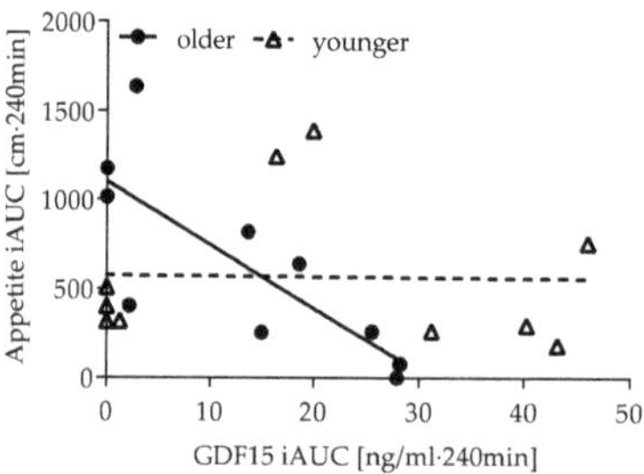

Figure 4. Correlation of appetite iAUC and GDF15 iAUC after dextrose ingestion in older (rho = −0.802, $p = 0.005$) and younger women (rho = −0.215, $p = 0.550$). iAUC: incremental area under the curve. Closed circles represent older women, open triangles younger women.

4. Discussion

Due to its many functions, GDF15 has been a target of pharmacological research to treat, e.g., cachexia and diabetes [18]. Moreover, GDF15 has recently gained attention as a potential biomarker for cellular senescence [19] and a key player in the aging process [2], but also as an important regulator of weight homeostasis and appetite [1,20]. To our knowledge, to date there are no studies investigating postprandial appetite and GDF15 concentrations in older adults compared to younger. In this analysis, we show that acute GDF15 response was different between older and younger women and dependent on the type of test meal in older women. Only in the older group and after dextrose ingestion was an increase in GDF15 concentrations found. Postprandial appetite increased over time in both age groups following dextrose ingestion; however, following protein ingestion, appetite remained low in the old, while increasing in the young. In older women, a higher GDF15 response to dextrose was associated with a lower increase in appetite.

The age difference regarding appetite after high protein ingestion might be due to the slower gastric emptying time in older age [10]. In addition, our results imply that GDF15 affects appetite in older women, since fasting GDF15 concentrations as well as their responses were negatively associated with appetite. However, these results also suggest that in younger women, GDF15 does not influence appetite. Studies in mice imply that the anorectic and even nauseating properties of GDF15 unfold only at high/pharmacological concentrations [21,22]. Yet, it is unclear what levels are required in humans for GDF15 to affect energy balance and exert anorexic effects. The association between GDF15 and appetite might therefore be even more prominent in older adults with strongly elevated concentrations, as seen in disease [16].

The regulation of GDF15 secretion is complex as various organs and tissues produce GDF15 [2]. Moreover, multiple stressors and stimuli are able to regulate its expression. One prominent regulator is the ISR, which is induced in response to, among other things, protein restriction [3]. As the dextrose meal was free of protein, the GDF15 increase after dextrose intake might also be interpreted as a response to a lack of protein. This is supported by the observation that GDF15 concentrations after protein ingestion do not change to the same extent as after dextrose intake. Furthermore, the different amounts of calories ingested (180 versus 450 kcal) might also have affected postprandial GDF15 concentrations. However, as short-term overfeeding studies in humans and rodents did not result in altered GDF15 concentrations [3], this effect appears to be negligible.

To date, only a few studies have investigated post-meal GDF15 concentrations. One study found increasing GDF15 concentrations after an oral glucose tolerance test (75 g dextrose) in younger to middle-aged adults with obesity [13]. In addition, it was shown that the ingestion of a high-carbohydrate or a high-fat mixed meal (protein content 12 E%) did not result in postprandial changes of GDF15. This was also seen in another study using mixed meals (protein content 15 E%) in a younger cohort [14]. Possibly, the protein content of the meals in these studies was sufficient, and therefore GDF15 expression was not induced. However, in none of these studies was postprandial appetite assessed.

Additionally, insulin resistance might play a role in the acute regulation of GDF15. In response to dextrose ingestion, glucose concentrations rose to higher levels at 60 min in older compared to younger women (8.27 versus 5.93 mmol/L), whereas postprandial insulin concentrations were not different between the age groups. This indicates that the older women in this analysis were more insulin-resistant than the younger women, which is a known age-associated effect. GDF15 and insulin iAUC were strongly positively correlated in younger women after protein ingestion, which is in line with the literature, wherein GDF15 was found to regulate insulin secretion [18]. This suggests a complex interplay between nutritional stimuli, GDF15 and insulin, as this is neither seen in response to dextrose nor in older women.

Our study is subject to limitations, such as the number of subjects and the subjective evaluation of the appetite. As a metabolic indicator of energy status as opposed to the subjective rating of appetite, we also analyzed postprandial GLP-1 concentrations. GLP-1 is known to enhance satiety and reduce energy intake [23], and overall exhibits similar actions to GDF15 (Figure 5) [24]. However, GLP-1 is associated differently with fasting and postprandial appetite compared to GDF15. This might imply a dissociation between the expected actions of GLP-1 on satiety and the subjective rating of appetite. A reason for this dissociation might be that humans can be more sensitive to external factors (such as meal size, company while eating, time of day) than internal biological stimuli [25]. Moreover, we were not able to control for all confounders that might have an effect on the findings.

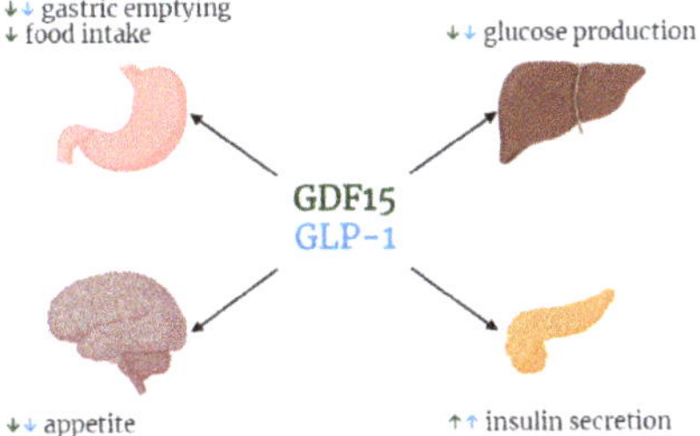

Figure 5. Overview of selected shared functions between GDF-15 and GLP-1. Arrows pointing up indicate enhancing of, arrows pointing down refer to a downregulation. Green refers to GDF15 functions, blue to GLP-1. Created with https://biorender.com/ (accessed on 23 September 2022).

In conclusion, we showed age-specific differences in the appetite's response to protein intake, and in the GDF15 response following dextrose ingestion. Possibly, studies on subjects with GDF15 concentrations higher than 1200 pg/mL, such as in clinical settings, might reveal a more prominent anorectic effect of GDF15 after meal ingestion.

Supplementary Materials: The following supporting information can be downloaded at: https://www.mdpi.com/article/10.3390/nu14194066/s1, Figure S1: Correlation of baseline GDF15 concentrations and (A) FMI and (B) BMI. BMI: body mass index, FMI: fat mass index; Figure S2: Incremental area under the curve (iAUC) for appetite (A), GDF15 (B), GLP-1 (C), glucose (D) and insulin (E) for each test meal and age group. OW: older women, YW: younger women, Dex: dextrose, HP: high protein.

Author Contributions: Data curation, C.H., S.L. and B.K.; writing—original draft preparation, C.H.; supervision, K.N.; writing—review and editing, B.K., M.K., S.K., U.M.-W. and K.N. All authors have read and agreed to the published version of the manuscript.

Funding: This research received no external funding.

Institutional Review Board Statement: The study was conducted in accordance with the Declaration of Helsinki, and was approved by the ethics committee of the University of Potsdam and registered at drks.de as DRKS00017090.

Informed Consent Statement: Informed consent was obtained from all subjects involved in the study.

Data Availability Statement: Data sharing not applicable. Data cannot be shared due to national data protection laws.

Conflicts of Interest: The authors declare no conflict of interest.

References

1. Breit, S.N.; Brown, D.A.; Tsai, V.W.W. The GDF15-GFRAL Pathway in Health and Metabolic Disease: Friend or Foe? *Annu. Rev. Physiol.* **2021**, *83*, 127–151. [CrossRef] [PubMed]
2. Conte, M.; Giuliani, C.; Chiariello, A.; Iannuzzi, V.; Franceschi, C.; Salvioli, S. GDF15, an emerging key player in human aging. *Ageing Res. Rev.* **2022**, *75*, 101569. [CrossRef] [PubMed]
3. Patel, S.; Alvarez-Guaita, A.; Melvin, A.; Rimmington, D.; Dattilo, A.; Miedzybrodzka, E.L.; Cimino, I.; Maurin, A.C.; Roberts, G.P.; Meek, C.L.; et al. GDF15 Provides an Endocrine Signal of Nutritional Stress in Mice and Humans. *Cell Metab.* **2019**, *29*, 707–718.e8. [CrossRef] [PubMed]
4. Hsu, J.Y.; Crawley, S.; Chen, M.; Ayupova, D.A.; Lindhout, D.A.; Higbee, J.; Kutach, A.; Joo, W.; Gao, Z.; Fu, D.; et al. Non-homeostatic body weight regulation through a brainstem-restricted receptor for GDF15. *Nature* **2017**, *550*, 255–259. [CrossRef]
5. Tsai, V.W.W.; Macia, L.; Johnen, H.; Kuffner, T.; Manadhar, R.; Jørgensen, S.B.; Lee-Ng, K.K.M.; Zhang, H.P.; Wu, L.; Marquis, C.P.; et al. TGF-b superfamily cytokine MIC-1/GDF15 is a physiological appetite and body weight regulator. *PLoS ONE* **2013**, *8*, e55174. [CrossRef]
6. Borner, T.; Wald, H.S.; Ghidewon, M.Y.; Zhang, B.; Wu, Z.; De Jonghe, B.C.; Breen, D.; Grill, H.J. GDF15 Induces an Aversive Visceral Malaise State that Drives Anorexia and Weight Loss. *Cell Rep.* **2020**, *31*, 107543. [CrossRef]
7. Xiong, Y.; Walker, K.; Min, X.; Hale, C.; Tran, T.; Komorowski, R.; Yang, J.; Davda, J.; Nuanmanee, N.; Kemp, D.; et al. Long-acting MIC-1/GDF15 molecules to treat obesity: Evidence from mice to monkeys. *Sci. Transl. Med.* **2017**, *9*, eaan8732. [CrossRef]
8. Hays, N.P.; Roberts, S.B. The anorexia of aging in humans. *Physiol. Behav.* **2006**, *88*, 257–266. [CrossRef]
9. Norman, K.; Haß, U.; Pirlich, M. Malnutrition in Older Adults-Recent Advances and Remaining Challenges. *Nutrients* **2021**, *13*, 2764. [CrossRef]
10. Di Francesco, V.; Zamboni, M.; Dioli, A.; Zoico, E.; Mazzali, G.; Omizzolo, F.; Bissoli, L.; Solerte, S.B.; Benini, L.; Bosello, O. Delayed postprandial gastric emptying and impaired gallbladder contraction together with elevated cholecystokinin and peptide YY serum levels sustain satiety and inhibit hunger in healthy elderly persons. *J. Gerontol. A Biol. Sci. Med. Sci.* **2005**, *60*, 1581–1585. [CrossRef]
11. Kleinert, M.; Bojsen-Møller, K.N.; Jørgensen, N.B.; Svane, M.S.; Martinussen, C.; Kiens, B.; Wojtaszewski, J.F.; Madsbad, S.; Richter, E.A.; Clemmensen, C. Effect of bariatric surgery on plasma GDF15 in humans. *Am. J. Physiol. Endocrinol. Metab.* **2019**, *316*, E615–E621. [CrossRef] [PubMed]
12. Galuppo, B.; Agazzi, C.; Pierpont, B.; Chick, J.; Li, Z.; Caprio, S.; Santoro, N. Growth differentiation factor 15 (GDF15) is associated with non-alcoholic fatty liver disease (NAFLD) in youth with overweight or obesity. *Nutr. Diabetes* **2022**, *12*, 9. [CrossRef] [PubMed]
13. Schernthaner-Reiter, M.H.; Kasses, D.; Tugendsam, C.; Riedl, M.; Peric, S.; Prager, G.; Krebs, M.; Promintzer-Schifferl, M.; Clodi, M.; Luger, A.; et al. Growth differentiation factor 15 increases following oral glucose ingestion: Effect of meal composition and obesity. *Eur. J. Endocrinol.* **2016**, *175*, 623–631. [CrossRef] [PubMed]
14. Tsai, V.W.W.; Macia, L.; Feinle-Bisset, C.; Manandhar, R.; Astrup, A.; Raben, A.; Lorenzen, J.K.; Schmidt, P.T.; Wiklund, F.; Pedersen, N.L.; et al. Serum Levels of Human MIC-1/GDF15 Vary in a Diurnal Pattern, Do Not Display a Profile Suggestive of a Satiety Factor and Are Related to BMI. *PLoS ONE* **2015**, *10*, e0133362. [CrossRef]
15. Herpich, C.; Haß, U.; Kochlik, B.; Franz, K.; Laeger, T.; Klaus, S.; Bosy-Westphal, A.; Norman, K. Postprandial dynamics and response of fibroblast growth factor 21 in older adults. *Clin. Nutr.* **2021**, *40*, 3765–3771. [CrossRef]
16. Herpich, C.; Franz, K.; Ost, M.; Otten, L.; Coleman, V.; Klaus, S.; Müller-Werdan, U.; Norman, K. Associations Between Serum GDF15 Concentrations, Muscle Mass, and Strength Show Sex-Specific Differences in Older Hospital Patients. *Rejuvenation Res.* **2021**, *24*, 14–19. [CrossRef]
17. Kyle, U.G.; Genton, L.; Slosman, D.O.; Pichard, C. Fat-free and fat mass percentiles in 5225 healthy subjects aged 15 to 98 years. *Nutrition* **2001**, *17*, 534–541. [CrossRef]

18. Wang, D.; Day, E.A.; Townsend, L.K.; Djordjevic, D.; Jørgensen, S.B.; Steinberg, G.R. GDF15: Emerging biology and therapeutic applications for obesity and cardiometabolic disease. *Nat. Rev. Endocrinol.* **2021**, *17*, 592–607. [CrossRef]
19. Basisty, N.; Kale, A.; Jeon, O.H.; Kuehnemann, C.; Payne, T.; Rao, C.; Holtz, A.; Shah, S.; Sharma, V.; Ferrucci, L.; et al. A proteomic atlas of senescence-associated secretomes for aging biomarker development. *PLoS Biol.* **2020**, *18*, e3000599. [CrossRef]
20. Baek, S.J.; Eling, T. Growth differentiation factor 15 (GDF15): A survival protein with therapeutic potential in metabolic diseases. *Pharmacol. Ther.* **2019**, *198*, 46–58. [CrossRef]
21. Borner, T.; Shaulson, E.D.; Ghidewon, M.Y.; Barnett, A.B.; Horn, C.C.; Doyle, R.P.; Grill, H.J.; Hayes, M.R.; De Jonghe, B.C. GDF15 Induces Anorexia through Nausea and Emesis. *Cell Metab.* **2020**, *31*, 351–362.e5. [CrossRef] [PubMed]
22. Klein, A.B.; Nicolaisen, T.S.; Ørtenblad, N.; Gejl, K.D.; Jensen, R.; Fritzen, A.M.; Larsen, E.L.; Karstoft, K.; Poulsen, H.E.; Morville, T.; et al. Pharmacological but not physiological GDF15 suppresses feeding and the motivation to exercise. *Nat. Commun.* **2021**, *12*, 1041. [CrossRef] [PubMed]
23. Aaboe, K.; Krarup, T.; Madsbad, S.; Holst, J.J. GLP-1: Physiological effects and potential therapeutic applications. *Diabetes Obes. Metab.* **2008**, *10*, 994–1003. [CrossRef] [PubMed]
24. Chiang, Y.A.; Ip, W.; Jin, T. The role of the Wnt signaling pathway in incretin hormone production and function. *Front. Physiol.* **2012**, *3*, 273. [CrossRef] [PubMed]
25. Levitsky, D.A. The non-regulation of food intake in humans: Hope for reversing the epidemic of obesity. *Physiol. Behav.* **2005**, *86*, 623–632. [CrossRef] [PubMed]

MDPI

Article

The Associations between Metalloestrogens, GSTP1, and SLC11A2 Polymorphism and the Risk of Endometrial Cancer

Kaja Michalczyk [1,*], Patrycja Kapczuk [2], Grzegorz Witczak [1], Mateusz Bosiacki [3], Mateusz Kurzawski [4], Dariusz Chlubek [2] and Aneta Cymbaluk-Płoska [1]

1 Department of Gynecological Surgery and Gynecological Oncology of Adults and Adolescents, Pomeranian Medical University, 70-111 Szczecin, Poland; grzegorzwitczak9@gmail.com (G.W.); anetac@data.pl (A.C.-P.)
2 Department of Biochemistry and Medical Chemistry, Pomeranian Medical University, Powstańców Wielkopolskich. 72, 70-111 Szczecin, Poland; patrycja.kapczuk@pum.edu.pl (P.K.); dchlubek@pum.edu.pl (D.C.)
3 Department of Functional Diagnostics and Physical Medicine, Pomeranian Medical University in Szczecin, 70-111 Szczecin, Poland; bosiacki.m@pum.edu.pl
4 Department of Experimental and Clinical Pharmacology, Pomeranian Medical University in Szczecin, Powstańców Wielkopolskich 72, 70-111 Szczecin, Poland; mateusz.kurzawski@pum.edu.pl
* Correspondence: kajamichalczyk@wp.pl

Citation: Michalczyk, K.; Kapczuk, P.; Witczak, G.; Bosiacki, M.; Kurzawski, M.; Chlubek, D.; Cymbaluk-Płoska, A. The Associations between Metalloestrogens, GSTP1, and SLC11A2 Polymorphism and the Risk of Endometrial Cancer. *Nutrients* **2022**, *14*, 3079. https://doi.org/10.3390/nu14153079

Academic Editors: Birgit-Christiane Zyriax and Nataliya Makarova

Received: 24 June 2022
Accepted: 25 July 2022
Published: 27 July 2022

Abstract: Background: The incidence of endometrial cancer (EC) is still rising. Numerous risk factors including patient characteristics and molecular instability have been identified for EC. The presence of specific molecular markers allows specific diagnostic and prognostic approaches. Several single nucleotide polymorphisms (SNPs) have been identified to influence endometrial cancer risk. Metalloestrogens are metal ions which can mimic estrogen activity; however, their role in uterine pathologies remains unknown. This study aimed to investigate total blood trace elements levels and evaluate the distribution of selected genotypes in GSTP1 and SLC11A2 genes. Methods: This retrospective case-control analysis was carried out in peripheral blood samples of 110 women with endometrial cancer (EC; $n = 21$), uterine fibroma ($n = 25$), endometrial polyp ($n = 48$), and normal endometrium ($n = 16$). Analysis included measurement of metals and phosphor in serum, and of genetic polymorphisms in GST (rs1695) and SLC11A2 (rs224589) in DNA from white blood cells. Serum trace elements were measured using ICP-OES spectrometry. SNPs were identified using Taq Man real-time PCR genotyping assays. Results: The study confirmed higher age (OR 2.19, 95% CI 1.69–2.24), post-menopausal status (OR 1.89, 95% CI 1.36–1.94), and diabetes type 2 (OR 1.54; 95% CI 0.97–1.72) as independent risk factors for EC. We also found a high level of Cd (OR 1.49; 95% CI 1.31–1.63) and a low level of Co (OR 0.76; 95% CI 0.53–0.59) to be independent risk factors of EC. None of the tested polymorphisms of GSTP1 and SLC11A2 were associated with EC risk. However, high Cd (OR 1.21, 95% CI 1.15–1.29) and Ni (OR 1.07, 95% CI 1.05–1.18) serum levels were significantly associated with a SLC1A2 TG genotype, and high Cd levels with GSTP1 (OR 1.05, 95% CI 1.01–1.13).

Keywords: metalloestrogens; nutrients; trace elements; GSTP1; SLC11A2; endometrial cancer

1. Introduction

Endometrial cancer (EC) is the world's most common gynecological malignancy and one of a few cancers with an increasing incidence [1]. Numerous factors have been identified to increase the risk of endometrial cancer, including patients' age, obesity, late menopause onset, and unbalanced estrogen [2,3]. The new molecular classification, based on the Cancer Genome Atlas Research Network, distinguishes four types of EC by their molecular characteristics: Polymerase Epsilon Mutation (POLE) ultramutated, microsatellite instability hypermutated (MSI), copy-number low (CNL), and copy-number high (CNH) [4]. The presence of specific molecular markers allows new, more specific diagnostic and prognostic possibilities and the use of targeted therapy. The most common alternations were found to occur in TP53, PTEN, PI3KCA, CTNNB1, ARID1A, and KRAS pathways [4]. Multiple studies have investigated the associations between diverse single nucleotide polymorphisms (SNPs) and endometrial cancer risk. Strong associations were found for HNF1B, KLF, EIF2AK, CYP19A1, SOX4, and MYC [5].

Endometrial hyperplasia is a term describing an excessive proliferation of endometrial cells. It is a benign condition; however, it is considered as a precursor state for endometrial carcinomas [6]. Endometrial hyperplasia is usually correlated with unopposed estrogen, which in the absence of abnormal endometrial proliferation in the progesterone, stimulates abnormal endometrial proliferation [7]. Additionally, other benign uterine lesions, including uterine fibroids, were found to be hormone-dependent and estrogen is considered to be the major mitogenic factor in the uterus [8].

There is limited information on the influence of trace elements on human endometrium and their role in carcinogenesis. Metals including Zn, Cu, and Fe are essential chemicals for human well-being in trace amounts. However, when in excess, they may cause adverse effects including the induction of cellular damage, alternation in cellular homeostasis, inflammation, production of ROS (reactive oxygen species), and finally carcinogenic activity [9,10]. Additionally, certain metals such as Aluminum, Cadmium, Copper, Cobalt, Nickel, Lead, Tin, and Chromium have been found to have the ability to mimic estrogen activity and to activate estrogen receptors and were therefore named metalloestrogens [11–13]. This is why we decided to measure serum trace elements concentration in patients diagnosed with malignant and benign uterine lesions.

Glutathione S-transferase (GST) is an enzyme that catalyzes the conjugation of glutathione into electrophilic compounds. The enzyme not only detoxifies endogenous and exogenous species but also participates in the activation of oxidative metabolites participating in carcinogenesis, including ROS, and regulates stress-induced signaling pathways [14,15]. Some studies have presented an association between GSTP1 gene polymorphisms and increased risk of cancers, including endometrial cancer [16–18]. A similar association was found for endometrial hyperplasia [19].

SLC11 is a family of integral membrane proteins that are divalent metal ions transporters that use H^+ -electrochemical gradient as a driving force to transport metal ions [20,21]. SLC11A2, also known as DMT1, and Nramp2, is widely distributed and expressed in the duodenum, erythroid cells, kidney, lung, brain, testis, thymus, and placenta [22]. The predominant substrates of SLC11A2 are Fe^{2+}, Cd^{2+}, Co^{2+}, Cu^{1+}, Mn^{2+}, Ni^{2+}, Pb^{2+}, and Zn^{2+} [22].

SLC11A2 has an important role in iron homeostasis and transport. Mutations in the SLC11A2 gene were found in patients suffering from hypochromic microcytic anemia with serum and liver iron overload [23,24], while its activation was found to lead to severe pathologies including autophagy and cell death in Parkinson's disease [25]. Additionally, overexpression of SLC11A2 was found to be associated with several cancers including esophageal, colorectal, and breast carcinomas [22,26,27]. They also correlated with an invasive form of the disease.

Single nucleotide polymorphism (SNP) are forms of DNA variation among individuals either caused by nucleotide transition or transversion. They may change the encoded amino acids into nonsynonymous, replacing one nucleotide with a different one, and can be silent (synonymous) or occur in the noncoding region. SNPs may result in gene expression changes, mRNA stability and protein coding. The identification of gene variations and their effect analysis may allow a betting understanding of their impact on gene function as they may be responsible for characteristics causing population diversity, genome evolution, familial or interindividual traits, and differences in disease prevalence and treatment response.

Genes and the genetic polymorphism of genes involved in metalloestrogen homeostasis (i.e., glutathione S-transferase P1 gene (GSTP1) and metal ions transport (i.e., the Solute Carrier 11 group A member 2 gene (SLC11A2) may be closely related to estrogen overstimulation and therefore serve as a potential risk factor of endometrial cancer. Genetic variation may explain the heterogeneity of patients and help identify those more susceptible to metalloestrogen stimulation. GSTP1 rs1695 and SLC11A2 rs224589 are widely studied polymorphisms of the genes coding regions of the mentioned enzymes that may alter enzyme activity among different genotypes.

This study aimed to investigate the association of the selected polymorphisms: rs1695 in GSTP1 and rs224589 in SLC11A2, together with serum and blood trace elements in different endometrial pathologies.

2. Materials and Methods

2.1. Study Participants

The study included 140 patients consecutively admitted to the Department of Gynecological Surgery and Gynecological Oncology of Adults and Adolescents, Pomeranian Medical University. This case-control study included patients with a confirmed diagnosis of endometrial cancer based on histopathological evaluation who were admitted for radical surgery. The control group consisted of patients admitted for hysteroscopy or laparoscopy/laparotomy with histopathologically confirmed benign uterine conditions or normal endometrium. The exclusion criteria included recurrence of endometrial cancer, previous cancer treatment or other types of primary care, and presence of unbalanced/untreated chronic diseases. In addition, patients with lost or incomplete data were removed from the study group. Finally, a total of 110 patients were included in the study analysis. The research was conducted in accordance with the Helsinki Declaration and with the consent of the Ethics Committee of Pomeranian Medical University in Szczecin under the number KB-0012/27/2020 on 9 March 2020. Patient characteristics are demonstrated in Table 1.

Table 1. Group characteristics.

		Number of Patients
Age	<50 years	55
	≥50–60 years	31
	≥60 years	33
BMI	<25	35
	≥25 <30	36
	≥30	25
Cigarette smoking	Yes	7
	No	101
Menopause status	Before	36
	After	64
Type 2 Diabetes	Yes	15
	No	93
Hypothyroidism	Yes	18
	No	90
Histopathological diagnosis	Endometrial cancer	21
	Uterine fibroma	25
	Endometrial polyp	48
	Normal endometrium	16

2.2. Laboratory Analyses

From each patient, two peripheral blood samples were collected for the study purpose: one was used to obtain serum for serum trace elements analysis, while the other was used for blood sample trace elements analysis and isolation of genomic DNA. The specimens were stored at a temperature of −80 degrees Celsius. The samples were obtained at the time of hospital admission for hysteroscopy/laparoscopy or laparotomy. Informed consent to participate in the study was obtained from all patients.

2.3. *Trace Elements Analysis*

For the purpose of elemental analysis, serum and whole blood samples underwent a microwave decomposition procedure using a microwave digestion system. After defrost and sample preparation, 65% HNO3 was added to the samples, which were then transferred into Teflon vessels and placed in the microwave. The process of sample digestion was composed of two stages: an initial of 15 min, at which the samples were gradually heated up to 180 °C, and the second of 20 min, at which the temperature was maintained at 180 °C. Samples were analyzed using inductively coupled plasma optical emission spectrometry (ICP-OES, ICAP 7400 Duo, Thermo Scientific, Waltham, MA, USA) equipped with a concentric nebulizer and cyclonic spray chamber to determine Zinc (Zn), Copper (Cu), Iron (Fe), Chromium (Cr), Cobalt (Co), Strontium (Sr), Phosphor (P), Magnesium (Mg), Cadmium (Cd), Nickel (Ni), and Manganese (Mn) content. The digested samples were further diluted 20-fold. For the analysis, 500 µL of yttrium was added with the final standard sample concentration at 0.5 mg/L and 1 mL of 1% Triton (Triton X-100, Sigma-Aldrich, Poland). The samples were further diluted with 0.075% HNO_3 (Suprapur, Merck, Poland) up to the volume of 10 mL and stored in the fridge at 4–8 °C final until analysis. The calibration curve was constructed using multielement standard solutions (ICP multielement standard solution IV, IX and XVI, Merck, Kenilworth, NJ, USA).

2.4. *Molecular Analysis*

For the study purpose, genomic DNA was isolated from 0.2 mL of a whole blood sample (all with blood cells) using a commercial kit for genomic DNA isolation using Genomic Mini AX Blood 1000 Spin (A&A Biotechnology). The genotyping of the selected SNPs was performed using pre-designed Genotyping Assays (TaqMan real-time PCR genotyping assays, Thermo Fisher Scientific, Assay IDs: C___2967992_1_, C___3237198_20). The following SNPs were genotyped: GSTP1 rs1695 A > G and SLC11A2 rs224589 T > G.

2.5. *Statistical Analysis*

The analysis was conducted using Statistica 10, StataSoft, Poland The comparison of patient characteristics between the groups was performed using the U-Mann Whitney test. The associations between the tested SNPs and cancer risk were calculated by the comparison of the frequencies of selected genotypes among the EC and control group. Odds ratios (OR) and the corresponding confidence intervals (95% CI) for each SNP were calculated using univariable regression models. *p*-value < 0.05 was adopted as the statistical significance threshold.

3. Results

We found no differences between total blood/serum trace elements concentration between endometrial cancer patients and the control group of patients diagnosed with benign uterine conditions or normal endometrium. When analyzed separately, we found a significant difference for Cd levels between patients diagnosed with endometrial cancer vs endometrial polyps as patients with endometrial polyps tended to have lower Cd concentrations ($p = 0.002$). We also found a significant difference in Cu levels between patients diagnosed with uterine fibromas vs. endometrial polyps ($p = 0.042$). A similar trend was noticed for Cd levels. Another association was found for Fe concentration as patients with endometrial polyps showed lower Fe expression than patients with normal endometrium ($p = 0.038$). All of the associations are demonstrated in Table 2.

Table 2. Associations between specific trace elements levels in selected groups.

		EC	Control	*p*-Value	EC	Polyp	*p*-Value	EC	Normal Endometrium	*p*-Value	EC	Fibroma	*p*-Value	Fibroma	Polyp	*p*-Value	Fibroma	Normal Endometrium	*p*-Value	Polyp	Normal Endometrium	*p*-Value
Cu	Below median	9	44	0.620	8	24	0.412	10	8	0.858	11	11	0.545	8	26	0.042	12	8	0.853	22	8	0.837
	Above median	11	42		11	21		10	9		9	13		16	18		12	9		22	9	
Zn	Below median	12	41	0.321	10	22	0.784	11	7	0.402	11	11	0.545	12	22	1.000	12	8	0.853	22	8	0.837
	Above median	8	45		9	23		9	10		9	13		12	22		12	9		22	9	
Pb	Below median	12	34	0.198	11	16	0.248	12	6	0.130	10	10	0.525	12	16	0.355	14	6	0.151	20	6	0.224
	Above median	7	39		7	20		7	10		8	12		9	20		9	10		14	9	
Cd	Below median	6	44	0.107	4	27	0.002	9	9	0.630	8	12	0.536	5	27	0.000	8	9	0.081	22	7	0.613
	Above median	12	37		14	14		11	8		9	9		19	14		15	5		21	9	
Co	Below median	5	18	0.710	5	9	0.686	4	3	0.782	5	5	0.653	6	9	0.705	5	3	0.858	9	5	0.228
	Above median	4	19		4	10		4	4		4	6		5	10		6	3		11	2	
Fe	Below median	12	41	0.321	12	20	0.281	10	8	0.858	12	10	0.226	11	23	0.612	9	11	0.086	18	12	0.038
	Above median	8	45		8	24		10	9		8	14		13	21		15	6		26	5	
P	Below median	7	45	0.205	7	25	0.106	8	10	0.254	11	11	0.545	9	25	0.128	10	10	0.279	22	8	0.837
	Above median	12	40		13	19		12	7		9	13		15	19		14	7		22	9	

Our study showed no correlation between patients' menopausal status and blood concentration of any of the investigated metalloestrogens (Table 3).

Table 3. Serum metalloestrogen concentration in pre- and postmenopausal patients.

Trace Element	Median Serum Concentration	N of Patients before Menopause	N of Patients after Menopause	OR	*p*-Value
Cu	0.875	34	39	1.88	0.182
Zn	0.989	34	39	0.70	0.442
Cd	0.003	9	19	1.72	0.505
Co	0.007	11	17	1.55	0.576
Fe	1.301	34	39	0.62	0.309
P	111.230	35	37	1.01	0.984

As a part of the study, we checked for any correlations between patients' characteristics and serum trace element. Patients' BMI and age did not influence either serum trace element concentration. Yet, we found some significant intracorrelations between selected trace elements levels. Serum Zn levels positively correlated with Cu, Fe, P, and Mg concentration. All of the correlations are presented in Table 4.

Table 4. Spearman correlation.

	BMI	Age	Zn	Cu	Fe	Cr	Co	Sr	P	Mg	Cd	Ni
BMI	1.000	0.178	0.027	0.198	−0.077	0.019	0.150	0.007	−0.099	−0.008	−0.165	−0.015
Age	0.178	1.000	−0.067	0.188	−0.031	0.051	−0.183	0.056	0.033	0.022	−0.078	−0.018
Zn	0.027	−0.067	1.000	**0.392**	**0.215**	0.008	0.066	0.048	**0.281**	**0.253**	−0.062	0.009
Cu	0.198	0.188	**0.392**	1.000	0.169	−0.068	0.032	0.138	**0.529**	**0.492**	−0.169	0.090
Fe	−0.077	−0.031	**0.215**	0.169	1.000	0.146	0.053	0.117	**0.314**	**0.339**	0.024	−0.068
Cr	0.019	0.051	0.008	−0.068	0.146	1.000	0.133	0.217	−0.291	0.110	0.032	0.331
Co	0.150	−0.183	0.066	0.032	0.053	0.133	1.000	**0.514**	**−0.402**	−0.021	0.237	0.034
Sr	0.007	0.056	0.048	0.138	0.117	0.217	**0.514**	1.000	−0.024	**0.264**	−0.049	0.038
P	−0.099	0.033	**0.281**	**0.529**	**0.314**	−0.291	**−0.402**	−0.024	1.000	**0.546**	−0.036	**−0.298**
Mg	−0.008	0.022	**0.253**	**0.492**	**0.339**	0.110	−0.021	**0.264**	**0.546**	1.000	0.096	−0.136
Cd	−0.165	−0.078	−0.062	−0.169	0.024	0.032	0.237	−0.049	−0.036	0.096	1.000	−0.372
Ni	−0.015	−0.018	0.009	0.090	−0.068	0.331	0.034	0.038	**−0.298**	−0.136	−0.372	1.000

The underlined variables are significant with $p < 0.05$.

We conducted a univariate logistic regression model to assess the risk factors for endometrial cancer. We found patients' age, BMI, menopausal status, and history of diabetes mellitus type 2 to influence the risk for endometrial cancer. However, trace elements concentration did not seem to influence the probability of cancer occurrence. The results are presented in Table 5.

Table 5. Univariate logistic regression model.

		Endometrial Cancer	Control Group	OR	Upper 95%CI	*p*-Value
Age	Above median	17	39	2.21	1.86–2.29	0.001
Menopause	Yes	19	45	1.48	1.20–1.49	0.001
Grading	1	1	0	-	-	0.733
	2–3	17	2	-		
Staging	1–2	11	1	-	-	0.773
	3–4	1	0	-		
Smoking	Yes	3	4	1.71	1.49–183	0.086
BMI		20	88	1.34	1.29–1.51	0.0032
Diabetes type 2	Yes	10	5	1.26	1.14–1.30	0.000
Hashimoto	Yes	1	17	0.53	0.44–0.65	0.121
Cu	Above median	11	42	1.28	1.17–1.31	0.620
Zn	Above median	8	45	0.71	0.70–0.86	0.321
Pb	Above median	7	39	0.63	0.60–0.71	0.198
Cd	Above median	12	37	1.38	1.32–1.44	0.107
Co	Above median	4	19	0.76	0.74–0.79	0.710
P	Above median	12	40	1.62	1.57–1.66	0.205

However, upon multivariate analysis, Cadmium and Cobalt levels were found to be associated with endometrial cancer risk, while above median Co levels were found to correlate with a decreased the risk of EC. The results are displayed in Table 6.

Table 6. Multivariate logistic regression model.

		Endometrial Cancer	Control Group	OR	95%CI	*p*-Value
Age	Above median	17	39	2.19	1.69–2.24	0.0028
Menopause		19	45	1.89	1.36–1.94	0.0002
Grading		1	0			ND
		17	2			
Staging		11	1			ND
		1	0			
Smoking		3	4	2.41	2.30–3.61	0.0824
BMI		20	88	1.22	1.18–1.38	0.0502
Diabetes		10	5	1.54	0.97–1.72	0.0029
Hashimoto		1	17	1.09	0.86–1.12	0.0943
Cu	Above median	11	42	1.61	1.29–1.80	0.2319
Zn	Above median	8	45	0.75	0.70–1.12	0.0540
Pb	Above median	7	39	0.66	0.59–0.66	0.2743
Cd	Above median	12	37	1.49	1.31–1.63	0.0367
Co	Above median	4	19	0.76	0.53–0.59	0.0423
P	Above median	12	40	1.22	1.16–1.23	0.2036

None of the tested polymorphisms revealed a correlation with endometrial cancer risk (Table 7). However, we found some non-significant differences in genotypes frequencies among endometrial cancer patients and controls as the TG genotype was more frequently expressed among EC patients compared with controls (35% vs. 25.5%, respectively).

Table 7. The associations between the analyzed SNPs and endometrial cancer risk.

	Genotype	Number of Cases	Endometrial Cancer	Control Group	OR	95%CI	*p*-Value
SLC11A2	GG	85	12 (60.0%)	73 (71.6%)	1	-	
	TG	33	7 (35.0%)	26 (25.5%)	1.638	1.235–1.702	0.348
	TT	4	1 (5.0%)	3 (2.9%)	2.027	1.649–2.094	0.464
GSTP1	AA	52	9 (45.0%)	43 (47.8%)	1	-	-
	AG	50	10 (50.0%)	40 (44.4%)	1.194	1.088–1.217	0.710
	GG	8	1 (5.0%)	7 (7.8%)	0.683	0.598–0.702	0.654

In our study, patients with high Cd (OR 1.21, 95% CI 1.15–1.29) and Ni (OR 1.07, 95% CI 1.05–1.18) serum levels were significantly associated with a SLC1A2 TG genotype, and high Cd levels with GSTP1 (OR 1.05, 95% CI 1.01–1.13). The prevalence of other genotypes did not seem to correlate with blood metalloestrogen levels (see Table 8).

Table 8. Polymorphisms prevalence with regard to metalloestrogen levels.

Studied Population (n)	Cu		*p*-Value	Studied Population (n)	Zn		*p*-Value	Studied Population (n)	Fe		*p*-Value	Studied Population (n)	Co		*p*-Value	Studied Population (n)	Cd		*p*-Value
	OR	95%CI			OR	95%CI			OR	95%CI			OR	95%CI			OR	95 %CI	
71	-	-	-	71	-	-	-	71	-	-	-	32	-	-	-	65	-	-	-
31	1.11	1.04–1.12	0.535	31	0.98	0.84–1.04	0.282	31	1.29	1.19–1.35	0.518	12	1.15		0.361	30	1.21	1.15–1.29	0.023
4	1.09	1.02–1.11	0.037	4	0.87	0.85–0.92	0.553	4	1.34	1.30–1.41	0.722	2	1.27	1.17–1.30	0.114	4	1.20	1.19–1.24	0.091
51	-	-	-	51	-	–	-	51	-	-	-	17	-	-	-	46	-	-	-
47	1.19	1.17–1.20	0.312	47	0.79	0.76–0.86	0.222	47	1.25	1.21–1.30	0.157	24	1.23	1.20–1.31	0.154	45	1.09	1.04–1.11	0.423
8	1.15	1.11–1.16	0.484	8	0.94	0.92–0.99	0.227	8	1.29	1.22–1.35	0.327	5	1.19	1.16–1.28	0.127	8	1.05	1.01–1.13	0.032
Studied Population (n)	**Sr**		***p*-Value**	**Studied Population (n)**	**Cr**		***p*-Value**	**Studied Population (n)**	**Mg**		***p*-Value**	**Studied Population (n)**	**Mn**		***p*-Value**	**Studied Population (n)**	**Ni**		***p*-Value**
	OR	**95%CI**			**OR**	**95%CI**			**OR**	**95%CI**			**OR**	**95%CI**			**OR**	**95%CI**	
63	-	-	-	27	-	-	-	71	-	-	-	55	-	-	-	36	-	-	-
29	0.89	0.82–0.90	0.400	11	0.76	0.74–0.88	0.355	31	1.24	1.19–1.28	0.518	23	0.86	0.84–1.02	0.058	20	1.07	1.05–1.18	0.040
4	0.77	0.74–0.81	0.237	1	-	-	-	4	1.11	1.09–1.21	0.722	2	-	-	-	2	-	-	-
47	-	-	-	17	-	-	-	51	-	-	-	37	0.97	0.93–0.99	0.087	34	-	-	-
43	0.91	0.89–0.94	0.196	22	1.01	0.97–1.08	0.624	47	1.15	1.12–1.20	0.145	36	0.90	0.87–0.97	0.869	21	1.10	1.08–1.21	0.827
6	0.83	0.80–0.86	0.274	0	-	-	-	8	1.27	1.22–1.28	0.058	7	0.86	0.85–0.93	0.680	3	1.16	1.15–1.19	0.564

4. Discussion

The etiology of most benign and malignant uterine pathologies is hormone-dependent. Estrogen is the major mitogenic factor in the uterus and is responsible for tissue remodeling during the menstrual cycle. However, abnormalities to the endometrial tissue including hormonal disbalance, changes in tissue microenvironment, abnormalities in cytokine or growth factors expression, or increased production of ROS can result in carcinogenesis. Metalloestrogens may mimic estrogen activity and activate estrogen receptors and thus can also be increased in various endometrial pathologies.

In the presented study, we analyzed the trace elements concentrations in different uterine pathologies. We did not find any significant correlations between the selected microelements and endometrial cancer when compared to the control group of benign uterine pathologies/normal endometrium as a whole; however, we discovered some significant correlations only when we divided the control group into separate categories for each histopathological diagnosis. We found some correlations for Cd, Cu, and Fe serum concentration. Higher Cd serum concentrations were observed in patients diagnosed with endometrial cancer when compared with patients diagnosed with endometrial polyps ($p = 0.002$). Additionally, higher Cd and Cu concentrations were found in patients diagnosed with uterine fibromas when compared with endometrial polyps; however, the median concentrations of Cd and Cu were still lower in patients diagnosed with uterine fibromas than in EC patients. Moreover, patients diagnosed with endometrial polyps also had significantly higher Fe serum expression when compared with patients with normal endometrium. Elevated copper levels (both serum and tissue) have been found in multiple cancers [28]. As copper is a co-factor in redox reactions of enzymes participating in basic biological reactions required for cell growth and development including superoxide dismutase and cytochrome c oxidase, its increased amount may correlate with an increased need for tissue proliferation. As uterine fibromas are benign tumors which are also characterized with a rapid growth, this may be the reason for an increased trace elements concentration. Further studies are required to deepen the knowledge on the role of trace elements and their distribution in different uterine pathologies.

So far, only a few studies have evaluated metalloestrogen concentrations in endometrial cancer and resulted in conflicting reports. Due to the high importance of metalloestrogens and their unexplained role in uterine pathologies, we decided to conduct a study to further evaluate their distribution in patients with different uterine conditions. Atakul et al. [29] found associations between serum Cu and Zn. In accordance with their study, lower Cu, Zn, and Cu/Zn ratio was found in patients diagnosed with endometrial cancer than in control group. Cu concentration also inversely corelated with myometrial invasion. Yaman et al. [30] investigated trace metal concentration in different cancerous and noncancerous endometrial, ovary, and cervical tissue samples. The authors found increased Fe, similar levels of Cu, and lower levels of Zn in endometrial cancer patients when compared with controls. Additionally, Rzymski et al. [31] investigated metal accumulation in uterine tissue samples. Compared with normal endometrium, endometrial cancer, hyperplasia, and CIN samples revealed significantly increased levels of Cd, Pb, Cu, Mn, and Cu/Zn ratio. Both current and former smoking status were associated with significantly higher Cd and Pb levels. Additionally, endometrial polyps, when compared with histologically normal endometrium, showed increased median concentrations of Al, Cd, Ni, and Pb. There was no significant association for Cu/Zn ratio. The study showed no correlation between patients' age, menopausal status and the concentration of any of the investigated elements in endometrial tissue sample. Additionally, in our study, the menopausal status did not influence any of the assessed metalloestrogen levels. As there are still few data on the role of metalloestrogens and their levels in patients with different gynecological conditions, further research is needed to evaluate their significance. It would also be interesting to assess endometrium/uterine tissue metalloestrogen levels and compare their expression with serum levels.

The limitation of our study is that we did not ask the patients about any recent use of dietary supplements. As the supplements containing elements such as Zn, Cu, or Fe are widely accessible, their use might have influenced their blood concentration. However, as the supplements in Eastern Europe are still not as popular and accessible as in the western countries, we believe that their use was very limited. There is no obvious explanation for the differences observed between the studies in serum microelement concentration; however, the discrepancies may be caused based on the study size and patient characteristics. The differences may be also caused by the methods used for serum microelement analyzes, e.g., colorimetry, spectrophotometry, ICP, which each has different sensitivity and selectivity. In this study, we used ICP-OES—a very accurate method to analyze the trace elements concentration, which was also previously described in other trace elements analyses.

As a part of the study, we also conducted univariate analysis to evaluate the influence of patient characteristics and the assessed variables on the endometrial cancer risk. We confirmed that a high patient age, BMI, post-menopausal status and diabetes type 2 were significant risk factors of EC (see Table 5). Upon multivariate analysis, we found cadmium and cobalt levels to be associated with endometrial cancer risk. Higher Cd levels were found to be associated with increased endometrial cancer risk (OR 1.49, $p = 0.0367$), while above median Co levels were found to correlate with lower EC risk (OR 0.76, $p = 0.0423$). In a study by Rzymski et al. [31], the author found both current and former smoking status to be associated with significantly higher Cd and Pb levels. In our study, we did not further evaluate the correlation between cigarette use and trace element levels, as smoking was found not to be associated with endometrial cancer risk in our analysis; however, the study findings by Rzymski may be a potential explanation for this observation and require further analysis. Cobalt is an essential component of vitamin B12. Vitamin B12 is essential for the maintenance of DNA methylation, repair, synthesis, and thus for cell development and proper function. Accumulating evidence suggests the role of increased levels of folate and B-vitamins to have a role in cancer formation. Supplemental use of vitamin B12 intake was found to be associated with type 2 EC. However, the study did not include multivariable analysis. There is limited and still conflicting evidence regarding the influence of vitamin B12 and folic acid on cancer incidence. Further studies are required to explain their effect on cancer risk and cancer formation.

GSTs have a particularly important detoxification capability that protects against environmental and oxidative stress. Recent studies reported alternated GST expression to be associated with increased cancer risk [32]. As genetic polymorphisms can alternate enzyme expression, we decided to measure selected SNPs. We analyzed whether polymorphisms rs1695 in GSTP1 and rs224589 in SLC11A2 genes are associated with the risk of endometrial cancer. In our study, we found no significant association between GSTP1 and SLC11A2 polymorphisms and endometrial cancer prevalence.

A meta-analysis by Zhao et al. [33] tried to evaluate the associations between GSTP1 Ile105Val polymorphism and gynecological cancer susceptibility; however, the researchers found no significant associations with any genetic model even when accounting for cancer type, ethnicity, and smoking status. The study included a limited population as only two studies discussed endometrial cancer patients. A study by Ozerkan et al. [34] found no associations between GSTP1 polymorphism and EC risk in Caucasian population. On the other hand, a study by Chan et al. [35] revealed GSTP1 Ile(105)Val polymorphism to be associated with an increased risk of endometrial cancer. In our study, we found no associations between GTSP1 polymorphism and EC risk; however, as there are contradictory results, further research is needed.

This was the first study to analyze functional polymorphism in SLC11A2 gene in endometrial cancer patients. Previous studies described SLCA11A2 overexpression in breast carcinomas [22,26]. As endometrial cancer is also hormone dependent, we wanted to determine if its gene polymorphism affects the risk of EC. Even though we found no correlation between SLC11A2 polymorphisms and endometrial cancer risk, we found a positive correlation between SCL11A2 TT genotype and Cu concentration ($p = 0.037$). On the other hand, patients with TG genotype demonstrated lower Cd and Ni levels ($p = 0.023$ and $p = 0.040$, respectively). Further research is needed to evaluate the correlations between selected genotypes and trace elements levels due to the limited number of patients included in the analysis.

5. Conclusions

The study confirmed higher patient age, post-menopausal status, presence of diabetes type 2, and higher BMI as independent risk factors for endometrial cancer. Menopausal status did not influence metalloestrogen levels. High serum cadmium and low cobalt concentrations were found to influence endometrial cancer risk. None of the tested genetic polymorphisms (rs1695 in GSTP1 and rs224589 in SLC11A2) were found to be associated with endometrial cancer risk. However, GTSP1 and SLC11A2 SNPs may correlate with selected trace elements concentrations as high Cd and Ni serum levels were significantly associated with the SLC1A2 TG genotype and high Cd levels with GSTP1. Analyses of a more extensive study group should be performed to confirm our findings.

Author Contributions: Conceptualization, K.M., A.C.-P. and D.C.; methodology, D.C. and M.K.; software, D.C. and M.B.; validation, M.K., D.C. and A.C.-P.; formal analysis, K.M. and P.K.; investigation, K.M. and P.K; resources, K.M.; data curation, K.M., G.W. and P.K.; writing—original draft preparation, K.M.; writing—review and editing, K.M., A.C.-P. and D.C.; visualization, K.M.; supervision, A.C.-P., M.K. and D.C.; project administration, A.C.-P. All authors have read and agreed to the published version of the manuscript.

Funding: This research received no external funding.

Institutional Review Board Statement: The study was conducted according to the guidelines of the Declaration of Helsinki, and approved by the Ethics Committee of Pomeranian Medical University in Szczecin (protocol code KB-0012/27/2020 of 9 March 2020).

Informed Consent Statement: Informed consent was obtained from all subjects involved in the study.

Data Availability Statement: The data presented in this study are available on request from the corresponding author. The data are not publicly available due to ethical restrictions

Conflicts of Interest: The authors declare no conflict of interest.

References

1. Rodriguez, A.C.; Blanchard, Z.; Maurer, K.A.; Gertz, J. Estrogen Signaling in Endometrial Cancer: A Key Oncogenic Pathway with Several Open Questions. *Horm. Cancer* **2019**, *10*, 51–63. [CrossRef] [PubMed]
2. McGonigle, K.F.; Karlan, B.Y.; Barbuto, D.A.; Leuchter, R.S.; Lagasse, L.D.; Judd, H.L. Development of endometrial cancer in women on estrogen and progestin hormone replacement therapy. *Gynecol. Oncol.* **1994**, *55*, 126–1332. [CrossRef] [PubMed]
3. Dossus, L.; Lukanova, A.; Rinaldi, S.; Allen, N.; Cust, A.E.; Becker, S.; Tjonneland, A.; Hansen, L.; Overvad, K.; Chabbert-Buffet, N.; et al. Hormonal, metabolic, and inflammatory profiles and endometrial cancer risk within the EPIC cohort—A factor analysis. *Am. J. Epidemiol.* **2013**, *177*, 787–799. [CrossRef]
4. Getz, G.; Gabriel, S.B.; Cibulskis, K.; Lander, E.; Sivachenko, A.; Sougnez, C.; Lawrence, M.; Kandoth, C.; Dooling, D.; Fulton, R.; et al. Integrated genomic characterization of endometrial carcinoma. *Nature* **2013**, *497*, 67–73.
5. Bafligil, C.; Thompson, D.J.; Lophatananon, A.; Smith, M.J.; Ryan, N.A.J.; Naqvi, A.; Evans, D.G.; Crosbie, E.J. Association between genetic polymorphisms and endometrial cancer risk: A systematic review. *J. Med. Genet.* **2020**, *57*, 591–600. [CrossRef]
6. O'Rourke, R.W. Endometrial hyperplasia, endometrial cancer, and obesity: Convergent mechanisms regulating energy homeostasis and cellular proliferation. *Surg. Obes. Relat. Dis.* **2014**, *10*, 926–928. [CrossRef] [PubMed]
7. Schindler, A.E. Progestogen deficiency and endometrial cancer risk. *Maturitas* **2009**, *62*, 334–337. [CrossRef] [PubMed]
8. Levy, B.S. Modern management of uterine fibroids. *Acta Obstet. Gynecol. Scand.* **2008**, *87*, 812–823. [CrossRef]
9. Salnikow, K.; Zhitkovich, A. Genetic and epigenetic mechanisms in metal carcinogenesis and cocarcinogenesis: Nickel, arsenic, and chromium. *Chem. Res. Toxicol.* **2008**, *21*, 28–44. [CrossRef] [PubMed]
10. Leonard, S.S.; Harris, G.K.; Shi, X. Metal-induced oxidative stress and signal transduction. *Free Radic. Biol. Med.* **2004**, *37*, 1921–1942. [CrossRef]
11. Byrne, C.; Divekar, S.D.; Storchan, G.B.; Parodi, D.A.; Martin, M.B. Metals and breast cancer. *J. Mammary Gland. Biol. Neoplasia* **2013**, *18*, 63–73. [CrossRef] [PubMed]
12. Martin, M.B.; Reiter, R.; Pham, T.; Avellanet, Y.R.; Camara, J.; Lahm, M.; Pentecost, E.; Pratap, K.; Gilmore, B.A.; Divekar, S.; et al. Estrogen-like activity of metals in MCF-7 breast cancer cells. *Endocrinology* **2003**, *144*, 2425–2436. [CrossRef]

13. Darbre, P.D. Metalloestrogens: An emerging class of inorganic xenoestrogens with potential to add to the oestrogenic burden of the human breast. *J. Appl. Toxicol.* **2006**, *26*, 191–197. [CrossRef] [PubMed]
14. Beckett, G.J.; Hayes, J.D. Glutathione S-Transferases: Biomedical Applications. *Adv. Clin. Chem.* **1993**, *30*, 281–380. [PubMed]
15. Singh, R.R.; Reindl, K.M. Glutathione s-transferases in cancer. *Antioxidants* **2021**, *10*, 701. [CrossRef] [PubMed]
16. Liu, P. Association of glutathione S-transferase P1 polymorphism and endometrial cancer risk in a chinese population. *Biomed. Res.* **2017**, 28.
17. Yu, Z.; Li, Z.; Cai, B.; Wang, Z.; Gan, W.; Chen, H.; Li, H.; Zhang, P.; Li, H. Association between the GSTP1 Ile105Val polymorphism and prostate cancer risk: A systematic review and meta-analysis. *Tumor Biol.* **2013**, *34*, 1855–1863. [CrossRef] [PubMed]
18. Harries, L.W.; Stubbins, M.J.; Forman, D.; Howard, G.C.W.; Wolf, C.R. Identification of genetic polymorphisms at the glutathione S-transferase Pi locus and association with susceptibility to bladder, testicular and prostate cancer. *Carcinogenesis* **1997**, *18*, 641–644. [CrossRef]
19. Elsaid, A.; Al-Kholy, W.; Ramadan, R.; Elshazli, R. Association of glutathione-S-transferase P1 (GSTP1)-313 A>G gene polymorphism and susceptibility to endometrial hyperplasia among Egyptian women. *Egypt. J. Med. Hum. Genet.* **2015**, *16*, 361–365. [CrossRef]
20. Nevo, Y.; Nelson, N. The NRAMP family of metal-ion transporters. *Biochim. Biophys. Acta Mol. Cell Res.* **2006**, *1763*, 609–620. [CrossRef] [PubMed]
21. Gunshin, H.; Mackenzie, B.; Berger, U.V.; Gunshin, Y.; Romero, M.F.; Boron, W.F.; Nussberger, S.; Gollan, J.L.; Hediger, M.A. Cloning and characterization of a mammalian proton-coupled metal-ion transporter. *Nature* **1997**, *388*, 482–488. [CrossRef] [PubMed]
22. Montalbetti, N.; Simonin, A.; Kovacs, G.; Hediger, M.A. Mammalian iron transporters: Families SLC11 and SLC40. *Mol. Aspects Med.* **2013**, *34*, 270–287. [CrossRef] [PubMed]
23. Lam-Yuk-Tseung, S.; Camaschella, C.; Iolascon, A.; Gros, P. A novel R416C mutation in human DMT1 (SLC11A2) displays pleiotropic effects on function and causes microcytic anemia and hepatic iron overload. *Blood Cells, Mol. Dis.* **2006**, *36*, 347–354.
24. Beaumont, C.; Delaunay, J.; Hetet, G.; Grandchamp, B.; De Montalembert, M.; Tchernia, G. Two new human DMT1 gene mutations in a patient with microcytic anemia, low ferritinemia, and liver iron overload. *Blood* **2006**, *107*, 4168–4170. [CrossRef] [PubMed]
25. Chew, K.C.M.; Ang, E.T.; Tai, Y.K.; Tsang, F.; Lo, S.Q.; Ong, E.; Ong, W.Y.; Shen, H.M.; Lim, K.L.; Dawson, V.L.; et al. Enhanced autophagy from chronic toxicity of iron and mutant A53T α-synuclein: Implications for neuronal cell death in parkinson disease. *J. Biol. Chem.* **2011**, *286*, 33380–33389. [CrossRef]
26. Brookes, M.J.; Hughes, S.; Turner, F.E.; Reynolds, G.; Sharma, N.; Ismail, T.; Berx, G.; McKie, A.T.; Hotchin, N.; Anderson, G.J.; et al. Modulation of iron transport proteins in human colorectal carcinogenesis. *Gut* **2006**, *55*, 1449–1460. [CrossRef]
27. Boult, J.; Roberts, K.; Brookes, M.J.; Hughes, S.; Bury, J.P.; Cross, S.S.; Anderson, G.J.; Spychal, R.; Iqbal, T.; Tselepis, C. Overexpression of cellular iron import proteins is associated with malignant progression of esophageal adenocarcinoma. *Clin. Cancer Res.* **2008**, *14*, 379–387. [CrossRef] [PubMed]
28. Gupte, A.; Mumper, R.J. Elevated copper and oxidative stress in cancer cells as a target for cancer treatment. *Cancer Treat. Rev.* **2009**, *35*, 32–46. [CrossRef]
29. Atakul, T.; Altinkaya, S.O.; Abas, B.I.; Yenisey, C. Serum Copper and Zinc Levels in Patients with Endometrial Cancer. *Biol. Trace Elem. Res.* **2020**, *195*, 46–54. [CrossRef]
30. Yaman, M.; Kaya, G.; Simsek, M. Comparison of trace element concentrations in cancerous and noncancerous human endometrial and ovary tissues. *Int. J. Gynecol. Cancer* **2007**, 17. [CrossRef]
31. Rzymski, P.; Niedzielski, P.; Rzymski, P.; Tomczyk, K.; Kozak, L.; Poniedziałek, B. Metal accumulation in the human uterus varies by pathology and smoking status. *Fertil. Steril.* **2016**, *105*, 1511–1518.e3. [CrossRef] [PubMed]
32. Okat, Z. Clinical importance of glutathione-s-transferase enzyme polymorphisms in Cancer. *Int. Phys. Med. Rehabil. J.* **2018**, *3*, 491–493. [CrossRef]
33. Zhao, E.; Hu, K.; Zhao, Y. Associations of the glutathione S-transferase P1 Ile105Val genetic polymorphism with gynecological cancer susceptibility: A meta-analysis. *Oncotarget* **2017**, *8*, 41734. [CrossRef]
34. Ozerkan, K.; Atalay, M.A.; Yakut, T.; Doster, Y.; Yilmaz, E.; Karkucak, M. Polymorphisms of glutathione-s-transferase M1, T1, and P1 genes in endometrial carcinoma. *Eur. J. Gynaecol. Oncol.* **2013**, *34*, 42–47. [PubMed]
35. Chan, Q.K.Y.; Khoo, U.S.; Ngan, H.Y.S.; Yang, C.Q.; Xue, W.C.; Chan, K.Y.K.; Chiu, P.M.; Ip, P.P.C.; Cheung, A.N.Y. Single nucleotide polymorphism of Pi-class glutathione S-transferase and susceptibility to endometrial carcinoma. *Clin. Cancer Res.* **2005**, *11*, 2981–2985. [CrossRef] [PubMed]

Article

Limiting Postpartum Weight Retention in Culturally and Linguistically Diverse Women: Secondary Analysis of the HeLP-her Randomized Controlled Trial

Mingling Chen [1], Siew Lim [2] and Cheryce L. Harrison [1,3,*]

1 Monash Centre for Health Research and Implementation, School of Public Health and Preventive Medicine, Monash University, Clayton, VIC 3168, Australia; mingling.chen@monash.edu
2 Eastern Health Clinical School, Monash University, Box Hill, VIC 3128, Australia; siew.lim1@monash.edu
3 Diabetes and Endocrine Unit, Monash Health, Clayton, VIC 3168, Australia
* Correspondence: cheryce.harrison@monash.edu; Tel.: +61-385-722-662

Abstract: Postpartum weight retention (PPWR) contributes to maternal obesity development and is more pronounced in culturally and linguistically diverse (CALD) women. Our antenatal healthy lifestyle intervention (HeLP-her) demonstrated efficacy in reducing PPWR in non-Australian-born CALD women compared with Australian-born women. In this secondary analysis, we aimed to examine differences in the intervention effect on behavioral and psychosocial outcomes between Australian-born and non-Australian-born women and explore factors associated with the differential intervention effect on PPWR. Pregnant women at risk of gestational diabetes (Australian-born $n = 86$, non-Australian-born $n = 142$) were randomized to intervention (four lifestyle sessions) or control (standard antenatal care). PPWR was defined as the difference in measured weight between 6 weeks postpartum and baseline (12–15 weeks gestation). Behavioral (self-weighing, physical activity (pedometer), diet (fat-related dietary habits questionnaire), self-perceived behavior changes), and psychosocial (weight control confidence, exercise self-efficacy, eating self-efficacy) outcomes were examined by country of birth. Multivariable linear regression analysis was conducted to assess factors associated with PPWR. The intervention significantly increased self-weighing, eating self-efficacy, and self-perceived changes to diet and physical activity at 6 weeks postpartum in non-Australian-born women, compared with no significant changes observed among Australian-born women. Intervention allocation and decreased intake of snack foods were predictors of lower PPWR in non-Australian-born women. Results indicate that the HeLP-her intervention improved dietary behaviors, contributing to the reduction of PPWR in CALD women. Future translations could prioritize targeting diet while developing more effective strategies to increase exercise engagement during pregnancy in this population.

Keywords: ethnicity; lifestyle intervention; postpartum weight retention; pregnancy

Citation: Chen, M.; Lim, S.; Harrison, C.L. Limiting Postpartum Weight Retention in Culturally and Linguistically Diverse Women: Secondary Analysis of the HeLP-her Randomized Controlled Trial. *Nutrients* **2022**, *14*, 2988. https://doi.org/10.3390/nu14142988

Academic Editor: Birgit-Christiane Zyriax

Received: 7 June 2022
Accepted: 18 July 2022
Published: 21 July 2022

1. Introduction

Overweight and obesity are significant global health challenges, affecting approximately 40% of women worldwide [1]. Pregnancy is recognized as a high-risk period for accelerated weight gain in women. Excessive gestational weight gain (GWG) contributes to postpartum weight retention (PPWR) and increases the risk of adverse outcomes in subsequent pregnancies, including preeclampsia, gestational diabetes mellitus (GDM), cesarean delivery, and large-for-gestational-age birth, as well as the development of long-term maternal obesity, cardiovascular disease, and metabolic syndrome in later life [2,3]. It is documented that PPWR is more pronounced among underserved populations, including women from culturally and linguistically diverse (CALD) backgrounds [4]. Observational studies in the US and Europe have shown women from South Asian, Middle Eastern, and African groups experience more weight retention postpartum than women of European background [5–7]. Contributory factors include lower socioeconomic status, higher psychosocial stress, and suboptimal lifestyle behaviors such as greater energy intake and less physical activity during and after pregnancy [5,6,8].

Lifestyle intervention in pregnancy comprising a healthy diet and/or physical activity optimizes GWG [9], thus promoting a return to pre-pregnancy weight after childbirth. Despite accumulated evidence on the benefits of such interventions during and following pregnancy, few studies have explored variation in intervention efficacy by ethnicity [10,11]. Given the increased risk of retaining weight postpartum in CALD women [4], understanding intervention effects on postpartum outcomes and associated factors that contribute to intervention effects in CALD groups is important to identify facilitating factors and effective strategies for weight management during pregnancy for these high-risk populations.

The Healthy Lifestyle in Pregnancy (HeLP-her) study is a low-intensity lifestyle intervention previously conducted in Australian antenatal care settings to optimize GWG and PPWR in women at increased risk of GDM [12,13]. The study population was ethnically diverse, with ~65% of women non-Australian-born. We have previously demonstrated the efficacy of the intervention, which was more effective in reducing weight retention at 6 weeks postpartum in non-Australian-born women (intervention 1.13 ± 4.11 kg vs. control 3.66 ± 5.47 kg, $p < 0.01$) compared with Australian-born women (−0.56 ± 4.93 kg vs. −0.74 ± 5.14 kg, $p = 0.87$) [13], yet exploratory analysis was not undertaken to elucidate this differential intervention effect. Here, this secondary analysis aims to: (1) examine differences in behavioral and psychosocial outcomes at 6 weeks postpartum between Australian-born and non-Australian-born women to further our understanding of ethnic variability in response to the intervention and (2) explore factors associated with the differential intervention effect on postpartum weight retention to provide insight into future intervention design and translation to benefit higher risk, CALD populations.

2. Materials and Methods

2.1. Study Design

Detailed study design and methods have been previously described [12,13]. In brief, women were recruited at three large metropolitan tertiary teaching hospitals in Victoria, Australia, between June 2008 and October 2010, combining over 8600 births per year. The hospitals serve a region of ethnically and culturally diverse populations with over one-third of residents non-Australian-born, comprising the largest refugee and migrant community in Victoria [14]. According to the Australian Bureau of Statistics, the region is classified of "average socioeconomic advantage" [15]. Inclusion criteria were: gestational age ≤ 15 weeks, singleton pregnancy, body mass index (BMI) ≥ 25 kg/m^2 (or BMI ≥ 23 kg/m^2 if high-risk ethnicity, i.e., Asian, African, and Polynesian [16]), and at increased risk of GDM as identified by a validated risk prediction tool [12,13,17]. Exclusion criteria included multiple pregnancies, BMI ≥ 45 kg/m^2, type 1 or 2 diabetes diagnosis, pre-existing chronic medical conditions, and non-English speaking women. Eligible women were randomized to intervention or control through computer-generated randomized sequencing. All women received standard antenatal care. Informed consent was obtained from all participants. The study was approved by the Southern Health Research Advisory and Ethics Committee. The trial was registered on the Australian New Zealand Clinical Trial Registry (ACTRN12608000233325) [12,13].

2.2. Intervention Group

Underpinned by the Social Cognitive Theory, the low-intensity behavior change program included four, 45 min individual lifestyle sessions delivered at 14–16, 20, 24, and 28 weeks gestation by a trained health coach [12,13]. The sessions delivered simple, pregnancy-specific healthy eating and physical activity messages based on National guidelines [18,19], as well as healthy GWG information according to the National Academy of Medicine (previously, Institute of Medicine [IOM]) guidelines [20]. Behavior change strategies were utilized to practice and increase self-management, including personal goal setting, problem-solving, action planning, self-monitoring, addressing barriers, and relapse prevention [12,13]. Ongoing support with mobile phone SMS messages was provided throughout the intervention. In addition, two healthy lifestyle postcards were distributed at 30 and 34 weeks gestation to reinforce behavior change and maintain engagement [12,13].

2.3. *Control Group*

The control group received one individual 15 min education session based on the population-based Australian Dietary and Physical Activity Guidelines [18,19] at baseline, as well as written pamphlets. No further support was provided during the study period [12,13].

2.4. *Data Sources*

Data collected at baseline (12–15 weeks gestation) and 6 weeks postpartum was used for this secondary analysis.

2.4.1. Demographics

Demographic information including age, country of birth, years lived in Australia, education, employment, household income, parity, and breastfeeding status, were collected using a structured self-administered questionnaire. Women were grouped by country of birth into two categories: Australian-born (born in Australia) and non-Australian-born (born outside of Australia). Those non-Australian-born are referred to as CALD women [21,22].

2.4.2. Anthropometrics

Anthropometric measurements (height, weight) were conducted by a registered nurse who was blinded to group allocation. Weight was measured to the nearest 0.1 kg on an electronic scale (Tanita model BWB-800 Digital Scale, Wedderburn Scales, Melbourne, Australia). The primary outcome was PPWR, calculated as the difference in measured weight (kg) between baseline and 6 weeks postpartum.

2.4.3. Health Behaviors

Physical Activity

The Yamax Digiwalker SW-700 Pedometer (Yamax Corporation, Tokyo, Japan) was used to assess the free-living step count per day as a tool with demonstrated accuracy in pregnancy, as previously reported [23]. Participants were asked to wear a sealed pedometer for a minimum of three to seven consecutive days during waking hours. Readings were processed to provide the average daily step count according to the total days worn.

Diet

Dietary behaviors related to adopting low-fat diets were assessed using a 20-item scale that was derived from the Fat-Related Dietary Habits Questionnaire [24]. The 20-item scale included five dimensions: "avoid fat as flavoring", "modify meats", "avoid frying", "substitute lower-fat products", and "replace foods with fruits and vegetables". Responses to the items were on a 3-point scale ("usually", "sometimes", and "rarely or never") and were coded through 1 to 3 to positively correlate with fat intake. An additional "don't eat this food" option was also provided for each item, coded as missing data. A summary mean score was calculated.

Self-Weighing

Self-reported frequency of self-weighing was collected via questionnaire. Data were dichotomized as frequent ("daily", "weekly" or "monthly") or not frequent ("occasionally" or "never") self-weighing.

2.4.4. Psychosocial Measures

Risk Perception

Risk perception for excess GWG and development of GDM was assessed on a 4-point scale adapted from the theory of health stage of change [25]. Data were dichotomized as no perceived risk ("definitely no risk" or "not really at risk") or perceived risk ("slight risk" or "high risk").

Weight Control Confidence

Confidence for weight control was assessed by asking "how confident are you that you can control your weight gain if you wished" and "how confident are you that you can control your weight gain if you experienced difficulties", which were adapted from the Chronic Disease Self-Efficacy Scale [26]. Responses were on a 10-point scale from "not at all confident" (1) to "totally confident" (10). A summary mean score was calculated, with higher scores indicating higher confidence.

Exercise and Eating Self-Efficacy

Self-efficacy for exercise and eating behaviors was assessed using a 14-item scale derived from a validated scale developed by Sallis et al. [27]. The 14 items consisted of four subscales: "sticking to it (exercise)", "making time for exercise", "sticking to it (eating)", and "reducing calories". Responses were on a 5-point scale from "not at all confident" (1) to "extremely confident" (5). The mean score of each subscale was calculated, with higher scores indicating higher self-efficacy.

2.4.5. Other Measures

At 6 weeks postpartum, participants were asked to report their self-perceived change in physical activity and diet since participating in the program on a 4-point scale; responses were dichotomized as perceived change ("lots of changes", "some changes", or "minor changes") or no perceived change ("no changes at all"). If a perceived change in physical activity or diet was reported, specific changes in these behaviors were collected. Participants were also asked to evaluate the program on assisting them in optimizing physical activity and healthy diet on a 4-point scale; responses were dichotomized as positive ("very helpful", "helpful", or "slightly helpful") or negative evaluation ("not helpful at all"). The number of sessions attended by participants was also recorded to assess the intervention compliance.

2.5. Statistical Analysis

As our primary analysis demonstrated a differential intervention effect on PPWR by country of birth (i.e., Australian-born vs. non-Australian-born) [13], here our secondary analysis was conducted by stratifying data according to country of birth to explore the differential responses in behavioral and psychosocial outcomes and factors associated with the effect of country of birth on PPWR based on the primary finding. Data are presented as mean $\pm$ standard deviation (continuous variables) and proportions (categorical variables). Differences in baseline or categorical outcome measures between groups (Australian-born vs. non-Australian-born, intervention vs. control) were compared using independent sample t tests for continuous variables and chi-square tests or Fisher's exact tests for categorical variables. Least squares means and their differences between groups (intervention vs. control) derived from multivariable linear regressions that adjusted for baseline variables (age, education, work, parity, BMI, self-weighing, risk perception of excess GWG, risk perception of GDM, eating self-efficacy) were used to present changes in continuous outcome measures. Due to the differing proportions of baseline BMI status (i.e., overweight or obesity) between Australian-born and non-Australian-born participants, we examined the potential interaction effect between intervention, baseline BMI status, and country of birth on the primary outcome of PPWR by conducting a three-way interaction term in linear regression analysis, with the intervention effect on PPWR stratified according to baseline BMI (i.e., Australian-born with overweight, Australian-born with obesity, non-Australian-born with overweight, non-Australian-born with obesity). To assess factors associated with PPWR, a multivariable linear regression model was constructed by including demographic, anthropometric, behavioral, and psychosocial variables with $p < 0.1$ on univariate analyses and using backward elimination to remove variables at $p > 0.05$. The model was adjusted for intervention allocation, age, baseline BMI and parity, the latter three as clinically relevant variables associated with PPWR. We used complete case analysis (i.e., all available data included in the analysis) as data were deemed missing at random, negating the need for multiple imputations [12,13]. Statistical significance was set at an α level of $p < 0.05$. All analyses were performed with SAS 9.4 (SAS Institute Inc., Cary, NC, USA).

3. Results

The HeLP-her study randomized 228 eligible women, with 86 Australian-born and 142 non-Australian-born. Over half (57%) of the non-Australian-born women were from South Asia (primarily India, Sri Lanka, Bangladesh, and Pakistan), 23% from East and Southeast Asia (primarily China, Vietnam, Malaysia, and Indonesia), and 12% from the Middle East, Africa, and Pacific Islands. Most (80%) of the non-Australian-born women had resided in Australia for less than 10 years. At 6 weeks postpartum, 202 women were followed up and completed primary outcome weight measurements (76 Australian-born and 126 non-Australian-born), with an overall attrition rate of 11.4% attributed to miscarriage or stillbirth, pre-term birth, and loss to contact (Figure S1).

3.1. Baseline Characteristics

Table 1 shows the baseline characteristics of the participants. Compared to Australian-born women, non-Australian-born women were younger, had a higher education level, were less likely to be employed, and were more likely to be primiparous. The mean BMI of non-Australian-born women was 28.0 ± 4.4 kg/m^2, lower than 34.3 ± 5.5 kg/m^2 in Australian-born women, with 22.5% and 73.3% being obese in the two groups, respectively. In addition, non-Australian-born women were less likely to weigh themselves frequently and had a lower risk perception for excess GWG and GDM at baseline than Australian-born women. Conversely, eating self-efficacy was higher in non-Australian-born compared with Australian-born women. There were no significant differences in the baseline levels of physical activity, fat-related dietary behaviors, weight control confidence, and exercise self-efficacy between the two groups.

Table 1. Baseline characteristics of participants.

Variables	Australian-Born (n = 86)	Non-Australian-Born (n = 142)	p Value
Demographics			
Age (years)	32.8 ± 4.4	31.5 ± 4.5	0.037
Education (%)			<0.001
High school or below	29.6	12.1	
Certificate/diploma	40.8	23.5	
Bachelor's degree or higher	29.6	64.4	
Work (%)			0.004
Full-time	23.5	28.0	
Part-time	43.2	22.0	
No paid work	33.3	50.0	
Household income (%)			0.050
<$40,000	22.2	36.2	
$40,000–80,000	35.8	33.1	
>$80,000	28.4	15.0	
Parity (%)			0.032
Primiparous	32.1	47.0	

Table 1. *Cont.*

Variables	Australian-Born (n = 86)	Non-Australian-Born (n = 142)	p Value
Anthropometrics			
Weight (kg)	90.8 ± 16.8	70.5 ± 13.3	<0.001
BMI (kg/m^2)	34.3 ± 5.5	28.0 ± 4.4	<0.001
BMI (kg/m^2) (%)			<0.001
Overweight (≤29.9)	26.7	77.5	
Obesity (≥30.0)	73.3	22.5	
Behavioral			
Physical activity (steps/day)	6201.0 ± 2921.4	5570.9 ± 3188.8	0.176
Fat-related dietary behaviors [a]	1.9 ± 0.3	1.9 ± 0.3	0.694
Frequent self-weighing (%)	62.9	40.9	0.006
Psychosocial			
Perceived risk of excess GWG (%)	85.7	71.4	0.036
Perceived risk of GDM (%)	63.5	40.0	0.004
Weight control confidence [b]	5.7 ± 2.0	5.9 ± 2.1	0.630
Exercise self-efficacy [c]			
Sticking to it	2.5 ± 0.9	2.4 ± 0.8	0.500
Making time for exercise	2.3 ± 0.9	2.3 ± 0.8	0.966
Eating self-efficacy [c]			
Sticking to it	2.6 ± 0.8	3.0 ± 0.9	0.015
Reducing calories	3.4 ± 0.9	3.6 ± 0.7	0.185

BMI, body mass index; GWG, gestational weight gain; GDM, gestational diabetes mellitus. Data are presented as mean ± SD or percentage. [a] 1 = usually choose low-fat; 3 = rarely or never choose low-fat. [b] 1 = not at all confident; 10 = totally confident. [c] 1 = not at all confident; 5 = extremely confident.

3.2. Intervention Effect

The intervention effects on behavioral and psychosocial measures are shown in Tables 2 and 3. Non-Australian-born women in the intervention group had significantly improved eating self-efficacy in "reducing calories" (intervention effect 0.68 (95%CI 0.15, 1.21), p = 0.013), and were more likely to report they had made changes to physical activity (intervention 89.4% vs. control 65.9%, p = 0.008) and diet (97.9% vs. 73.2%, p < 0.001) at 6 weeks postpartum, compared to the control group. A higher proportion of non-Australian-born women in the intervention group reported specific changes to diet than those in the control group, including increased fruit and vegetable consumption, increased low-fat dairy products, decreased intake of snack foods, and decreased takeaway and convenience foods (all p < 0.05). Frequent self-weighing in non-Australian-born women increased from 48.3% at baseline to 72.3% at 6 weeks postpartum (p = 0.013) in the intervention group, with no significant change from baseline (31.9%) to 6 weeks postpartum (36.6%) in the control group (p = 0.645). No significant differences between intervention and control were found in changes in physical activity, fat-related dietary behaviors, weight control confidence, and exercise self-efficacy in non-Australian-born women (all p > 0.05). In Australian-born women, none of the measures significantly differed between intervention and control. Frequent self-weighing in Australian-born women was similar over time (baseline to 6 weeks postpartum) irrespective of intervention allocation.

Table 2. Changes in behavioral and psychosocial measures from baseline to 6 weeks postpartum.

Variables	Australian-Born (n = 76)				Non-Australian-Born (n = 126)			
	Intervention (n = 39)	Control (n = 37)	Intervention Effect	p Value	Intervention (n = 65)	Control (n = 61)	Intervention Effect	p Value
Behavioral change in								
Physical activity (steps/day)	5135.0 (−14,637.0, 24,907.0)	243.4 (−14,874.5, 15,361.4)	4891.6 (−11,427.0, 21,210.1)	0.537	912.1 (−1473.8, 3298.1)	−1711.2 (−4683.5, 1261.1)	2623.3 (−670.7, 5917.3)	0.114
Fat-related dietary behaviors [a]	0.07 (−0.19, 0.32)	0.11 (−0.11, 0.33)	−0.04 (−0.25, 0.17)	0.695	−0.07 (−0.24, 0.10)	−0.02 (−0.22, 0.19)	−0.05 (−0.28, 0.17)	0.646
Psychosocial change in								
Weight control confidence [b]	−0.61 (−3.21, 2.00)	−0.52 (−2.74, 1.70)	−0.09 (−2.24, 2.07)	0.934	0.30 (−0.87, 1.47)	0.55 (−0.88, 1.98)	−0.25 (−1.82, 1.31)	0.745
Exercise self-efficacy [c]								
Sticking to it	0.49 (−0.19, 1.18)	0.49 (−0.09, 1.08)	0.00 (−0.57, 0.57)	0.997	0.05 (−0.38, 0.48)	0.06 (−0.46, 0.57)	0.00 (−0.57, 0.56)	0.991
Making time for exercise	0.41 (−0.54, 1.37)	0.34 (−0.47, 1.15)	0.08 (−0.71, 0.87)	0.838	0.19 (−0.24, 0.62)	0.00 (−0.51, 0.52)	0.19 (−0.38, 0.75)	0.504
Eating self-efficacy [c]								
Sticking to it	−0.08 (−0.91, 0.76)	0.38 (−0.33, 1.09)	−0.46 (−1.15, 0.23)	0.179	0.06 (−0.30, 0.42)	−0.16 (−0.57, 0.25)	0.22 (−0.21, 0.66)	0.306
Reducing calories	0.69 (0.03, 1.35)	0.68 (0.12, 1.24)	0.01 (−0.54, 0.56)	0.968	0.25 (−0.15, 0.66)	−0.43 (−0.91, 0.06)	0.68 (0.15, 1.21)	0.013

Data are presented as least squares means with 95% CIs from linear regression models with adjustment for age, education, work, parity, baseline BMI, baseline self-weighing, risk perception of excess GWG, risk perception of GDM, and baseline eating self-efficacy (sticking to it). p value for the intervention effect. [a] 1 = usually choose low-fat; 3 = rarely or never choose low-fat. [b] 1 = not at all confident; 10 = totally confident. [c] 1 = not at all confident; 5 = extremely confident.

Table 3. Other behavioral measures at 6 weeks postpartum.

Variables	Australian-Born (n = 76)			Non-Australian-Born (n = 126)		
	Intervention (n = 39)	Control (n = 37)	p Value	Intervention (n = 65)	Control (n = 61)	p Value
Frequent self-weighing	57.7	53.3	0.744	72.3	36.6	<0.001
Perceived change to physical activity	65.4	66.7	0.920	89.4	65.9	0.008
Increased number of regular physical activity sessions	30.8	36.7	0.642	31.1	24.4	0.488
Increased time spent on physical activity sessions	3.9	16.7	0.200 [a]	6.7	9.8	0.704 [a]
Increased physical intensity of exercise sessions	0.0	10.0	0.240 [a]	2.2	9.8	0.188 [a]
Perceived change to diet	88.5	86.7	1.000 [a]	97.9	73.2	<0.001
Increased fruit and vegetable consumption	42.3	44.8	0.851	77.8	55.0	0.026
Increased low-fat dairy products	38.5	24.1	0.251	60.0	35.0	0.021
Decreased fruit juice, cordial and soft drink consumption	19.2	27.6	0.467	42.2	37.5	0.657
Decreased intake of snack foods	46.2	41.4	0.722	53.3	30.0	0.030
Decreased takeaway and convenience foods	38.5	51.7	0.324	55.6	20.0	<0.001

Data are presented as percentages. [a] Fisher's exact test due to expected cell frequencies of <5.

Most Australian-born (96.4%) and non-Australian-born women (95.4%) had a positive evaluation of the program at 6 weeks postpartum. Of the women allocated to the intervention, 94.9% Australian-born and 90.8% non-Australian-born women attended all four sessions, with no significant differences in the intervention compliance ($p = 0.707$).

3.3. Interaction Analysis

Table S1 shows the intervention effect according to country of birth and baseline BMI. There was a significant intervention effect on reducing PPWR in non-Australian-born women with overweight (intervention 1.97 ± 3.92 kg vs. control 3.98 ± 5.53 kg, $p = 0.040$), as well as a trend toward less PPWR in non-Australian-born women with obesity (−1.42 ± 3.77 kg vs. 2.03 ± 5.03 kg, $p = 0.052$). In contrast, no significant intervention effect was revealed among Australian-born women regardless of BMI status ($p > 0.05$). No interaction effect was found between intervention, baseline BMI status, and country of birth ($p = 0.237$).

3.4. Regression Analysis

The regression results are presented in Tables S2 and S3. The multivariable analysis showed intervention allocation and decreased intake of snack foods were independent predictors of lower PPWR in non-Australian-born women. Other demographic, anthropometric, behavioral, and psychosocial factors did not significantly impact their PPWR. In Australian-born women, increased time spent on physical activity sessions was the only factor significantly associated with PPWR.

4. Discussion

We previously reported a greater effect of the HeLP-her intervention on reducing weight retention at 6 weeks postpartum in non-Australian-born women compared with Australian-born women [13]. In this study, we expand on these findings by examining factors that may be related to the differential intervention effect, including demographic and anthropometric characteristics as well as behavioral and psychosocial factors targeted by the intervention. Here, we found that non-Australian-born women receiving the intervention had improved self-efficacy for dietary behaviors, were more likely to self-weigh frequently, and reported changes to diet and physical activity at 6 weeks postpartum, compared to standard antenatal care. In contrast, we did not observe any significant changes in behavioral or psychosocial measures among Australian-born women following the intervention. Exploratory analysis showed intervention allocation and decreased intake of snack foods were predictors of lower weight retention postpartum in non-Australian-born women.

The HeLP-her intervention is non-prescriptive, utilizing simple messages on healthy eating and physical activity aligned with national dietary and physical activity recommendations, underpinned by practicing skills in self-management. Weight self-monitoring has been identified as an essential component in weight management, enabling immediate adjustment to weight-related behaviors as well as reinforcement [28]. Yet previous studies in the US have shown ethnic minority groups are less likely to report frequent self-weighing compared to their white counterparts [29,30], as consistent with the baseline findings in non-Australian-born women in the HeLP-her study. This could be partly attributable to decreased risk perception related to weight gain during pregnancy and lower awareness of GDM risk, as observed here at baseline. Encouragingly, our results showed the HeLP-her intervention significantly improved the self-weighing behaviors in non-Australian-born women, despite no improvement observed in Australian-born women. The greater improvement in self-weighing frequency following the intervention among non-Australian-born women suggests receptiveness to the intervention messaging with amenability to behavior change in this group.

The intervention focused on small, sustainable behavior adjustments using individualized goal setting and action planning. Therefore, it is not surprising that we were unable to detect significant changes in physical activity measured by daily step counts or fat-related dietary behaviors measured in dietary scores in non-Australian-born women, per our previous findings [12,13]. It is possible that the measurement tools used were less sensitive to detect the small behavioral changes encouraged as part of the intervention. Despite no differences in quantitatively measured physical activity or dietary behaviors, non-Australian-born women were more likely to report perceived changes in diet and physical activity. Particularly, changes in several specific dietary behaviors were reported, among which decreased intake of snack foods was found to be associated with less PPWR on the multivariable analysis. In line with this, non-Australian-born women had higher eating self-efficacy in "sticking to it" at baseline, which persisted into 6 weeks postpartum. Furthermore, they had greater improvement in self-efficacy of "reducing calories" after the intervention. These are reflective of improved confidence in the ability to make behavior changes as well as increased commitment and practicing behavior change towards dietary behaviors. In contrast to the positive changes seen in diet, we did not find any changes in specific physical activity behaviors related to the increasing number, time, or intensity of exercise, nor the improvement in exercise self-efficacy over time among non-Australian-born women. Previous studies have found a lower level of physical activity during pregnancy among Asians, Middle Easterners, and Africans compared to white populations [31–33]. It is shown that women's beliefs, attitudes, barriers, and intentions towards exercise during pregnancy differ between cultures [32]. For women from CALD backgrounds, safety concerns about exercising during pregnancy is a significant barrier, reflected by cultural beliefs [32,34]. For example, in traditional Chinese culture, pregnant women are advised to restrict exercise due to concern of miscarriage [35]. For this reason, it is plausible that dietary behaviors may be more readily modified, with fewer barriers to behavior change, than physical activity behaviors in CALD women during pregnancy. However, given physical activity is safe and associated with optimized weight and reduction in complications during pregnancy [36,37], it is imperative to find ways to address barriers and improve health knowledge towards physical activity in this population.

In contrast to non-Australian-born women, we did not find significant intervention effects on behavioral, psychosocial, and weight outcomes among Australian-born women irrespective of BMI status (i.e., overweight or obesity). This is consistent with previous Australian-based trials, including the large LIMIT randomized trial conducted in predominantly white women with overweight or obesity, which reported no differences in GWG between antenatal care and lifestyle advice following six intervention sessions throughout pregnancy [38]. In our study, women born in Australia had a higher baseline risk perception, potentially reflecting a higher level of health literacy and confidence or familiarity in access to healthcare services and information [39]. Therefore, the low-intensity intervention format utilizing simple health messaging may not have been sufficient to influence further behavior change towards diet, physical activity and self-management behaviors. Future research, potentially including more tailored or prescriptive intervention, is needed to evaluate outcomes with different intervention types and intensities in this population.

Strengths and Limitations

Here, we used a rigorous study design, utilized robust measures (including a validated GDM screening tool, objective measurement of weight, and pedometer for assessing physical activity), and reported high compliance and high fidelity to the intervention delivery [12,13]. Limitations include the absence of measured dietary intake (e.g., energy intake) and health literacy that may have elucidated findings further [40]. Also, despite a relatively high retention rate (88.6%) at follow-up of 6 weeks postpartum, up to 30% of questionnaire data were missing. Furthermore, the non-Australian-born participants in our study were of moderate socioeconomic advantage and predominantly Asian, reflective of the broader Australian demographic data on migrant populations [41]. As a secondary analysis, our results may not be fully generalizable to all populations, which remained to be confirmed in larger, population-based studies.

5. Conclusions

Women from CALD backgrounds experience greater health inequity with an increased risk of adverse health outcomes during pregnancy [42]. Strategies that are accessible, relevant, culturally acceptable, and cost-effective are needed to support health improvement for these women during pregnancy. Our results suggest a low-intensity intervention based on simple health messages alongside routine antenatal care is acceptable and relevant to diverse cultures with demonstrated efficacy in reducing postpartum weight retention among high-risk CALD groups. The improvement in weight outcome appears to derive from small individually driven changes to dietary behaviors during pregnancy, potentially reflective of increased receptiveness to dietary, compared with physical activity, behavior change. Further research is needed to address barriers to exercise in this population to maximize exercise engagement during pregnancy and promote broader health benefits.

Supplementary Materials: The following supporting information can be downloaded at: https://www.mdpi.com/article/10.3390/nu14142988/s1, Figure S1: CONSORT diagram; Table S1: Intervention effect on weight change from baseline to 6 weeks postpartum according to country of birth and baseline BMI; Table S2: Univariate regression analysis for predictors of weight change from baseline to 6 weeks postpartum; Table S3: Multivariable regression analysis for predictors of weight change from baseline to 6 weeks postpartum.

Author Contributions: C.L.H. conceptualized and contributed to the original study design. C.L.H. and M.C. contributed to the secondary study design. C.L.H. conducted the study and collected the data. M.C. performed the data analysis and wrote the manuscript. C.L.H. and S.L. critically reviewed and revised the manuscript. All authors have read and agreed to the published version of the manuscript.

Funding: This project was supported by a BRIDGES grant from the International Diabetes Federation. BRIDGES, an International Diabetes Federation project, was supported by an educational grant from Lilly Diabetes (Project Number: LT07-121). The Jack Brockhoff Foundation also provided funding for this study. This study was also supported by the National Health and Medical Research Council Centre for Research Excellence for Health in Preconception and Pregnancy (CRE-HiPP; APP1171142; C.L.H.); National Medical Health and Research Council Fellowship (S.L.); and Australian Government Research Training Program Scholarship (M.C.).

Institutional Review Board Statement: The study was conducted in accordance with the Declaration of Helsinki, and approved by the Southern Health Research Advisory and Ethics Committee (project number 07216C; approval date 1 April 2008).

Informed Consent Statement: Informed consent was obtained from all subjects involved in the study.

Data Availability Statement: The data presented in the current study are available from the corresponding author on reasonable request.

Acknowledgments: The authors would like to acknowledge Amanda Hulley, Lauren Snell, and Melanie Gibson-Helm for recruitment and data collection, Deborah Thompson and Nicole Ng for data entry, and Helena Teede and posthumously, Catherine Lombard, for funding support and initial input into the study concept and design.

Conflicts of Interest: The authors declare no conflict of interest.

References

1. World Health Organization. Facts about Overweight and Obesity. Available online: https://www.who.int/news-room/fact-sheets/detail/obesity-and-overweight (accessed on 25 January 2022).
2. Catalano, P.M.; Shankar, K. Obesity and pregnancy: Mechanisms of short term and long term adverse consequences for mother and child. *BMJ* **2017**, *356*, j1. [CrossRef]
3. Villamor, E.; Cnattingius, S. Interpregnancy weight change and risk of adverse pregnancy outcomes: A population-based study. *Lancet* **2006**, *368*, 1164–1170. [CrossRef]
4. Headen, I.E.; Davis, E.M.; Mujahid, M.S.; Abrams, B. Racial-ethnic differences in pregnancy-related weight. *Adv. Nutr.* **2012**, *3*, 83–94. [CrossRef] [PubMed]

5. Boardley, D.J.; Sargent, R.G.; Coker, A.L.; Hussey, J.R.; Sharpe, P.A. The relationship between diet, activity, and other factors, and postpartum weight change by race. *Obstet. Gynecol.* **1995**, *86*, 834–838. [CrossRef]
6. Van Poppel, M.N.; Hartman, M.A.; Hosper, K.; van Eijsden, M. Ethnic differences in weight retention after pregnancy: The ABCD study. *Eur. J. Public Health* **2012**, *22*, 874–879. [CrossRef]
7. Waage, C.W.; Falk, R.S.; Sommer, C.; Mørkrid, K.; Richardsen, K.R.; Baerug, A.; Shakeel, N.; Birkeland, K.I.; Jenum, A.K. Ethnic differences in postpartum weight retention: A Norwegian cohort study. *BJOG* **2016**, *123*, 699–708. [CrossRef]
8. Endres, L.K.; Straub, H.; McKinney, C.; Plunkett, B.; Minkovitz, C.S.; Schetter, C.D.; Ramey, S.; Wang, C.; Hobel, C.; Raju, T.; et al. Postpartum weight retention risk factors and relationship to obesity at 1 year. *Obstet. Gynecol.* **2015**, *125*, 144–152. [CrossRef] [PubMed]
9. Teede, H.J.; Bailey, C.; Moran, L.J.; Bahri Khomami, M.; Enticott, J.; Ranasinha, S.; Rogozinska, E.; Skouteris, H.; Boyle, J.A.; Thangaratinam, S.; et al. Association of Antenatal Diet and Physical Activity-Based Interventions With Gestational Weight Gain and Pregnancy Outcomes: A Systematic Review and Meta-analysis. *JAMA Intern. Med.* **2021**, *182*, 106–114. [CrossRef] [PubMed]
10. Effect of diet and physical activity based interventions in pregnancy on gestational weight gain and pregnancy outcomes: Meta-analysis of individual participant data from randomised trials. *BMJ* **2017**, *358*, j3119. [CrossRef]
11. Cantor, A.G.; Jungbauer, R.M.; McDonagh, M.; Blazina, I.; Marshall, N.E.; Weeks, C.; Fu, R.; LeBlanc, E.S.; Chou, R. Counseling and Behavioral Interventions for Healthy Weight and Weight Gain in Pregnancy: Evidence Report and Systematic Review for the US Preventive Services Task Force. *JAMA* **2021**, *325*, 2094–2109. [CrossRef]
12. Harrison, C.L.; Lombard, C.B.; Strauss, B.J.; Teede, H.J. Optimizing healthy gestational weight gain in women at high risk of gestational diabetes: A randomized controlled trial. *Obesity* **2013**, *21*, 904–909. [CrossRef] [PubMed]
13. Harrison, C.L.; Lombard, C.B.; Teede, H.J. Limiting postpartum weight retention through early antenatal intervention: The HeLP-her randomised controlled trial. *Int. J. Behav. Nutr. Phys. Act.* **2014**, *11*, 134. [CrossRef] [PubMed]
14. *Monash Health Annual Report 2018–2019*; Monash Health: Melbourne, Australia, 2019.
15. *Information Paper: An Introduction to Socio-Economic Indexes for Areas (SEIFA) 2006*; Australian Bureau of Statistics: Canberra, Australia, 2008.
16. WHO/IASO/IOTF. *The Asia-Pacific Perspective: Re-Defining Obesity and Its Treatment*; Health Communications Australia: Melbourne, Australia, 2000.
17. Teede, H.J.; Harrison, C.L.; Teh, W.T.; Paul, E.; Allan, C.A. Gestational diabetes: Development of an early risk prediction tool to facilitate opportunities for prevention. *Aust. N. Z. J. Obstet. Gynaecol.* **2011**, *51*, 499–504. [CrossRef] [PubMed]
18. *Dietary Guidelines for Australian Adults*; National Health and Medical Research Council: Canberra, Australia, 2003.
19. *National Physical Activity Guidelines for Adults*; Australian Government Department of Health and Ageing: Canberra, Australia, 2009.
20. Rasmussen, K.; Yaktine, A.L. Institute of Medicine and National Research Council Committee to Reexamine IOM Pregnancy Weight Guidelines. In *Weight Gain during Pregnancy: Reexamining the Guidelines*; National Academies Press: Washington, DC, USA, 2009.
21. Ethnic Communities' Council of Victoria. ECCV Glossary of Terms. Available online: https://eccv.org.au/glossary-of-terms/ (accessed on 12 July 2022).
22. Pham, T.T.L.; Berecki-Gisolf, J.; Clapperton, A.; O'Brien, K.S.; Liu, S.; Gibson, K. Definitions of Culturally and Linguistically Diverse (CALD): A Literature Review of Epidemiological Research in Australia. *Int. J. Environ. Res. Public Health* **2021**, *18*, 737. [CrossRef] [PubMed]
23. Harrison, C.L.; Thompson, R.G.; Teede, H.J.; Lombard, C.B. Measuring physical activity during pregnancy. *Int. J. Behav. Nutr. Phys. Act.* **2011**, *8*, 19. [CrossRef]
24. Kristal, A.R.; White, E.; Shattuck, A.L.; Curry, S.; Anderson, G.L.; Fowler, A.; Urban, N. Long-term maintenance of a low-fat diet: Durability of fat-related dietary habits in the Women's Health Trial. *J. Am. Diet. Assoc.* **1992**, *92*, 553–559. [CrossRef]
25. Prochaska, J.O.; DiClemente, C.C. Transtheoretical therapy: Toward a more integrative model of change. *Psychother. Theory Res. Pract.* **1982**, *19*, 276–288. [CrossRef]
26. Lorig, K.R.; Sobel, D.S.; Ritter, P.L.; Laurent, D.; Hobbs, M. Effect of a self-management program on patients with chronic disease. *Eff. Clin. Pract.* **2001**, *4*, 256–262.
27. Sallis, J.F.; Pinski, R.B.; Grossman, R.M.; Patterson, T.L.; Nader, P.R. The development of self-efficacy scales for health-related diet and exercise behaviors. *Health Educ. Res.* **1988**, *3*, 283–292. [CrossRef]
28. Vanwormer, J.J.; French, S.A.; Pereira, M.A.; Welsh, E.M. The impact of regular self-weighing on weight management: A systematic literature review. *Int. J. Behav. Nutr. Phys. Act.* **2008**, *5*, 54. [CrossRef]
29. Gavin, K.L.; Linde, J.A.; Pacanowski, C.R.; French, S.A.; Jeffery, R.W.; Ho, Y.Y. Weighing frequency among working adults: Cross-sectional analysis of two community samples. *Prev. Med. Rep.* **2015**, *2*, 44–46. [CrossRef] [PubMed]
30. Linde, J.A.; Jeffery, R.W.; French, S.A.; Pronk, N.P.; Boyle, R.G. Self-weighing in weight gain prevention and weight loss trials. *Ann. Behav. Med.* **2005**, *30*, 210–216. [CrossRef] [PubMed]
31. Catov, J.M.; Parker, C.B.; Gibbs, B.B.; Bann, C.M.; Carper, B.; Silver, R.M.; Simhan, H.N.; Parry, S.; Chung, J.H.; Haas, D.M.; et al. Patterns of leisure-time physical activity across pregnancy and adverse pregnancy outcomes. *Int. J. Behav. Nutr. Phys. Act.* **2018**, *15*, 68. [CrossRef] [PubMed]

32. Guelfi, K.J.; Wang, C.; Dimmock, J.A.; Jackson, B.; Newnham, J.P.; Yang, H. A comparison of beliefs about exercise during pregnancy between Chinese and Australian pregnant women. *BMC Pregnancy Childbirth* **2015**, *15*, 345. [CrossRef] [PubMed]
33. Richardsen, K.R.; Falk, R.S.; Jenum, A.K.; Mørkrid, K.; Martinsen, E.W.; Ommundsen, Y.; Berntsen, S. Predicting who fails to meet the physical activity guideline in pregnancy: A prospective study of objectively recorded physical activity in a population-based multi-ethnic cohort. *BMC Pregnancy Childbirth* **2016**, *16*, 186. [CrossRef] [PubMed]
34. Moore, A.P.; Flynn, A.C.; Adegboye, A.R.A.; Goff, L.M.; Rivas, C.A. Factors Influencing Pregnancy and Postpartum Weight Management in Women of African and Caribbean Ancestry Living in High Income Countries: Systematic Review and Evidence Synthesis Using a Behavioral Change Theoretical Model. *Front. Public Health* **2021**, *9*, 637800. [CrossRef]
35. Lee, D.T.; Ngai, I.S.; Ng, M.M.; Lok, I.H.; Yip, A.S.; Chung, T.K. Antenatal taboos among Chinese women in Hong Kong. *Midwifery* **2009**, *25*, 104–113. [CrossRef]
36. Physical Activity and Exercise During Pregnancy and the Postpartum Period: ACOG Committee Opinion, Number 804. *Obstet. Gynecol.* **2020**, *135*, e178–e188. [CrossRef]
37. Harrison, C.L.; Brown, W.J.; Hayman, M.; Moran, L.J.; Redman, L.M. The Role of Physical Activity in Preconception, Pregnancy and Postpartum Health. *Semin. Reprod. Med.* **2016**, *34*, e28-37. [CrossRef]
38. Dodd, J.M.; Turnbull, D.; McPhee, A.J.; Deussen, A.R.; Grivell, R.M.; Yelland, L.N.; Crowther, C.A.; Wittert, G.; Owens, J.A.; Robinson, J.S. Antenatal lifestyle advice for women who are overweight or obese: LIMIT randomised trial. *BMJ* **2014**, *348*, g1285. [CrossRef]
39. Beauchamp, A.; Buchbinder, R.; Dodson, S.; Batterham, R.W.; Elsworth, G.R.; McPhee, C.; Sparkes, L.; Hawkins, M.; Osborne, R.H. Distribution of health literacy strengths and weaknesses across socio-demographic groups: A cross-sectional survey using the Health Literacy Questionnaire (HLQ). *BMC Public Health* **2015**, *15*, 678. [CrossRef] [PubMed]
40. Garad, R.; McPhee, C.; Chai, T.L.; Moran, L.; O'Reilly, S.; Lim, S. The Role of Health Literacy in Postpartum Weight, Diet, and Physical Activity. *J. Clin. Med.* **2020**, *9*, 2463. [CrossRef] [PubMed]
41. Australia's Population by Country of Birth. Available online: https://www.abs.gov.au/statistics/people/population/australias-population-country-birth/latest-release (accessed on 1 May 2022).
42. Karger, S.; Bull, C.; Enticott, J.; Callander, E.J. Options for improving low birthweight and prematurity birth outcomes of indigenous and culturally and linguistically diverse infants: A systematic review of the literature using the social-ecological model. *BMC Pregnancy Childbirth* **2022**, *22*, 3. [CrossRef]

Article

Association of Dietary Pattern with Cardiovascular Risk Factors among Postmenopausal Women in Taiwan: A Cross-Sectional Study from 2001 to 2015

Sabrina Aliné [1], Chien-Yeh Hsu [2,3], Hsiu-An Lee [4], Rathi Paramastri [1] and Jane C.-J. Chao [1,3,5,*]

1 School of Nutrition and Health Sciences, College of Nutrition, Taipei Medical University, 250 Wu-Hsing Street, Taipei 11031, Taiwan; sabrinaline14@gmail.com (S.A.); rara.paramastri@gmail.com (R.P.)
2 Department of Information Management, National Taipei University of Nursing and Health Sciences, 365 Ming-Te Road, Peitou District, Taipei 11219, Taiwan; cyhsu@ntunhs.edu.tw
3 Master Program in Global Health and Development, College of Public Health, Taipei Medical University, 250 Wu-Hsing Street, Taipei 11031, Taiwan
4 National Health Research Institutes, 35 Keyan Road, Zhunan Town, Miaoli County 35053, Taiwan; billy72325@gmail.com
5 Nutrition Research Center, Taipei Medical University Hospital, 252 Wu-Hsing Street, Taipei 11031, Taiwan
* Correspondence: chenjui@tmu.edu.tw; Tel.: +886-2-2736-1661 (ext. 6548); Fax: +886-2-2737-3112

Abstract: Unhealthy diet and inappropriate lifestyle contribute to an imbalance in cardiometabolic profiles among postmenopausal women. This research aimed to analyze the association between dietary pattern and changes in cardiovascular risk factors among postmenopausal Taiwanese women using binary logistic regression. This cross-sectional study involved 5689 postmenopausal Taiwanese women aged 45 years and above, and the data were obtained from Mei Jau Health Management Institution database between 2001 and 2015. The cardiovascular risk dietary pattern characterized by high intakes of processed food, rice/flour products, organ meat, and sauce was derived by reduced rank regression. Participants in the highest quartile of the cardiovascular risk dietary pattern were more likely to have high levels of systolic blood pressure (OR = 1.29, 95% CI 1.08–1.53), diastolic blood pressure (OR = 1.28, 95% CI 1.01–1.62), atherogenic index of plasma (OR = 1.26, 95% CI 1.06–1.49), triglycerides (OR = 1.38, 95% CI 1.17–1.62), and fasting blood glucose (Q3: OR = 1.45, 95% CI 1.07–1.97). However, this dietary pattern was not correlated with total cholesterol, low-density lipoprotein cholesterol, high-density lipoprotein cholesterol, and C-reactive protein. Therefore, adherence to the cardiovascular risk dietary pattern increases the risk of having higher levels of blood pressure, triglycerides, fasting blood glucose in postmenopausal Taiwanese women.

Keywords: postmenopausal women; dietary pattern; cardiovascular risk factors; reduced rank regression

Citation: Aliné, S.; Hsu, C.-Y.; Lee, H.-A.; Paramastri, R.; Chao, J.C.-J. Association of Dietary Pattern with Cardiovascular Risk Factors among Postmenopausal Women in Taiwan: A Cross-Sectional Study from 2001 to 2015. *Nutrients* **2022**, *14*, 2911. https://doi.org/10.3390/nu14142911

Academic Editors: Birgit-Christiane Zyriax and Nataliya Makarova

Received: 15 June 2022
Accepted: 14 July 2022
Published: 15 July 2022

1. Introduction

Menopause is defined as the cessation of menstruation owing to a decrease in ovarian follicles and the further reduction of estradiol levels. It occurs mostly at a median age of 51 years [1]. The diagnosis of menopause is based on no menstrual period for 12 consecutive months in women [2,3]. Around 467 million postmenopausal women were registered in the world in the 1990s, and by 2030 the number of postmenopausal women is expected to be 1.2 billion with 47 million new postmenopausal women each year [4]. Several studies revealed that postmenopause was associated with increased inflammatory markers such as C-reactive protein (CRP), interleukin-1α (IL-1α), and tumor necrosis factor-α (TNF-α) and an imbalance in cardiometabolic profiles such as low levels of high-density lipoprotein cholesterol (HDL-C) and elevated levels of total cholesterol (TC), low-density lipoprotein cholesterol (LDL-C), triglycerides (TG), visceral fat, and blood glucose [5–8]. These imbalanced cardiometabolic profiles were favorable for the progression of atherosclerosis and an increased risk of cardiovascular disease (CVD) [9]. Cardiovascular disease was known as the leading cause of mortality worldwide between 1990 and 2019. This scourge claimed around 18.6 million individuals' lives in 2019 [10]. According to the report by the Ministry of Health and Welfare, Taiwan, heart disease was the second leading cause of death following malignant neoplasms in 2020 [11]. Shen et al. found that in Taiwanese women, early age at menopause between 45 and 49 years was linked to higher CVD death rate and all-cause mortality [12]. In 2018, the percentage of the elderly aged 65 years and above in Taiwan surpassed 14% and has become an aged society [13]. The prevalence of cardiovascular disease, diabetes, and cancer was high among the Taiwanese elderly during the past decade [14]. Aging and atherosclerosis can cause vascular wall damage and estrogen receptor loss, and a decrease in circulating estrogen also reduces estrogen receptors in both vascular endothelium and vascular smooth muscle cells [3]. Additionally, women with vasomotor symptoms have significantly higher blood pressure, elevated circulating total cholesterol levels, and greater body mass index (BMI) than women without vasomotor symptoms [3].

Evidence showed that postmenopausal Chinese women increased the risk of dyslipidemia after multiple adjustment as compared to premenopausal women probably due to the loss of endogenous estrogen after menopause [15]. Some studies also supported that menopause was associated with adverse changes of cardiometabolic profiles and increased risk and mortality of CVD [5,15,16]. Research conducted by Lin et al. demonstrated that compared to premenopausal women in North Taiwan, postmenopausal women had considerably greater odds of having central obesity, metabolic syndrome, high blood pressure, and high blood triglycerides [17]. In addition, diet has been associated with cardiovascular risk factors and other health-related outcomes. A healthy balanced diet plays a significant role in the prevention and mortality reduction of chronic diseases [18]. However, postmenopausal women consuming an unhealthy diet such as high intake of sodium, added sugar, trans fats, and red meat but low intake of fruit, whole grains, fibers, fish, nuts, and legumes were correlated with abnormal fasting blood glucose, high BMI, hypertension, and high blood cholesterol which are considered as risk factors of CVD among postmenopausal women [19]. Brazilian postmenopausal women who consumed a low-quality diet with an excessive intake of sodium and low intakes of vegetables and fruit had central obesity, higher blood pressure, and increased levels of blood lipids and fasting blood glucose [20].

The dietary pattern is considered as a new approach applied in nutritional epidemiology to assess the relationship between dietary factors and disease risk [21]. However, little is known about the outcomes resulting from the association between dietary patterns and CVD risk factors among postmenopausal Taiwanese women. Hence, the aim of this study was to analyze the association between dietary patterns and changes in cardiovascular risk factors such as blood pressure, blood lipids, blood glucose, and CRP among postmenopausal Taiwanese women.

2. Materials and Methods

2.1. Study Population and Data Source

This cross-sectional study was conducted using the database from 2001 to 2015, and the data were collected by the Mei Jau (MJ) Health Screening Centers which are located in Taipei, Taoyuan, Taichung, and Kaohsiung cities in Taiwan. All the subjects signed the consent form and agreed their data only for research use without their identity before their health check-up at the MJ Health Screening Center. While visiting the MJ Health Screening Center, all the subjects filled the questionnaires to collect information about their socio-demographic status, lifestyle, and dietary habits by the self-reported questionnaires. Blood samples were analyzed for biochemical parameters. The study was approved by the Joint Institutional Review Board of Taipei Medical University (TMU-JIRB N202007075). There were 377,124 subjects who visited the MJ Health Screening Center between 2005 and 2015. We included women aged ≥45 years who self-reported menopausal status after missing their menstrual period for at least 12 consecutive months using a questionnaire. We excluded 299,450 participants who were male, had disease conditions such as cancer, cystic fibrosis, lung disease, cirrhosis, kidney disease, or infectious disease, or used any forms of lipid-lowering drugs. In addition, we excluded 68,985 women who were non-postmenopausal, aged less than 45 years, or failed to complete the questionnaire about their dietary habits. After excluding 3000 participants who had multiple entries between 2005 and 2015, a total 5689 postmenopausal women were retained in this study (Figure 1).

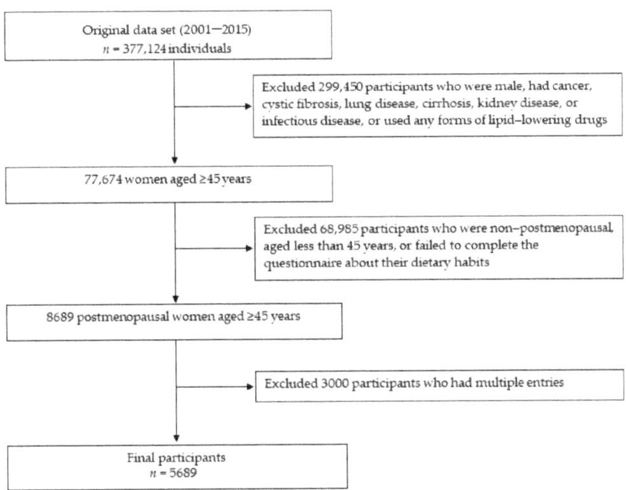

Figure 1. Flowchart of study participants.

2.2. Dietary Assessment and Other Covariates

A semi-quantitative food frequency questionnaire (FFQ) was developed, standardized, and validated by the MJ Health Management Institution, and used to assess dietary habits of the subjects. The FFQ questionnaire contained the closed-ended questions about 22 non-overlapping food groups with a total of 85 individual food items consumed by the participants in the past month [22]. The intake frequency was assessed in accordance with daily and weekly consumption. Each question was given the definition about one serving size of the food item, and presented 5 frequency response options as described previously [22]. Dietary data were collected for further frequency response options as described previously [22]. Dietary data were collected for further analysis to derive the dietary pattern using a reduced rank regression (RRR) model. The RRR model as a multi-variable linear function was performed to derive the dietary pattern related to the disease of interest by a priori and a posteriori approaches based on the response variables for identifying a linear combination of the predictor variables [23].

Demographic data such as age, education (≤high school or >high school), and marital status (never married, married, or divorced/widowed) were collected. We also evaluated lifestyle data including smoking status (no or yes), drinking alcohol (no or yes), physical activity frequency (<150 min/week or ≥150 min/week), and sleep duration (<6 h, 6–8 h, or >8 h). Medical history regarding hypertension, diabetes mellitus, and CVD was recorded. All covariates were assessed using a self-reported questionnaire.

2.3. Anthropometric, Clinical, and Biochemical Data

Anthropometric parameters such as height, weight, waist circumference (WC), and waist-to-hip ratio (WHR) were assessed using an anthropometer with electronic scale at the MJ Health Screening Center. The values of BMI were calculated using weight (kg) divided by height (m^2) [24]. To identify central obesity among the participants, WC (≥80 cm) [25] and WHR (≥0.85) [26] were measured and calculated. Blood pressure was measured twice at 10 min intervals using a standardized sphygmomanometer. Biochemical data such as total cholesterol (TC), low-density lipoprotein cholesterol (LDL-C), high-density lipoprotein cholesterol (HDL-C), triglycerides (TG), fasting blood glucose (FBG), and C-reactive protein (CRP) were assessed after overnight fasting for 12–14 h by the central laboratory at the MJ Health Management Institution. Blood TC, HDL-C, TG, and FBG were evaluated using the commercial kits (Randox Laboratories Ltd., Antrim, UK). The levels of LDL-C were determined by Friedewald formula (LDL-C (mg/dL) = TC-HDL-C-TG/5) [27]. Atherogenic index of plasma (AIP) as an indicator for CVD risk was calculated by the following formula: AIP = log(TG/HDL-C) [28]. Inflammatory marker CRP was diagnosed by the reagent from Fortress Diagnostics (Antrim, UK). Cardiovascular disease risk factors were defined as: systolic blood pressure (SBP) $\geq$140 mmHg and/or diastolic blood pressure (DBP) $\geq$ 90 mmHg [20,28], AIP $\geq$ 0.24 with high risk of CVD [29], TC $\geq$ 5.17 mmol/L (200 mg/dL) [20], LDL-C $\geq$ 2.59 mmol/L (100 mg/dL) [20], HDL-C $\leq$ 1.29 mmol/L (50 mg/dL) [20], TG $\geq$ 1.69 mmol/L (150 mg/dL) [20], FBG $\geq$ 7.0 mmol/L (126 mg/dL) [28], and CRP $\geq$ 28.6 nmol/L (3 mg/L) [30].

2.4. Statistical Analysis

Statistical analysis was performed using SAS version 9.4 (SAS Institute Inc., Chicago, IL, USA) and IBM SPSS 20 (IBM Corp., Armonk, NY, USA). Kolmogorov–Smirnov test was used to determine the distribution of the data. To compare the differences between two groups, Mann–Whitney U test and chi-square test were used for categorical data. To compare data among multiple groups, one-way analysis of variance (ANOVA) and Kruskal–Wallis test were performed. We used binary logistic regression expressed as odds ratios (ORs) and 95% confidence intervals (CIs) to determine the association between the dietary pattern and cardiovascular risk factors. The dietary pattern was derived by RRR using PROC PLS function in SAS 9.4, and 22 food groups were considered as the predictors. After performing Pearson's correlation coefficient, triglycerides, systolic blood pressure, fasting blood glucose, and AIP were retained as the response variables (Figure 2). In compliance with previous investigation, to obtain the dietary pattern linked to CVD risk, the value of factor loading was set at $\geq$0.20 [31]. The dietary factor score for each food group was calculated by summing food frequency intake weighed by their factor loadings. Finally, we only retained the first dietary factor for further analysis because it explained the maximum variation of the response variables. The derived dietary pattern was then divided into quartiles according to the dietary factor score. The reference group for the cardiovascular risk dietary pattern was quartile 1 (Q1) which was the lowest quartile of the dietary factor score, and quartile 4 (Q4) represented the highest quartile of the dietary factor score. In binary logistic regression analysis, model 1 was unadjusted, model 2 was adjusted for age, BMI, WC, and WHR, and model 3 was adjusted for model 2 variables plus education, family income, smoking, drinking alcohol, physical activity frequency, and sleep duration. The p-value < 0.05 was considered statistically significant.

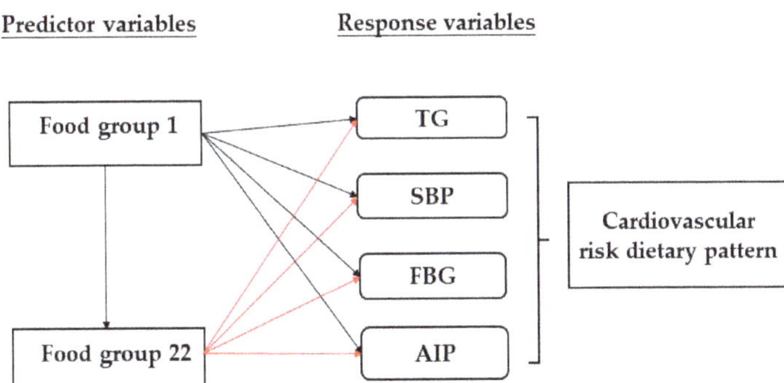

Figure 2. Cardiovascular risk dietary pattern derived from reduced rank regression model. TG: triglycerides, SBP: systolic blood pressure, FBG: fasting blood glucose, AIP: atherogenic index of plasma.

3. Results

3.1. Characteristics of Study Participants

Table 1 presents the demographic and lifestyle characteristics of the participants. The majority of postmenopausal women in this study had education below high school (81.5%), non-professional occupation (63.9%), low annual income (<NTD800,000: 69.1%), married status (70.3%), no smoking (97.9%), no drinking alcohol (95.5%), less physical activity frequency (<150 min/week: 55.5%), and sleep duration for 6–8 h (58.6%). The anthropometric, clinical, and biochemical data are shown in Table 2. The majority of postmenopausal women had normal BMI (44.2%), waist circumference (56.1%), and waist-to-hip ratio (68.9%). However, 31.3% postmenopausal women were overweight, 22.7% subjects were obese, 43.9% subjects had central obesity, and 31.1% subjects had abnormal waist-to-hip ratio. The prevalence of hypertension, diabetes, and CVD was 11.3%, 17.5%, and 10.7%, respectively. The mean value of AIP (0.3 ± 0.3) was higher than 0.24 defined as a CVD risk factor. The mean values of TC (5.9 ± 0.8 mmol/L) and LDL-C (3.7 ± 0.8 mmol/L) were abnormal among postmenopausal women. Among 5689 participants, only 7.3% subjects had normal FBG level (<7.0 mmol/L, data not shown).

Table 1. Demographic and lifestyle characteristics of postmenopausal women aged ≥45 years (n = 5689) [1].

Variables	Participants (n = 5689)
Age (years)	60.6 ± 7.6
Education	
<High school	4636 (81.5)
≥High school	1053 (18.5)
Occupation	
Non-professional	3637 (63.9)
Professional	1269 (22.3)
Unemployed/retired	783 (13.8)
Annual family income (NTD)	
<800,000	3929 (69.1)
810,000–1,600,000	1347 (23.7)
>1,600,000	413 (7.2)
Marital status	
Never married	83 (1.5)
Married	4002 (70.3)
Widows/divorced	1604 (28.2)

Table 1. *Cont.*

Variables	Participants (n = 5689)
Smoking	
No	5567 (97.9)
Yes	122 (2.1)
Drinking alcohol	
No	5433 (95.5)
Yes	256 (4.5)
Physical activity frequency	
<150 min/week	3160 (55.5)
≥150 min/week	2529 (44.5)
Sleep duration	
<6 h	1917 (33.7)
6–8 h	3333 (58.6)
>8 h	439 (7.7)

[1] Continuous data are presented as mean ± SD and categorical data are expressed as numbers (percentage).

Table 2. Demographic, clinical, and biochemical data of postmenopausal women aged ≥45 years (n = 5689) [1].

Variables	Participants (n = 5689)
Body mass index (kg/m^2)	
<18.5	100 (1.8)
18.5–23.9	2516 (44.2)
24–26.9	1780 (31.3)
≥27	1293 (22.7)
Waist circumference	
<80 cm	3191 (56.1)
≥80 cm	2498 (43.9)
Waist-to-hip ratio	
<0.85	3922 (68.9)
≥0.85	1767 (31.1)
Prevalence of chronic disease	
Hypertension	642 (11.3)
Diabetes mellitus	994 (17.5)
Cardiovascular disease	609 (10.7)
Systolic blood pressure (mmHg)	133 ± 20
Diastolic blood pressure (mmHg)	75 ± 12
Atherogenic index of plasma	0.3 ± 0.3
Total cholesterol (mmol/L)	5.9 ± 0.8
Low-density lipoprotein cholesterol (mmol/L)	3.7 ± 0.8
High-density lipoprotein cholesterol (mmol/L)	1.5 ± 0.4
Triglycerides (mmol/L)	1.6 ± 0.8
Fasting blood glucose (mmol/L)	6.6 ± 1.9
C-reactive protein (nmol/L)	26.8 ± 47.2

[1] Continuous data are presented as mean ± SD and categorical data are expressed as numbers (percentage).

3.2. Cardiovascular Risk Dietary Pattern

A dietary pattern identified as a "cardiovascular risk dietary pattern" was derived using the RRR model. Four food groups including processed food, rice/flour products, organ meat, and sauce showed a positive correlation (factor loading ≥0.20) with the cardiovascular risk dietary pattern, meanwhile food groups such as dairy products, fruits, whole grains, and sweet bread had a negative correlation with this dietary pattern (factor loading

≤−0.20) (Figure 3). The cardiovascular risk dietary pattern explained 6.6% cumulative percentage of variation and 1.7% of the total variation for the four response variables. The explained variation was 1.5% for TG, 1.6% for AIP, and 1.8% for both SBP and FBG.

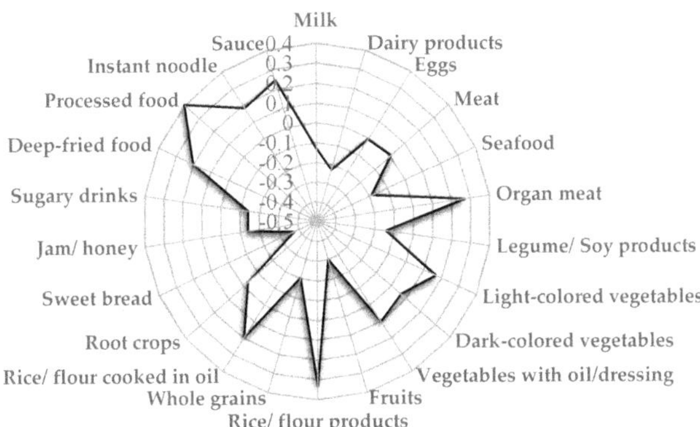

Figure 3. Spider-web diagram of factor loadings for cardiovascular risk dietary pattern.

3.3. *Association between the Dietary Pattern and Cardiovascular Risk Factors*

The unadjusted and adjusted models for the association of the cardiovascular risk dietary pattern with SBP, DBP, and AIP among postmenopausal Taiwanese women are presented in Table 3. Model 1 was unadjusted, model 2 was adjusted for age, BMI, WC, and WHR, and model 3 was adjusted for model 2 variables plus education, family income, smoking, drinking alcohol, physical activity frequency, and sleep duration. The results showed that participants in the higher quartiles (Q3 and Q4) of the cardiovascular risk dietary pattern were more likely to increase the odds of having high SBP (OR = 1.40–1.84), high DBP (OR = 1.28–1.69), and high AIP (OR = 1.43–1.69) compared to those in the reference group (Q1) before adjustment. After adjusting variables in models 2 and 3, participants in the highest quartile (Q4) of the cardiovascular risk dietary pattern were still more likely to increase the odds of having high SBP (model 2: OR = 1.42, 95% CI 1.20–1.68, model 3: OR = 1.29, 95% CI 1.08–1.53), high DBP (model 2: OR = 1.43, 95% CI 1.13–1.79, model 3: OR = 1.28, 95% CI 1.01–1.62), and high AIP (model 2: OR = 1.29, 95% CI 1.09–1.52, model 3: OR = 1.26, 95% CI 1.06–1.49).

The association of the cardiovascular risk dietary pattern with TC, LDL-C, and HDL-C among postmenopausal Taiwanese women in the unadjusted and adjusted models is shown in Table 4. The cardiovascular risk dietary pattern was not correlated with the odds of high TC in all models. Participants in the Q2 quartile of the cardiovascular risk dietary pattern were more likely to decrease the odds of having high LDL-C in all models (model 1: OR = 0.63, 95% CI 0.47–0.83, model 2: OR = 0.68, 95% CI 0.51–0.91, model 3: OR = 0.71, 95% CI 0.53–0.94) compared to those in the Q1 quartile. Participants in the higher quartiles (Q3 and Q4) of the cardiovascular risk dietary pattern were more likely to decrease the odds of having low HDL-C in the unadjusted model; however, no association was found after adjustment in models 2 and 3.

Table 5 demonstrates the association of the cardiovascular risk dietary pattern with TG, FBG, and CRP in the unadjusted and adjusted models among postmenopausal Taiwanese women. Participants in the higher quartiles (Q2–Q4) of the cardiovascular risk dietary pattern were more likely to increase the odds of having high TG in all models (model 1: OR = 1.39–1.79, model 2: OR = 1.21–1.43, model 3: OR = 1.18–1.38) compared to those in the lowest quartile (Q1). Participants in the higher quartile (Q3) of the cardiovascular risk dietary pattern were more likely to increase the odds of having high FBG in all models (model 1: OR = 1.75, 95% CI 1.30–2.35, model 2: OR = 1.54, 95% CI 1.14–2.07, model 3: OR = 1.45, 95% CI 1.07–1.97). Participants in the higher quartiles (Q3 and Q4) of the cardiovascular risk dietary pattern were more likely to increase the odds of having high

CRP (Q3: OR = 1.38, 95% CI 1.13–1.67, Q4: OR = 1.51, 95% CI 1.25–1.83) only in the unadjusted model.

Table 3. Binary logistic regression for the association between the dietary pattern, systolic blood pressure (SBP), diastolic blood pressure (DBP), and atherogenic index of plasma (AIP) (n = 5689).

Dietary Pattern	Cardiovascular Disease Risk Factors [1]		
	High SBP	High DBP	High AIP
	Odds Ratio (95% Confidence Interval)		
Model 1 [2]			
Q1 (reference)	1	1	1
Q2	1.29 (1.09–1.52) ***	1.22 (0.96–1.54)	1.41 (1.21–1.64) ***
Q3	1.40 (1.19–1.65) ***	1.28 (1.01–1.63) *	1.43 (1.23–1.67) ***
Q4	1.84 (1.58–2.16) ***	1.69 (1.35–2.12) ***	1.69 (1.45–1.98) ***
p for trend	0.000	0.000	0.000
Model 2 [3]			
Q1 (reference)	1	1	1
Q2	1.15 (0.97–1.36)	1.13 (0.89–1.44)	1.29 (1.09–1.51) **
Q3	1.19 (1.00–1.41) *	1.14 (0.89–1.46)	1.18 (1.01–1.39) *
Q4	1.42 (1.20–1.68) ***	1.43 (1.13–1.79) **	1.29 (1.09–1.52) **
p for trend	0.000	0.016	0.005
Model 3 [4]			
Q1 (reference)	1	1	1
Q2	1.09 (0.92–1.29)	1.07 (0.84–1.37)	1.28 (1.09–1.50) **
Q3	1.10 (0.92–1.31)	1.05 (0.82–1.34)	1.16 (0.99–1.37)
Q4	1.29 (1.08–1.53) **	1.28 (1.01–1.62) *	1.26 (1.06–1.49) **
p for trend	0.030	0.144	0.013

[1] High SBP, high DBP, and high AIP were defined as SBP ≥ 140 mmHg, DBP ≥ 90 mmHg, and AIP ≥ 0.24, respectively. [2] Model 1 was unadjusted. [3] Model 2 was adjusted for age, body mass index, waist circumference, and waist-to-hip ratio. [4] Model 3 was adjusted for model 2 variables plus education, family income, smoking, drinking alcohol, physical activity frequency, and sleep duration. * $p < 0.05$, ** $p < 0.01$, *** $p < 0.001$, significantly different from the reference group.

Table 4. Binary logistic regression for the association between the dietary pattern, total cholesterol (TC), low-density lipoprotein cholesterol (LDL-C), and high-density lipoprotein cholesterol (HDL-C) (n = 5689).

Dietary Pattern	Cardiovascular Disease Risk Factors [1]		
	High TC	High LDL-C	Low HDL-C
	Odds Ratio (95% Confidence Interval)		
Model 1 [2]			
Q1 (reference)	1	1	1
Q2	0.84 (0.67–1.05)	0.63 (0.47–0.83) **	0.88 (0.75–1.04)
Q3	0.97 (0.77–1.22)	0.82 (0.60–1.10)	0.81 (0.69–0.96) *
Q4	0.87 (0.70–1.09)	0.78 (0.58–1.05)	0.73 (0.62–0.86) ***
p for trend	0.349	0.013	0.002
Model 2 [3]			
Q1 (reference)	1	1	1
Q2	0.92 (0.73–1.15)	0.68 (0.51–0.91) **	0.95 (0.81–1.13)
Q3	1.11 (0.88–1.39)	0.92 (0.68–1.24)	0.93 (0.79–1.11)
Q4	1.08 (0.86–1.35)	0.94 (0.69–1.27)	0.90 (0.76–1.06)
p for trend	0.334	0.022	0.665

Table 4. *Cont.*

Dietary Pattern	Cardiovascular Disease Risk Factors [1]		
	High TC	High LDL-C	Low HDL-C
	Odds Ratio (95% Confidence Interval)		
Model 3 [4]			
Q1 (reference)	1	1	1
Q2	0.92 (0.74–1.16)	0.71 (0.53–0.94) *	0.98 (0.83–1.16)
Q3	1.11 (0.87–1.40)	0.99 (0.73–1.35)	0.98 (0.82–1.16)
Q4	1.08 (0.85–1.37)	1.04 (0.77–1.42)	0.96 (0.81–1.14)
p for trend	0.381	0.013	0.971

[1] High TC, high LDL-C, and low HDL-C were defined as TC $\geq$ 5.17 mmol/L (200 mg/dL), LDL-C $\geq$ 2.59 mmol/L (100 mg/dL), and HDL-C $\leq$ 1.29 mmol/L (50 mg/dL), respectively. [2] Model 1 was unadjusted. [3] Model 2 was adjusted for age, body mass index, waist circumference, and waist-to-hip ratio. [4] Model 3 was adjusted for model 2 variables plus education, family income, smoking, drinking alcohol, physical activity frequency, and sleep duration. * $p < 0.05$, ** $p < 0.01$, *** $p < 0.001$, significantly different from the reference group.

Table 5. Binary logistic regression for the association between the dietary pattern, triglycerides (TG), fasting blood glucose (FBG), and C-reactive protein (CRP) (n = 5689).

Dietary Pattern	Cardiovascular Disease Risk Factors [1]		
	High TG	High FBG	High CRP
	Odds Ratio (95% Confidence Interval)		
Model 1 [2]			
Q1 (reference)	1	1	1
Q2	1.39 (1.19–1.63) ***	1.19 (0.92–1.55)	1.12 (0.92–1.37)
Q3	1.42 (1.21–1.66) ***	1.75 (1.30–2.35) ***	1.38 (1.13–1.67) **
Q4	1.79 (1.54–2.09) ***	1.42 (1.08–1.86) *	1.51 (1.25–1.83) ***
p for trend	0.000	0.002	0.000
Model 2 [3]			
Q1 (reference)	1	1	1
Q2	1.29 (1.10–1.51) **	1.10 (0.84–1.43)	0.99 (0.81–1.22)
Q3	1.21 (1.03–1.43) *	1.54 (1.14–2.07) **	1.14 (0.93–1.39)
Q4	1.43 (1.22–1.68) ***	1.16 (0.87–1.53)	1.14 (0.93–1.39)
p for trend	0.000	0.040	0.322
Model 3 [4]			
Q1 (reference)	1	1	1
Q2	1.27 (1.09–1.50) **	1.07 (0.82–1.39)	0.99 (0.80–1.21)
Q3	1.18 (1.00–1.40) *	1.45 (1.07–1.97) *	1.11 (0.90–1.36)
Q4	1.38 (1.17–1.62) ***	1.05 (0.79–1.41)	1.09 (0.89–1.34)
p for trend	0.001	0.971	0.569

[1] High TG, high FBG, and high CRP were defined as TG $\geq$ 1.69 mmol/L (150 mg/dL), FBG $\geq$ 7.0 mmol/L (126 mg/dL), and CRP $\geq$ 28.6 nmol/L (3 mg/L), respectively. [2] Model 1 was unadjusted. [3] Model 3 was adjusted for age, body mass index, waist circumference, and waist-to-hip ratio. [4] Model 4 was adjusted for model 2 variables plus education, family income, smoking, drinking alcohol, physical activity frequency, and sleep duration. * $p < 0.05$, ** $p < 0.01$, *** $p < 0.001$, significantly different from the reference group.

4. Discussion

4.1. Association between the Dietary Pattern and Cardiovascular Risk Factors

In this cross-sectional study of 5689 postmenopausal Taiwanese women, we derived the cardiovascular risk dietary pattern and found a positive association with several CVD risk factors such as SBP, DBP, AIP, TG, and FBG. Among the participants in the highest quartile of the cardiovascular risk dietary pattern, 60.4% of postmenopausal women were overweight or obese, and 64.6% were physically inactive (<150 min/week) (data not shown). The cardiovascular risk dietary pattern was recognized by high consumption of processed food, rice/flour products, organ meat, and sauce, but low intakes of dairy products, fruit, whole grains, and sweet bread. The cardiovascular risk dietary pattern reflected similar characteristics as the western dietary pattern recognized by high intakes of processed food, meat, organ meat, rice/flour products, but low consumption of fruit, dark-colored vegetables, bread, and legume/soy products among Taiwanese middle-aged and elderly with chronic kidney disease [23]. Processed food and organ meat are often rich in calories, cholesterol, and/or saturated fat, and all of which could contribute to excessive energy consumption [23].

Low fiber and excessive salt and/or sugar in processed food as well as unbalanced saturated and unsaturated fats in animal food could be correlated with abnormal blood pressure, blood lipids, and blood glucose among Taiwanese middle-aged adults and elderly [23]. Highly refined carbohydrate in rice/flour products, a dietary component for high intake in the cardiovascular risk dietary pattern, could be associated with increases in cardiovascular risk and the development of atherosclerosis among middle-aged adults [32].

Our results revealed that the cardiovascular risk dietary pattern was positively associated with blood pressure. We found that the prevalence of hypertension was only 11.3% among 5689 postmenopausal Taiwanese women. Unlike our results, the previous studies conducted among postmenopausal women reported that the prevalence of hypertension was 31.6% and 56.0% in Brazilian and Chinese postmenopausal women, respectively [20,33]. Weight gain and increased sensitivity to salt in the diet might occur due to hormonal changes after menopause and age-associated metabolic changes, which could lead to a raise in blood pressure [34]. Weight status and physical activity could also contribute to abnormal blood pressure. Postmenopausal women aged <65 years with overweight (33.3%) or obesity (42.9%) also had higher prevalence of high blood pressure (130 mmHg/85 mmHg) compared to those who had normal weight (18.8%), and those who did not do aerobic exercise tended to have higher prevalence of high blood pressure compared to those who did aerobic exercise actively (44.0% vs. 14.3%, $p = 0.06$) [20].

Our findings showed that the cardiovascular risk dietary pattern was correlated with an increase in AIP among postmenopausal women. Numerous studies demonstrated that AIP was an important cardiovascular risk factor and a better predictor for CVD [33,35,36]. The previous studies have reported that AIP was a better predictor of the fractional esterification rate of HDL-C which is a powerful predictor of CVD [35], and a more sensitive diagnostic marker for studies of CVD [35], and a more sensitive diagnostic marker for CVD among postmenopausal women compared to traditional lipid parameters [35,36].

Our results revealed that participants in the highest quartile (Q4) of the cardiovascular risk dietary pattern were more likely to increase the odds of having high CRP before adjustment, even the association between dietary pattern and CRP was not significant after adjustment. A previous study conducted in Southern Brazil among postmenopausal women also observed that participants with high CRP were positively correlated with BMI, WC, body fat, TG, glucose, sedentary lifestyle, and excessive dietary carbohydrate intake (>55% of total energy) [37].

Although the association between aberrant lipid profiles and certain nutrients or food groups has been established, few have demonstrated the association between dietary pattern or quality and blood lipids in postmenopausal women [32]. We found that the cardiovascular risk dietary pattern was positively correlated with increased odds of high TG among postmenopausal women after full adjustment. Brazilian postmenopausal women with a low-quality inadequate diet characterized by an excessive intake of sodium (>2400 mg/day) had increases in the prevalence of high TC and high LDL-C known as cardiovascular risk factors [20]. However, Tardivo and co-workers [32] showed that there was no significant association between diet quality determined by healthy eating index scores and blood lipids in Brazilian postmenopausal women. A study conducted among Korean women showed that postmenopausal women who consumed the western dietary pattern with high intakes of oil and fats, meat, eggs, fast food, and sweets but low intake of grains were correlated with hyper LDL-C [38]. Other studies conducted among Chinese women and Japanese women consuming a western dietary pattern with high intakes of milk, dairy products, and fast food but a low intake of rice or vegetables revealed an imbalance in lipid profile, especially increases in TC and LDL-C [39,40]. The abnormality of various serum lipids was linked to hormonal changes, such as the rise in circulating androgen and the reduction in estrogen, during the menopausal transition period [39].

Adherence to a western type dietary pattern could be associated with the status of being overweight or obese and having high WC, which might contribute to metabolic alteration. The metabolic changes in postmenopausal women could explain the imbalance of CVD-related biochemical variables [41]. Because of estrogen deficiency, postmenopausal women could increase CVD risk factors including central obesity, elevated blood pressure, increased blood lipids, decreased glucose tolerance, and increased vascular inflammation [42]. Compared to premenopausal women, postmenopausal women were more prone to increase blood lipids, which could lead to increase the risk for the development of atherogenesis [16]. In addition, the dietary components could be correlated with abnormal CVD-related biochemical variables in postmenopausal women. High consumption of energy [43], saturated fatty acids [44,45], trans fats [45], cholesterol [46], and eggs [46] was associated with an increased risk of CVD or abnormal CVD-related biochemical variables among postmenopausal women. In contrast, a low-fat dietary pattern [45] or the dietary pattern with high consumption of plant food such as whole grains, vegetables, fruits, legumes, and nuts or seeds, but low intakes of processed food, red meat, sugar, and sodium [47] were correlated with a reduced risk of CVD among postmenopausal women. The cardiovascular risk dietary pattern identified in our study was characterized by high intakes of processed food, rice/flour products, organ meat, and sauce which were accompanied by a high amount of energy, saturated fats, trans fats, cholesterol, added sugar, and sodium. Although the underlying mechanism for the effects of dietary patterns or dietary components on CVD risk factors among postmenopausal women has not been fully understood, changes in lipid metabolism and the increased accumulation of visceral fat related to estrogen deficiency in postmenopausal women could partially contribute to the effect of the dietary pattern on CVD risk factors.

4.2. Strengths and Limitations

To our knowledge, the present study is the first one to identify the cardiovascular risk dietary pattern in postmenopausal Taiwanese women using the RRR model as a novel and powerful method. Additionally, the RRR model gave more explanation about the association between the dietary pattern and the disease of interest. Since the RRR-derived dietary pattern was generated by a disease-specific response, the response variables were correlated to the disease of concern [48]. Instead of explaining the variation in significant biomarkers, principal component analysis only provided the explanation of the overall variation in food group intake [48]. Meanwhile, by maximizing the explained variation in the biomarkers for diet-related disorders, the RRR model could be able to predict dietary pattern scores. Researchers can also determine the percentage variance using the RRR approach from the predictor variables and response variables, and both of which contributed to the dietary component [48]. Both the corresponding response scores and the explained variation in the predictor variables could be used to evaluate the extracted factor scores [48]. The large study population collected for 15 years could be representative of postmenopausal Taiwanese women. We also included demographic, anthropometric, clinical, biochemical, and dietary data to explore the association between these variables. However, a number of methodological limitations need to be addressed. First, our study was a cross-sectional study which provided features of eating habits and other characteristics at a specific time point and could raise the possibility of reverse causation bias. Second, the information for FFQ used to identify dietary habits could have self-reported bias. Additionally, the FFQ could be used for an estimate of habitual food intake but not for actual nutrient consumption. Even though the analysis was adjusted for the majority of known confounding variables, the residual confounding bias due to unknown or unmeasured covariates could not be completely ruled out. A longitudinal study is needed to explore the association between dietary patterns and CVD risk factors among postmenopausal Taiwanese women. Further research should be conducted to compare the association in premenopausal versus postmenopausal Taiwanese women.

5. Conclusions

The cardiovascular risk dietary pattern with a high intake of processed food, rice/flour products, organ meat, and sauce is associated with increased odds of high blood pressure, AIP, TG, and FBG among postmenopausal women. Our study suggests that choosing a healthier dietary pattern with a lower intake of processed food, rice/flour products, organ meat, and sauce could reduce the risk of CVD in postmenopausal Taiwanese women.

Author Contributions: Conceptualization, S.A. and J.C.-J.C.; data curation, C.-Y.H. and H.-A.L.; formal analysis, S.A. and R.P.; writing—original draft preparation, S.A. and J.C.-J.C.; writing—review and editing, J.C.-J.C.; supervision, C.-Y.H. and J.C.-J.C. All authors have read and agreed to the published version of the manuscript.

Funding: This research received no external funding.

Institutional Review Board Statement: The study was conducted in accordance with the Declaration of Helsinki, and approved by the Taipei Medical University–Joint Institutional Review Board (N202007075 and date of approval 11 September 2020).

Informed Consent Statement: All the participants signed the consent form authorized by the Mei Jau Health Management Institution.

Data Availability Statement: The data that support the findings of this study are available from the Mei Jau Health Management Institution, but restricted for research use only. The data are not publicly available.

Acknowledgments: The authors thank the Mei Jau Health Management Institution for collecting and providing their database available for our study.

Conflicts of Interest: The authors declare no conflict of interest.

References

1. Stamatelopoulos, K.; Papavagelis, C.; Augoulea, A.; Armeni, E.; Karagkouni, I.; Avgeraki, E.; Georgiopoulos, G.; Yannakoulia, M.; Lambrinoudaki, I. Dietary patterns and cardiovascular risk in postmenopausal women: Protocol of a cross-sectional and prospective study. *Maturitas* **2018**, *116*, 59–65. [CrossRef]
2. Dalal, P.; Agarwal, M. Postmenopausal syndrome. *Indian J. Psychiatry* **2015**, *57* (Suppl. S2), 222–232. [CrossRef]
3. Newson, L. Menopause and cardiovascular disease. *Post Reprod. Health* **2018**, *24*, 44–49. [CrossRef] [PubMed]
4. Hill, K. The demography of menopause. *Maturitas* **1996**, *23*, 113–127. [CrossRef]
5. Taleb-Belkadi, O.; Chaib, H.; Zemour, L.; Fatah, A.; Chafi, B.; Mekki, K. Lipid profile, inflammation, and oxidative status in peri- and postmenopausal women. *Gynecol. Endocrinol.* **2016**, *32*, 982–985. [CrossRef]
6. Gaspard, U.J.; Gottal, J.-M.; van den Brûle, F.A. Postmenopausal changes of lipid and glucose metabolism: A review of their main aspects. *Maturitas* **1995**, *21*, 171–178. [CrossRef]
7. Yamatani, H.; Takahashi, K.; Yoshida, T.; Takata, K.; Kurachi, H. Association of estrogen with glucocorticoid levels in visceral fat in postmenopausal women. *Menopause* **2013**, *20*, 437–442. [CrossRef] [PubMed]
8. Inaraja, V.; Thuissard, I.; Andreu-Vazquez, C.; Jodar, E. Lipid profile changes during the menopausal transition. *Menopause* **2020**, *27*, 780–787. [CrossRef] [PubMed]
9. Boukhris, M.; Tomasello, S.D.; Marzà, F.; Bregante, S.; Pluchinotta, F.R.; Galassi, A.R. Coronary Heart Disease in Postmenopausal Women with Type II Diabetes Mellitus and the Impact of Estrogen Replacement Therapy: A Narrative Review. *Int. J. Endocrinol.* **2014**, *2014*, 413920. [CrossRef] [PubMed]
10. Roth, G.A.; Mensah, G.A.; Johnson, C.O.; Addolorato, G.; Ammirati, E.; Baddour, L.M.; Barengo, N.C.; Beaton, A.Z.; Benjamin, E.J.; Benziger, C.P.; et al. Global Burden of Cardiovascular Diseases and Risk Factors, 1990–2019: Update From the GBD 2019 Study. *J. Am. Coll. Cardiol.* **2020**, *76*, 2982–3021. [CrossRef] [PubMed]
11. Ministry of Health and Welfare, Taiwan. 2020 Cause of Death Statistics. Available online: https://www.mohw.gov.tw/cp-5256-63399-2.html (accessed on 22 January 2022).
12. Shen, T.-Y.; Strong, C.; Yu, T. Age at menopause and mortality in Taiwan: A cohort analysis. *Maturitas* **2020**, *136*, 42–48. [CrossRef] [PubMed]
13. Landscape, NTU Research and Development. Fast Population Aging: A Challenge of National Health Insurance in Taiwan. Available online: http://research.ord.ntu.edu.tw/landscape/inner.aspx?id=260&chk=4dd6e804-41d7-420d-a4a2-beacba644b44 (accessed on 10 June 2022).

14. Hsu, C.-C.; Chang, H.-Y.; Wu, I.-C.; Chen, C.-C.; Tsai, H.-J.; Chiu, Y.-F.; Chuang, S.-C.; Hsiung, W.-C.; Tsai, T.-L.; Liaw, W.-J.; et al. Cohort Profile: The Healthy Aging Longitudinal Study in Taiwan (HALST). *Int. J. Epidemiol.* **2017**, *46*, 1106. [CrossRef] [PubMed]
15. He, L.; Tang, X.; Li, N.; Wu, Y.; Wang, J.; Li, J.; Zhang, Z.; Dou, H.; Liu, J.; Yu, L.; et al. Menopause with cardiovascular disease and its risk factors among rural Chinese women in Beijing: A population-based study. *Maturitas* **2012**, *72*, 132–138. [CrossRef] [PubMed]
16. Pardhe, B.D.; Ghimire, S.; Shakya, J.; Pathak, S.; Shakya, S.; Bhetwal, A.; Khanal, P.R.; Parajuli, N.P. Elevated Cardiovascular Risks among Postmenopausal Women: A Community Based Case Control Study from Nepal. *Biochem. Res. Int.* **2017**, *2017*, 3824903. [CrossRef] [PubMed]
17. Lin, W.-Y.; Yang, W.-S.; Lee, L.-T.; Chen, C.-Y.; Liu, C.-S.; Lin, C.-C.; Huang, K.-C. Insulin resistance, obesity, and metabolic syndrome among non-diabetic pre- and post-menopausal women in North Taiwan. *Int. J. Obes.* **2006**, *30*, 912–917. [CrossRef] [PubMed]
18. Banna, J.C.; Gilliland, B.; Keefe, M.; Zheng, D. Cross-cultural comparison of perspectives on healthy eating among Chinese and American undergraduate students. *BMC Public Health* **2016**, *16*, 1015. [CrossRef]
19. Ferreira, L.L.; Silva, T.R.; Maturana, M.A.; Spritzer, P.M. Dietary intake of isoflavones is associated with a lower prevalence of subclinical cardiovascular disease in postmenopausal women: Cross-sectional study. *J. Hum. Nutr. Diet.* **2019**, *32*, 810–818. [CrossRef]
20. Ventura, D.D.A.; Fonseca, V.D.M.; Ramos, E.G.; Marinheiro, L.P.F.; De Souza, R.A.G.; Chaves, C.R.M.D.M.; Peixoto, M.V.M. Association between quality of the diet and cardiometabolic risk factors in postmenopausal women. *Nutr. J.* **2014**, *13*, 121. [CrossRef]
21. Zhang, F.; Tapera, T.M.; Gou, J. Application of a new dietary pattern analysis method in nutritional epidemiology. *BMC Med. Res. Methodol.* **2018**, *18*, 119. [CrossRef]
22. Muga, M.A.; Owili, P.; Hsu, C.-Y.; Rau, H.-H.; Chao, J.C.-J. Association between Dietary Patterns and Cardiovascular Risk Factors among Middle-Aged and Elderly Adults in Taiwan: A Population-Based Study from 2003 to 2012. *PLoS ONE* **2016**, *11*, e0157745. [CrossRef]
23. Kurniawan, A.L.; Hsu, C.-Y.; Rau, H.-H.; Lin, L.-Y.; Chao, J.C.-J. Association of kidney function-related dietary pattern, weight status, and cardiovascular risk factors with severity of impaired kidney function in middle-aged and older adults with chronic kidney disease: A cross-sectional population study. *Nutr. J.* **2019**, *18*, 27. [CrossRef] [PubMed]
24. Chin, C.-C.; Kuo, Y.-H.; Yeh, C.-Y.; Chen, J.-S.; Tang, R.; Changchien, C.-R.; Wang, J.-Y.; Huang, W.-S. Role of body mass index in colon cancer patients in Taiwan. *World J. Gastroenterol.* **2012**, *18*, 4191–4198. [CrossRef] [PubMed]
25. Syauqy, A.; Hsu, C.-Y.; Rau, H.-H.; Chao, J.C.-J. Association of dietary patterns, anthropometric measurements, and metabolic parameters with C-reactive protein and neutrophil-to-lymphocyte ratio in middle-aged and older adults with metabolic syndrome in Taiwan: A cross-sectional study. *Nutr. J.* **2018**, *17*, 106. [CrossRef] [PubMed]
26. Ahmad, N.; Adam, S.I.M.; Nawi, A.M.; Hassan, M.R.; Ghazi, H.F. Abdominal obesity indicators: Waist circumference or waist-to-hip ratio in Malaysian adults population. *Int. J. Prev. Med.* **2016**, *7*, 82. [CrossRef] [PubMed]
27. Friedewald, W.T.; Levy, R.I.; Fredrickson, D.S. Estimation of the Concentration of Low-Density Lipoprotein Cholesterol in Plasma, Without Use of the Preparative Ultracentrifuge. *Clin. Chem.* **1972**, *18*, 499–502. [CrossRef] [PubMed]
28. Wang, C.; Du, Z.; Ye, N.; Liu, S.; Geng, D.; Wang, P.; Sun, Y. Using the Atherogenic Index of Plasma to Estimate the Prevalence of Ischemic Stroke within a General Population in a Rural Area of China. *BioMed Res. Int.* **2020**, *2020*, 7197054. [CrossRef]
29. Li, Y.-W.; Kao, T.-W.; Chang, P.-K.; Chen, W.-L.; Wu, L.-W. Atherogenic index of plasma as predictors for metabolic syndrome, hypertension and diabetes mellitus in Taiwan citizens: A 9-year longitudinal study. *Sci. Rep.* **2021**, *11*, 9900. [CrossRef]
30. Pearson, T.A.; Mensah, G.A.; Alexander, R.W.; Anderson, J.L.; Cannon, R.O., III; Criqui, M.; Fadl, Y.Y.; Fortmann, S.P.; Hong, Y.; Myers, G.L.; et al. Markers of inflammation and cardiovascular disease: Application to clinical and public health practice: A statement for healthcare professionals from the Centers for Disease Control and Prevention and the American Heart Association. *Circulation* **2003**, *107*, 499–511. [CrossRef]
31. Sun, Q.; Wen, Q.; Lyu, J.; Sun, D.; Ma, Y.; Man, S.; Yin, J.; Jin, C.; Tong, M.; Wang, B.; et al. Dietary Pattern Derived by Reduced Rank Regression and Cardiovascular Disease: A Cross-Sectional Study. *Nutr. Metab. Cardiovasc. Dis.* **2022**, *32*, 337–345. [CrossRef]
32. Tardivo, A.P.; Nahas-Neto, J.; Nahas, E.A.P.; Maesta, N.; Rodrigues, M.A.H.; Orsatti, F. Associations between healthy eating patterns and indicators of metabolic risk in postmenopausal women. *Nutr. J.* **2010**, *9*, 64. [CrossRef]
33. Wu, T.-T.; Gao, Y.; Zheng, Y.-Y.; Ma, Y.-T.; Xie, X. Atherogenic index of plasma (AIP): A novel predictive indicator for the coronary artery disease in postmenopausal women. *Lipids Health Dis.* **2018**, *17*, 197. [CrossRef] [PubMed]
34. Taddei, S. Blood pressure through aging and menopause. *Climacteric* **2009**, *12* (Suppl. S1), 36–40. [CrossRef] [PubMed]
35. Bo, M.S.; Cheah, W.L.; Lwin, S.; Nwe, T.M.; Win, T.T.; Aung, M. Understanding the Relationship between Atherogenic Index of Plasma and Cardiovascular Disease Risk Factors among Staff of an University in Malaysia. *J. Nutr. Metab.* **2018**, *2018*, 7027624. [CrossRef]
36. Nwagha, U.I.; Ikekpeazu, E.J.; Ejezie, F.E.; Neboh, E.E.; Maduka, I.C. Atherogenic index of plasma: A significant indicator for the onset of atherosclerosis during menopause in hypertensive females of Southeast Nigeria. *Afr. Health Sci.* **2010**, *10*, 248–252. [PubMed]
37. Alves, B.C.; Silva, T.R.; Spritzer, P.M. Sedentary lifestyle and high-carbohydrate intake are associated with low-grade chronic inflammation in post-menopause: A cross-sectional study. *Rev. Bras. Ginecol. Obstet.* **2016**, *38*, 317–324. [CrossRef]

38. Lee, J.; Hoang, T.; Lee, S.; Kim, J. Association Between Dietary Patterns and Dyslipidemia in Korean Women. *Front. Nutr.* **2022**, *8*, 756257. [CrossRef] [PubMed]
39. Zhang, J.; Wang, Z.; Wang, H.; Du, W.; Su, C.; Zhang, J.; Jiang, H.; Jia, X.; Huang, F.; Zhai, F.; et al. Association between dietary patterns and blood lipid profiles among Chinese women. *Public Health Nutr.* **2016**, *19*, 3361–3368. [CrossRef]
40. Htun, N.C.; Suga, H.; Imai, S.; Shimizu, W.; Takimoto, H. Food intake patterns and cardiovascular risk factors in Japanese adults: Analyses from the 2012 National Health and nutrition survey, Japan. *Nutr. J.* **2017**, *16*, 61. [CrossRef]
41. Grundy, S.M. Obesity, metabolic syndrome, and cardiovascular disease. *J. Clin. Endocrinol. Metab.* **2004**, *89*, 2595–2600. [CrossRef]
42. Rosano, G.M.C.; Vitale, C.; Marazzi, G.; Volterrani, M. Menopause and cardiovascular disease: The evidence. *Climacteric* **2007**, *10* (Suppl. S1), 19–24. [CrossRef]
43. Zheng, C.; Beresford, S.A.; Van Horn, L.; Tinker, L.F.; Thomson, C.A.; Neuhouser, M.L.; Di, C.; Manson, J.E.; Mossavar-Rahmani, Y.; Seguin, R.; et al. Simultaneous Association of Total Energy Consumption and Activity-Related Energy Expenditure With Risks of Cardiovascular Disease, Cancer, and Diabetes Among Postmenopausal Women. *Am. J. Epidemiol.* **2014**, *180*, 526–535. [CrossRef] [PubMed]
44. Rathnayake, K.M.; Weech, M.; Jackson, K.G.; Lovegrove, J.A. Meal Fatty Acids Have Differential Effects on Postprandial Blood Pressure and Biomarkers of Endothelial Function but Not Vascular Reactivity in Postmenopausal Women in the Randomized Controlled Dietary Intervention and VAScular function (DIVAS)-2 Study. *J. Nutr.* **2018**, *148*, 348–357. [CrossRef] [PubMed]
45. Howard, B.V.; Van Horn, L.; Hsia, J.; Manson, J.E.; Stefanick, M.I.; Wassertheil-Smoller, S.; Kuller, L.H.; LaCroix, A.Z.; Langer, R.D.; Lasser, N.I.; et al. Low-Fat Dietary Pattern and Risk of Cardiovascular Disease: The Women's Health Initiative Randomized Controlled Dietary Modification Trial. *Obstet. Gynecol. Surv.* **2006**, *61*, 451–453. [CrossRef]
46. Chen, G.-C.; Chen, L.-H.; Mossavar-Rahmani, Y.; Kamensky, V.; Shadyab, A.H.; Haring, B.; Wild, R.A.; Silver, B.; Kuller, L.H.; Sun, Y.; et al. Dietary cholesterol and egg intake in relation to incident cardiovascular disease and all-cause and cause-specific mortality in postmenopausal women. *Am. J. Clin. Nutr.* **2020**, *113*, 948–959. [CrossRef] [PubMed]
47. Hirahatake, K.M.; Jiang, L.; Wong, N.D.; Shikany, J.M.; Eaton, C.; Allison, M.A.; Martin, L.; Garcia, L.; Zaslavsky, O.; Odegaard, A.O. Diet Quality and Cardiovascular Disease Risk in Postmenopausal Women With Type 2 Diabetes Mellitus: The Women's Health Initiative. *J. Am. Heart Assoc.* **2019**, *8*, e013249. [CrossRef]
48. Manios, Y.; Kourlaba, G.; Grammatikaki, E.; Androutsos, O.; Ioannou, E.; Roma-Giannikou, E. Comparison of two methods for identifying dietary patterns associated with obesity in preschool children: The Genesis study. *Eur. J. Clin. Nutr.* **2010**, *64*, 1407–1414. [CrossRef]

Article

Nutritional Risk Factors Associated with Vasomotor Symptoms in Women Aged 40–65 Years

Alexandra Tijerina [1,†], Yamile Barrera [1,†], Elizabeth Solis-Pérez [1], Rogelio Salas [1], José L. Jasso [1], Verónica López [1], Erik Ramírez [1], Rosario Pastor [2,3], Josep A. Tur [3,4,5,*] and Cristina Bouzas [2,3,4,5]

1 Facultad de Salud Pública y Nutrición, Universidad Autónoma de Nuevo León, Monterrey 64460, Mexico; alexandra.tijerinas@uanl.mx (A.T.); yamile.barreracrnz@uanl.edu.mx (Y.B.); elizabeth.solis@uanl.mx (E.S.-P.); rogelio.salasg@uanl.mx (R.S.); jose.jassom@uanl.mx (J.L.J.); veronica.lopezg@uanl.mx (V.L.); erik.ramirezl@uanl.mx (E.R.)
2 Faculty of Health Sciences, Catholic University of Avila, 05005 Avila, Spain; rosario.pastor@ucavila.es (R.P.); cristina.bouzas@uib.es (C.B.)
3 Research Group on Community Nutrition and Oxidative Stress, University of Balearic Islands–IUNICS, 07122 Palma de Mallorca, Spain
4 Health Institute of the Balearic Islands (IDISBA), 07120 Palma de Mallorca, Spain
5 CIBER Physiopathology of Obesity and Nutrition (CIBEROBN), Institute of Health Carlos III (ISCIII), 28029 Madrid, Spain
* Correspondence: pep.tur@uib.es
† These authors contributed equally to this work.

Abstract: Vasomotor symptoms (VMS) are the most common symptoms among menopausal women; these include hot flashes and night sweats, and palpitations often occur along with hot flashes. Some studies in Mexico reported that around 50% of women presented with VMS mainly in the menopausal transition. It has been proven that VMS are not only triggered by an estrogen deficiency, but also by nutritional risk factors. Evidence of an association between nutritional risk factors and VMS is limited in Mexican women. The aim of this study is to identify nutritional risk factors associated with VMS in women aged 40–65 years. This is a comparative cross-sectional study, undertaken in a retrospective way. A sample group (n = 406 women) was divided into four stages according to STRAW+10 (Stages of Reproductive Aging Workshop): late reproductive, menopausal transition, early postmenopause, and late postmenopause. Hot flashes were present mainly in the early postmenopause stage (38.1%, $p \leq 0.001$). Two or more VMS were reported in 23.2% of women in the menopausal transition stage and 29.3% in the early postmenopause stage ($p < 0.001$). The presence of VMS was associated with different nutritional risk factors (weight, fasting glucose levels, cardiorespiratory fitness, and tobacco use) in women living in the northeast of Mexico.

Keywords: vasomotor symptoms; risk factors; reproductive aging; menopause; women; Mexico

Citation: Tijerina, A.; Barrera, Y.; Solis-Pérez, E.; Salas, R.; Jasso, J.L.; López, V.; Ramírez, E.; Pastor, R.; Tur, J.A.; Bouzas, C. Nutritional Risk Factors Associated with Vasomotor Symptoms in Women Aged 40–65 Years. *Nutrients* **2022**, *14*, 2587. https://doi.org/10.3390/nu14132587

Academic Editors: Birgit-Christiane Zyriax and Nataliya Makarova

Received: 4 May 2022
Accepted: 20 June 2022
Published: 22 June 2022

1. Introduction

Vasomotor symptoms (VMS) are the most common symptoms among menopausal women. These symptoms are short-term menopausal disorders, which include hot flashes and night sweats [1,2], and are often accompanied by palpitations [2]. According to STRAW+10 (Stages of Reproductive Aging Workshop) criteria, which is a "gold standard" that classifies women into stages of reproductive aging, VMS can appear in late menopausal transition but are more common in the early postmenopause [3,4]. Some studies in Mexico reported that VMS in women appear mainly in the menopausal transition, with a prevalence of 45% in women aged 45–55 years [5] and 47.63% in women aged 51–63 years [6].

It has been proven that VMS are not only triggered by estrogen deficiency, but also by different nutritional risk factors that can be modifiable and non-modifiable. Clinical practice guidelines for the menopausal stage [7–9], along with several other studies, delimitate that a body mass index higher or equal to 25, hypertension, and tobacco use are nutritional risk factors for VMS [10–12]. Their mechanism seems to relate to impaired heat conductance, blood flow, and hormones [12,13]. However, there is limited evidence concerning other factors such as fasting glucose ≥100 mg/dL, poor cardiorespiratory fitness, and excessive total fat intake.

There is also limited evidence concerning the nutritional factors associated with VMS in Mexican women. In Mexico, the prevalence of VMS has been reported but the associated risk factors have not [5,6]. A previous study has evaluated the social and nutritional factors associated with menopausal symptoms in Mexican women; however, there was no distinction of VMS, as nine different symptoms were arranged into one climacteric group [14]. These studies were carried out in central Mexico and there is limited evidence for the northeast region. The aim of this study is to identify nutritional risk factors associated with VMS in women aged 40–65 years in northeast Mexico.

2. Materials and Methods

2.1. Design and Subjects

This is a cross-sectional study which was carried out from 2015 to 2017. Women enrolled in this study were 40–65 years of age, living in the metropolitan area of Monterrey, in Nuevo León state, Mexico. They were apparently healthy, they voluntarily agreed to participate, and all provided written informed consent. Exclusion criteria included illnesses that affected their habitual eating habits and having undergone a hysterectomy. Women with incomplete data were eliminated from the study analysis (Figure 1). A total of $n = 406$ women were included in the study, representative of the study population (560,115 women aged 40–65 years) of Nuevo León state in the year 2020. This was according to a finite population equation, with 5% error and 95% confidence intervals, considering a VMS proportion of 50% [15].

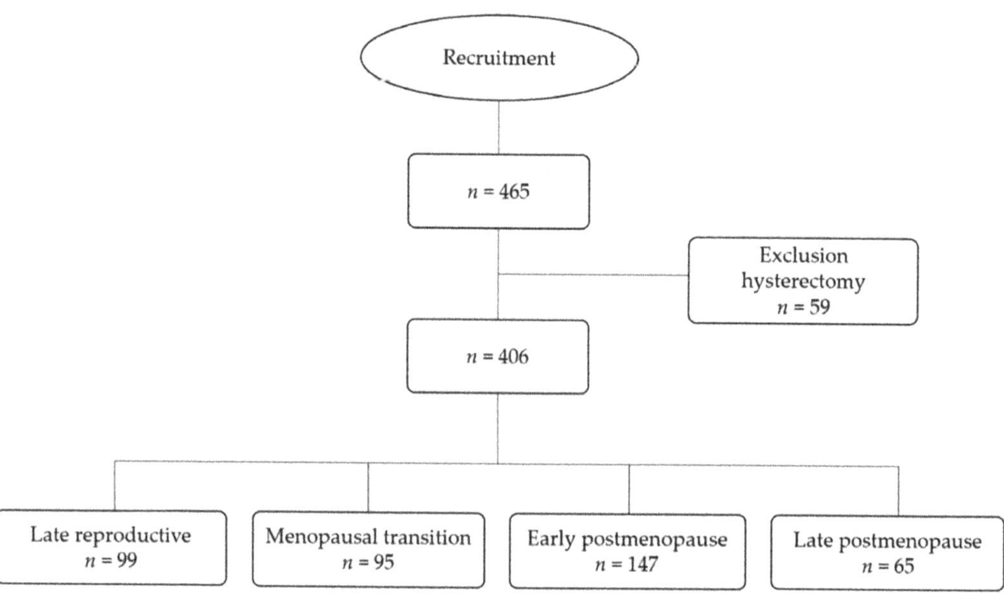

Figure 1. Flowchart of the recruitment of the study population.

Women were invited to participate via physical flyers and social media. They were screened for the inclusion criteria via a telephone call and scheduled for an appointment at the Center for Research in Nutrition and Public Health of the Facultad de Salud Pública y

Nutrición, Universidad Autónoma de Nuevo León (School of Public Health and Nutrition, Autonomous University of Nuevo León, translated into English).

This study followed the Declaration of Helsinki, and it was approved by the ethics committee of the Facultad de Salud Pública y Nutrición with protocol ID: 15–FaSPyN–SA–11.

2.2. *Stages of Reproductive Aging (STRAW+10 Criteria)*

A questionnaire was used to obtain data on menstrual cycles, including previous menstrual period, presence of blood flow or amenorrhea, and changes in length between cycles. According to the STRAW+10 criteria [3], women were classified into 4 stages of reproductive aging (Figure 1). Late reproductive was defined as "the presence of blood flow without changes or with short cycles." Menopausal transition was defined as "the presence of blood flow with long cycles or with at least one interval of amenorrhea ≥60 days," "amenorrhea <60 days and without changes," or "amenorrhea ≥60 days." Early postmenopause was defined as "amenorrhea ≥12 months, but ≤8 years," and late postmenopause was defined as "amenorrhea >8 years" [3].

2.3. *Vasomotor Symptoms*

Vasomotor symptoms (VMS) were reported in a questionnaire and included hot flashes and night sweats. Palpitations were also registered, as they are often present along with hot flashes [2]. Women had the option to answer yes (presence) or no (absence). After the information was collected, 4 categories were established: (1) absence, (2) presence of hot flashes, (3) presence of night sweats or palpitations, and (4) presence of 2 or more VMS, including hot flashes, night sweats, and palpitations.

2.4. *Nutrition Assessment*

A nutrition assessment of participants was performed according to the Nutrition Care Process of the Academy of Nutrition and Dietetics [16] as follows:

2.4.1. Anthropometric Measurements

Body mass index (BMI) was determined by the formula BMI = weight (kg)/height^2 (m^2), using a scale (Seca 874, ± 0.1 kg, Azcapotzalco, Mexico) for the weight and a digital stadiometer (Seca 274, ± 2 mm, Azcapotzalco, Mexico) for the height. BMI was classified as obese ≥30 kg/m^2, overweight 25–29.9 kg/m^2, or normal weight 18.5–24.9 kg/m^2 [17].

2.4.2. Biochemical Data

Venous blood samples were collected at fast, centrifuged at 3500 rpm for 12 min, and serum was obtained. The serum was frozen at −80 °C until assays were performed with the glucose oxidase/peroxidase method. Fasting glucose was obtained using A25 autoanalyzer (software version 4.1.1) (CV = 1.2%) (BioSystems S.A, Barcelona, Spain), according to the Norma Oficial Mexicana NOM–253–SSA1–2012 [9]. Fasting glucose was classified as high ≥100 mg/dL or normal <100 mg/dL [18]. Women on treatments using hypoglycemic drugs were also considered to have high fasting glucose levels.

2.4.3. Nutrition-Focused Physical Exam Findings

Systolic and diastolic blood pressure measurements were performed to the nearest 1 mmHg using a digital sphygmomanometer, according to the Norma Oficial Mexicana NOM–030–SSA2–2009 [19]. Two readings were taken 5 min apart and the average was calculated. Blood pressure was classified as hypertensive between ≥130 and ≥80 mmHg, elevated between 120–129 and <80 mmHg, or normal between <120 and <80 mmHg [20]. Women on treatments using antihypertensive drugs were considered to be hypertensive.

The cardiorespiratory fitness (CF) of the women was obtained by measuring the walking distance achieved in meters (m) during a six-minute (min) test, in a 15 m × 28 m field. The CF value was reported as meters per minute, using the formula CF = m/6 min. CF was determined as poor at <400 m/6 min or excellent at ≥400 m/6 min [21].

Prevalence of symptoms are reported as four categories (absence, hot flashes, night sweats or palpitations, and 2 or more VMS) (Figure 2). The study reveals that of all participants (n = 406), 41.4% experienced an absence of VMS and 29.6% presented with hot flashes only. An absence of VMS prevailed in women in the late reproductive stage (67.7%). Hot flashes were mainly reported in the early postmenopause stage (38.1%) followed by late postmenopause (33.8%). The presence of either night sweats or palpitations was reported mainly in women in the late reproductive stage (11.1%), although the presence of this category did not differ between different stages (p = 0.202). A combination of two or more VMS was reported by 23.2% of women in menopausal transition and 29.3% in early postmenopause; only 6.1% of women in the late reproductive stage presented with two or more VMS.

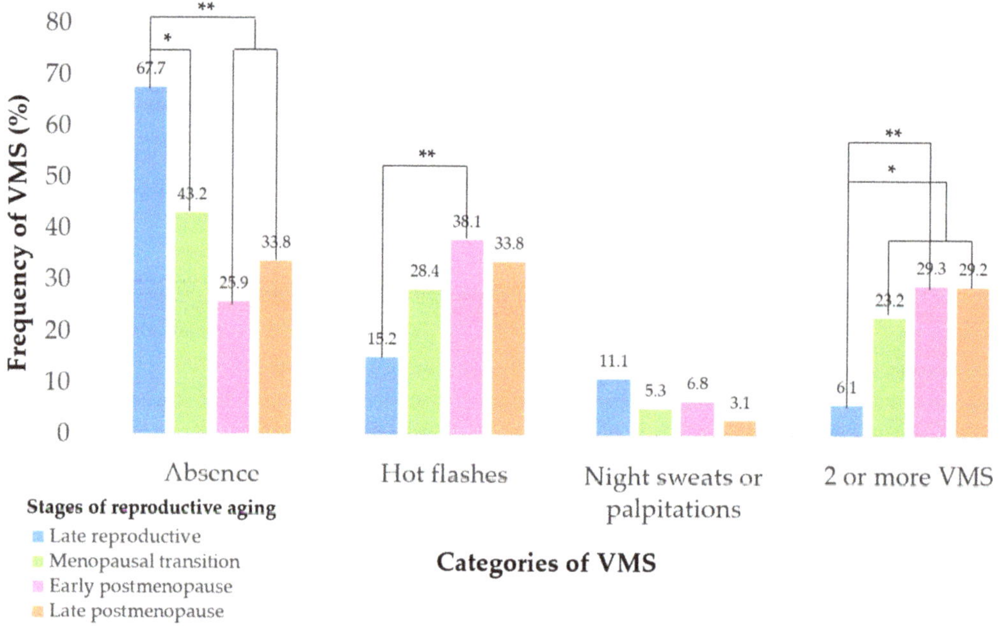

Figure 2. Prevalence of vasomotor symptoms. Data were analyzed using a chi-square test with the Marascuilo procedure (post hoc test). They are expressed as percentages (%). * $p < 0.01$, ** $p < 0.001$.

The association between nutritional risk factors and VMS is shown in Table 2 for unadjusted and adjusted models. In Model 1 (unadjusted model), an overweight BMI (25–29.9 kg/m^2) denoted a risk for the presence of hot flashes (PR 2.92, 95% CI: 1.66–6.32) (OR 3.24, 95% CI: 1.66–6.33, p = 0.001). The presence of either night sweats or palpitations was associated with high levels of fasting glucose (≥100 mg/dL) (OR 2.63, 95% CI: 1.09–6.37, p = 0.031) (PR 2.49, 95% CI: 1.09–6.33) and poor cardiorespiratory fitness (OR 15.01, 95% CI: 1.94–115.62, p = 0.009) (PR 8.03, 95% CI: 1.95–115.62). In addition, the presence of two or more VMS was associated with high fasting glucose levels (≥100 mg/dL) (OR 2.27, 95% CI: 1.29–3.99, p = 0.004) (PR 1.98, 95% CI: 1.29–3.98) and tobacco use (OR 3.19, 95% CI: 1.25–8.11, p = 0.015) (PR 2.25, 95% CI: 1.26–8.10).

Table 2. Association between nutritional risk factors and vasomotor symptoms in women aged 40–65 years (n = 406).

	Vasomotor Symptoms								
	Hot Flashes			Night Sweats or Palpitations			2 or More VMS		
Nutritional Factors	**OR (95% CI)**	**PR (95% CI)**	***p***	**OR (95% CI)**	**PR (95% CI)**	***p***	**OR (95% CI)**	**PR (95% CI)**	***p***
BMI Obese, ≥30 kg/m^2									
Model 1, unadjusted [a]	1.32 (0.65–2.67)	1.30 (0.66–2.65)	0.434	1.34 (0.38–4.70)	1.33 (0.39–4.62)	0.643	1.17 (0.57–2.40)	1.16 (0.59–2.31)	0.651
Model 2, adjusted [b]	1.46 (0.70–3.05)	1.43 (0.71–3.01)	0.306	1.35 (0.38–4.78)	1.34 (0.39–4.65)	0.635	1.27 (0.60–2.68)	1.25 (0.61–2.65)	0.524
Model 3, adjusted [c]	1.30 (0.62–2.72)	1.28 (0.62–2.73)	0.487	1.24 (0.34–4.49)	1.24 (0.35–4.41)	0.740	1.16 (0.54–2.47)	1.15 (0.55–2.45)	0.697
BMI Overweight, 25–29.9 kg/m^2									
Model 1, unadjusted [a]	3.24 (1.66–6.33)	2.92 (1.66–6.32)	0.001	2.49 (0.75–8.26)	2.45 (0.75–8.25)	0.135	1.12 (0.52–2.38)	1.11 (0.53–2.38)	0.768
Model 2, adjusted [b]	3.26 (1.63–6.52)	2.93 (1.63–6.51)	0.001	2.49 (0.74–8.31)	2.45 (0.75–8.27)	0.136	1.12 (0.52–2.45)	1.11 (0.54–2.31)	0.760
Model 3, adjusted [c]	3.26 (1.63–6.54)	2.93 (1.63–6.52)	0.001	2.23 (0.65–7.65)	2.20 (0.65–7.61)	0.200	1.08 (0.49–2.39)	1.08 (0.52–2.24)	0.836
BMI Normal 18.5–24.9 kg/m^2 was reference in all models									
Fasting glucose High, ≥100 mg/dL									
Model 1, unadjusted [a]	1.47 (0.87–2.51)	1.35 (0.87–2.48)	0.149	2.63 (1.09–6.37)	2.49 (1.09–6.33)	0.031	2.27 (1.29–3.99)	1.98 (1.29–3.98)	0.004
Model 2, adjusted [b]	1.30 (0.75–2.25)	1.23 (0.75–2.25)	0.347	2.50 (1.02–6.11)	2.38 (1.02–6.11)	0.045	2.03 (1.13–3.63)	1.81 (1.13–3.63)	0.017
Model 3, adjusted [c]	1.29 (0.74–2.25)	1.23 (0.75–2.23)	0.361	2.67 (1.04–6.84)	2.53 (1.05–6.80)	0.039	1.94 (1.08–3.50)	1.75 (1.08–3.48)	0.027
Fasting glucose Normal <100 mg/dL was reference in all models									
Cardiorespiratory fitness Poor, <400 m/6 min									
Model 1, unadjusted [a]	1.53 (0.18–12.53)	1.33 (0.19–12.17)	0.689	15.01 (1.94–115.62)	8.03 (1.95–115.62)	0.009	d	d	d
Model 2, adjusted [b]	1.05 (0.11–9.32)	1.03 (0.14–8.06)	0.963	13.57 (1.63–113.01)	7.63 (1.63–112.92)	0.016	d	d	d
Model 3, adjusted [c]	1.06 (0.12–8.70)	1.04 (0.14–8.23)	0.956	16.17 (2.02–129.11)	8.33 (2.03–129.05)	0.009	d	d	d
Cardiorespiratory fitness Excellent ≥400 m/6 min was reference in all models									
Blood pressure Hypertensive, systolic ≥130 mmHg or diastolic ≥80 mmHg									
Model 1, unadjusted [a]	1.47 (0.86–2.50)	1.37 (0.86–2.51)	0.157	1.85 (0.76–4.50)	1.80 (0.77–4.46)	0.170	1.33 (0.73–2.40)	1.29 (0.74–2.39)	0.340
Model 2, adjusted [b]	1.20 (0.68–2.10)	1.17 (0.69–2.09)	0.519	1.66 (0.67–4.10)	1.63 (0.68–4.06)	0.267	1.11 (0.60–2.07)	1.10 (0.62–2.00)	0.727
Model 3, adjusted [c]	1.26 (0.72–2.22)	1.21 (0.73–2.17)	0.405	1.86 (0.75–4.63)	1.81 (0.75–4.58)	0.178	1.08 (0.58–2.03)	1.07 (0.52–2.24)	0.792

Table 2. *Cont.*

Nutritional Factors	Vasomotor Symptoms								
	Hot Flashes			Night Sweats or Palpitations			2 or More VMS		
	OR (95% CI)	PR (95% CI)	p	OR (95% CI)	PR (95% CI)	p	OR (95% CI)	PR (95% CI)	p
Blood pressure Elevated, systolic 120–129 mmHg and diastolic 80 mmHg									
Model 1, unadjusted [a]	0.65 (0.22–1.89)	0.69 (0.22–1.93)	0.437	[d]	[d]	0.979	1.56 (0.60–4.04)	1.47 (0.60–4.03)	0.358
Model 2, adjusted [b]	0.47 (0.15–1.43)	0.51 (0.15–1.45)	0.189	[d]	[d]	0.986	1.20 (0.44–3.25)	1.18 (0.44–3.24)	0.719
Model 3, adjusted [c]	0.58 (0.19–1.75)	0.62 (0.19–1.76)	0.336	[d]	[d]	0.989	1.50 (0.54–4.10)	1.43 (0.55–4.10)	0.429
Blood pressure Normal <120 and <80 mmHg was reference in all models									
Total fat intake Excessive >30% of total energy intake									
Model 1, unadjusted [a]	1.10 (0.60–2.02)	1.09 (0.61–1.97)	0.749	2.50 (0.65–9.49)	2.47 (0.66–9.50)	0.178	1.48 (0.72–3.03)	1.46 (0.72–3.03)	0.283
Model 2, adjusted [b]	1.41 (0.74–2.69)	1.38 (0.74–2.68)	0.295	2.78 (0.71–10.81)	2.75 (0.72–10.79)	0.140	1.85 (0.87–3.90)	1.80 (0.88–3.90)	0.106
Model 3, adjusted [c]	1.37 (0.72–2.61)	1.34 (0.73–2.57)	0.328	2.24 (0.57–8.69)	2.22 (0.58–8.67)	0.243	1.97 (0.92–4.21)	1.91 (0.92–4.20)	0.079
Total fat intake Adequate ≤30% of total energy intake was reference in all models									
Tobacco use Yes									
Model 1, unadjusted [a]	2.14 (0.84–5.42)	1.64 (0.85–5.40)	0.107	2.38 (0.55–10.30)	2.19 (0.55–10.24)	0.244	3.19 (1.25–8.11)	2.25 (1.26–8.10)	0.015
Model 2, adjusted [b]	1.96 (0.76–5.05)	1.56 (0.76–5.05)	0.163	2.18 (0.49–9.66)	2.03 (0.50–9.60)	0.303	2.89 (1.10–7.57)	2.12 (1.11–7.55)	0.030
Model 3, adjusted [c]	2.04 (0.78–5.31)	1.60 (0.79–5.26)	0.140	2.31 (0.52–10.29)	2.14 (0.52–10.22)	0.270	3.05 (1.14–8.13)	2.19 (1.15–8.11)	0.025
Tobacco use *No* was reference in all models									

VMS: vasomotor symptoms. BMI: body mass index. OR: odds ratio. PR: prevalence ratio. 95%CI: 95% confidence interval. [a] Model 1: unadjusted. [b] Model 2: adjusted for age, hormone use, and physical activity level. [c] Model 3: adjusted for stage of reproductive aging, hormone use, physical activity level, alcohol intake, and caffeine intake. [d] Insufficient cases for a statistical result. $p < 0.05$ denotes statistical significance.

After adjusting for age, hormone use, and physical activity level, Model 2 showed very similar results to Model 1 (unadjusted). Excessive weight (BMI 25–29.9 kg/m^2) was a nutritional risk for the presence of hot flashes (PR 2.93, 95% CI: 1.63–6.51) (OR 3.26, 95% CI: 1.63–6.52, $p = 0.001$). Women were at risk of presenting with either night sweats or palpitations if fasting glucose levels were $\geq$100 mg/dL (PR 2.38, 95% CI: 1.02–6.11) (OR 2.50, 95% CI: 1.02–6.11, $p = 0.045$) and if they had poor cardiorespiratory fitness levels (PR 7.63, 95% CI: 1.63–112.92) (OR 13.57, 95% CI: 1.63–113.01, $p = 0.016$). Presence of two or more VMS was also associated with high fasting glucose levels ($\geq$100 mg/dL) (PR 1.81, 95% CI: 1.13–3.63) (OR 2.03, 95% CI: 1.13–3.63, $p = 0.017$) and smoking habits (PR 2.12, 95% CI: 1.11–7.55) (OR 2.89, 95% CI: 1.10–7.57, $p = 0.030$).

Model 3 was adjusted to include the stage of reproductive aging, hormone use, physical activity level, alcohol intake, and caffeine intake (Table 2), denoting consistency in results from Models 1 and 2. The presence of hot flashes was associated with an overweight BMI (25–29.9 kg/m^2) (PR 2.93, 95% CI: 1.63–6.52) (OR 3.26, 95% CI: 1.63–6.54, $p = 0.001$). Levels of fasting glucose $\geq$100 mg/dL (PR 2.53, 95% CI: 1.05–6.80) (OR 2.67, 95% CI: 1.04–6.84, $p = 0.039$) and poor cardiorespiratory fitness (PR 8.33, 95% CI: 2.03–129.05) (OR 16.17, 95% CI: 2.02–129.11, $p = 0.009$) are risk factors associated with the presence of night sweats or palpitations. High fasting glucose $\geq$100 mg/dL (PR 1.75, 95% CI: 1.08–3.48) (OR 1.94, 95% CI: 1.08–3.50, $p = 0.027$) and tobacco use (PR 2.19, 95% CI: 1.15–8.11) (OR 3.05, 95% CI: 1.14–8.13, $p = 0.025$) denoted a risk for the presence of two or more VMS.

There was no association between blood pressure and total fat intake and the presence of VMS in any proposed model (unadjusted or adjusted).

4. Discussion

This observational study in women aged 40–65 years from Nuevo León state, in northeast Mexico, determined the association between nutritional risk factors and the presence of vasomotor symptoms. Menopausal onset was at 47.0 years; this is younger than women living in central Mexico, as previous studies have found that women from Queretaro started menopause at 49.1 years [5] and those from Mexico City at 50.0 years [6]. The onset of menopause at earlier ages, defined as early menopause (<45 years old), [31] could be due to some of the trigger factors that women reported in this study, such as tobacco use and an overweight or obese BMI [32]. Moreover, it has been demonstrated that menopause at early ages is related to a higher risk of cardiovascular disease and mortality, especially in women aged 50–78 years [33].

The experience of VMS is similar among Mexican women and follows a pattern that predominates in the early postmenopause stage. In this study, 29.6% of all women showed hot flashes, while 38.1% of women in the early postmenopause stage reported having this specific VMS. In a previous study in central Mexico, 21.3% of women aged 40–60 years reported having hot flashes, mainly during menopausal transition [34]. In other countries, such as India, the United Kingdom, Australia, and the United States of America, hot flashes were more frequently reported during the postmenopausal stage [35,36].

The presence of two or more VMS was reported in 22.2% of all women in the current study and was most frequently reported by early postmenopausal women (29.3%). However, previous studies have shown greater incidence of two VMS (hot flashes and night sweats) at the postmenopausal stage; this was found in 50.0% of women aged 24–44 years from the United States of America [11] and in 53.3% of women aged 40–65 years from India [35]. Therefore, in this study, the higher frequency of two or more VMS agrees with the STRAW+10 criteria, which suggests greater a likelihood of symptoms occurring during the early postmenopausal stage [3].

In this study, several women in the late reproductive stage also reported the presence of VMS; 15.2% had hot flashes and 11.1% had either night sweats or palpitations. Previous studies have reported night sweats in 40.0% and hot flashes in 29.0% of late reproductive women aged 35–55 years [37]. It is important to note that VMS can begin at earlier stages, such as the late reproductive stage, although in a low proportion due to estrogen reduction through constant aging and death of follicles [38]. Thus, assessment and diagnosis of VMS is relevant at earlier stages.

The association between nutritional factors and the presence of VMS was analyzed using three regression models to obtain the odds ratio (OR), and a formula using OR values to obtain the prevalence ratio (PR). These demonstrate the risks of women being overweight, having fasting glucose levels above 100 mg/mL, having a cardiorespiratory fitness below 400 m/6 min, and being a smoker. Adjustment of Models 2 and 3 for covariates did not affect the association. Poor cardiorespiratory fitness was the only risk factor with subtle changes in Model 3 that resulted in slightly higher risk and a wider 95% CI.

Hot flashes were associated with being overweight in this study population (BMI 25–29.9 kg/m^2) at PR 2.92–2.93 and OR 3.24–3.26 ($p = 0.001$ in all models). Overweight and obese BMIs have previously been associated with the presence of VMS, as reported in the clinical practice guidelines [7,8] and in some studies from Scotland [10], North America [27,39], Australia [40], and South Korea [26]. Women showing a higher BMI tend to have an excess of body fat; visceral fat increases by up to 20% during the postmenopausal stage [41]. In this study, obesity was not associated with hot flashes, which may be because more obese women were at later stages, such as the late postmenopause stage, in which hot flashes were less frequently reported, similar to a previous study [39]. Excessive fat does not allow heat conduction through the skin; therefore, the body tries to release it by maximizing vasodilation, which increases the central body temperature beyond the sweating threshold [13,42]. However, our findings may also suggest that the mechanisms of estrone, associated with a decrease in hot flashes, are naturally occurring in our study of postmenopausal women [39,43].

Fasting glucose levels above 100 mg/dL were associated with either night sweats or palpitations and with two or more VMS. A Swedish study in women aged 50–64 years reported night sweats as the only VMS associated with high glucose levels ($p < 0.05$) [44]. In a longitudinal study of Australian women aged 45–50 years, it was stated that there was a significant association between night sweats and diabetes (OR 1.91, 99% CI: 1.08–3.35, $p < 0.001$) in an adjusted model including similar covariates as this study: age, educational level, length of time, BMI, physical activity level, tobacco use, alcohol intake, menopausal status, and hormone use [45]. Therefore, chronic hyperglycemia could be a strong associated factor. High levels of glucose have been associated with insomnia, because the hypothalamic-pituitary-adrenal axis is altered [46]. This axis also regulates steroid and adrenal secretions [47], so its alteration impacts VMS occurrence.

The association between high blood pressure and VMS could be due to the increased activity of the sympathetic nervous system [48], and thus the increased activity of the adrenalin and sweat glands [49]. Although models in this study showed non-significant results when systolic and diastolic blood pressures were elevated or at hypertension levels, other authors have reported significant associations [44,50]. A study from Sweden in postmenopausal women, aged 50–64 years, reported an association between the presence of night sweats and systolic blood pressure (OR 2.07, $p < 0.001$) [44]. In addition, women in the United States of America aged 45–54 years who were under antihypertensive treatment had 1.80 times greater risk of presenting with hot flashes ($p < 0.05$) [50].

Fat intake was not associated with the presence of VMS in any model (unadjusted or adjusted). Intake of foods high in fat, especially saturated fat obtained from animal sources such as red or white meat, can increase levels of LDL cholesterol (low-density lipoprotein) [51], and therefore increase the presence of VMS [44]. This is supported by findings of studies in Australian women aged 45–50 years, in which those who followed a high-fat diet presented a significant risk for the presence of both hot flashes and night sweats (unadjusted model, OR 1.16, $p = 0.002$) [52]. Additionally, a study of women aged 40–85 years in the United States of America, who were in menopausal transition and postmenopause, VMS were higher among those who consumed more red and white meat, seafood, and dairy, while an absence of VMS was presented among women consuming a plant-based diet [53].

It has been suggested that a decrease in VMS may be observed with improving cardiorespiratory fitness [54,55]. Cardiorespiratory fitness was associated with improved health ($p < 0.001$), emotions ($p = 0.05$), and occupational quality of life ($p = 0.03$), suggesting a positive effect on reduced menopausal symptoms in women aged 45–60 years in the United States of America [55]. However, there has been no reported association with hot flashes and night sweats in Spanish women aged 45–60 years [56]. In the present study, poor cardiorespiratory fitness was the nutritional factor with the highest association to either night sweats or palpitations (PR 7.63–8.33; OR 13.57–16.17, $p < 0.05$). Women who had been physically active since menopausal transition had more protection from VMS. A higher cardiorespiratory fitness level decreases the activation of the sympathetic nervous system, which narrows the blood vessels of the body; in turn, the thermoneutral zone maintains homeostasis [57].

Tobacco use has been reported by different studies as a nutritional risk factor for VMS. Cigarettes are composed of different metals that serve as endocrine disruptors which alter the hormonal balance, so estrogens can be affected until they trigger VMS [12,58]. Women that have been smoking for many years may be at increased risk of having VMS. In a longitudinal study, current smokers aged 24–44 years showed 2.5 times greater risk of having hot flashes and night sweats (95% CI: 1.5–5.3, p <0.05) after adjusting for age, hormone level, BMI, hormone use, marital status, and parity [11]. A study of women in the late reproductive stage and the menopausal transition stage from the United States of America, aged 42–52 years, showed an association between the presence of VMS (hot flashes and night sweats), the number of cigarettes smoked (unadjusted model, OR 1.6, 95% CI: 1.3–1.9, $p < 0.05$), and passive smoke exposure (unadjusted model, OR 1.3, 95% CI: 1.2–1.4, 0.05) [59].

Strengths and Limitations

The main strength of the current study is that it has determined the association between nutritional risk factors and the presence of vasomotor symptoms in women aged 40–65 years from Nuevo León state, in northeast Mexico. The analysis of prevalence and risk is also a strength of this study as it avoids over- or underestimation of this association; however, data could not be compared against other publications. A limitation of this study is the lack of precision in the responses of presence of VMS, as women were questioned in a dichotomous way (yes or no) without considering frequency or intensity. The analysis of some nutritional variables, such as blood pressure and cardiorespiratory function, was limited due to an insufficient number of cases when assessing the categories of VMS. Intakes of subtypes of fat were not analyzed. For future research, it is recommended that women present a similar intake of alcohol and caffeine for more uniformity in those covariates, which may achieve greater precision in adjusted models. In addition, longitudinal studies are needed to infer specific causes or to determine strong risk factors for this specific population.

5. Conclusions

The presence of VMS was associated with different nutritional risk factors (weight, fasting glucose levels, cardiorespiratory fitness, and tobacco use) in women living in northeast Mexico. This association was independent of covariates including age, stage of reproductive aging, hormone use, reported physical activity, alcohol intake, and caffeine intake. This evidence supports the need for updating these nutritional risk factors in clinical practice guidelines and its application as an instrument in primary health care services, which assist women at different stages of reproductive aging, at local and national levels.

Author Contributions: A.T., Y.B., E.S.-P. and J.A.T. designed the study; A.T. and Y.B. wrote the protocol; A.T. recruited the participants and collected samples; Y.B. and J.L.J. conducted the statistical analysis; A.T., Y.B. and J.A.T. wrote the first draft of the manuscript, and all authors (A.T., Y.B., E.S.-P., R.S., J.L.J., V.L., E.R., R.P., C.B. and J.A.T.) commented on previous versions of the manuscript. All authors have read and agreed to the published version of the manuscript.

Funding: C.B. and J.A.T. are funded by the Instituto de Salud Carlos III through the CIBEROBN CB12/03/30038, which is co-funded by the European Regional Development Fund. The funding sponsors had no role in the design of the study, in the collection, analysis, or interpretation of the data, in the writing of the manuscript, or in the decision to publish the results.

Institutional Review Board Statement: The study was conducted according to the guidelines of the Declaration of Helsinki and approved by the ethics committee of the Facultad de Salud Pública y Nutrición of the protocol ID: 15–FaSPyN–SA–11.

Informed Consent Statement: Informed consent was obtained from all subjects involved in the study. The results and writing of this manuscript followed the Committee on Publication Ethics (COPE) guidelines on how to deal with potential acts of misconduct, maintaining integrity of the research and its presentation following the rules of good scientific practice, the trust in the journal, the professionalism of scientific authorship, and the entire scientific endeavor. Written informed consent has been obtained from the patient(s) to publish this paper.

Data Availability Statement: There are restrictions on the availability of data for this trial, due to the signed consent agreements around data sharing, which only allow access to external researchers for studies following the project purposes. Those wishing to access the trial data used in this study can make a request to pep.tur@uib.es.

Acknowledgments: The authors especially thank the participants for their enthusiastic collaboration and the personnel for outstanding support and exceptional effort. CIBEROBN is an initiative of Instituto de Salud Carlos III, Spain.

Conflicts of Interest: The authors declare no conflict of interest.

References

1. Biglia, N.; Cagnacci, A.; Gambacciani, M.; Lello, S.; Maffei, S.; Nappi, R. Vasomotor symptoms in menopause: A biomarker of cardiovascular disease risk and other chronic diseases? *Climacteric* **2017**, *20*, 306–312. [CrossRef] [PubMed]
2. Hoffman, B.; Schorge, J.; Schaffer, J.; Halvorson, L.; Bradshaw, K.; Cunningham, F. *Williams Ginecología*, 2nd ed.; McGraw-Hill Interamericana Editores: Mexico City, Mexico, 2014.
3. Harlow, S.; Gass, M.; Hall, J.; Lobo, R.; Maki, P.; Rebar, R.; Sherman, S.; Sluss, P.; De Villiers, T. Executive summary of the Stages of Reproductive Aging Workshop + 10: Addressing the unfinished agenda of staging reproductive aging. *Menopause* **2012**, *19*, 387–395. [CrossRef] [PubMed]
4. Woods, N.; Mitchell, E. Symptoms during the perimenopause: Prevalence, severity, trajectory, and significance in women's lives. *Am. J. Med.* **2005**, *118*, 14S–24S. [CrossRef] [PubMed]
5. Carranza, S.; Sandoval, C. Comparación de la frecuencia y magnitud de los síntomas vasomotores en mujeres pre y posmenopáusicas de la Ciudad de México. *Ginecol. Obstet. Méx.* **2013**, *81*, 127–132.
6. Ortiz, G.; Arellano, A.; Sánchez, A.; Salazar, C.; Escobar, L.; Zavala, A. Descripción demográfica, bioquímica y sintomática según los estadios reproductivos STRAW+10 en mujeres mexicanas en la peri y postmenopausia. *Ginecol. Obstet. Méx.* **2020**, *88*, 29–40. [CrossRef]
7. Ministerio de Sanidad, Servicios Sociales e Igualdad (MSSSI). Guía de Práctica Clínica Sobre el Abordaje de Síntomas Vasomotores y Vaginales Asociados a la Menopausia y la Postmenopausia. Available online: https://portal.guiasalud.es/wp-content/uploads/2018/12/GPC_571_Menopausia_AETSA_herram.pdf (accessed on 29 March 2021).
8. Secretaría de Salud (SSA). Guía de Práctica Clínica: Atención al Climaterio y Menopausia. Available online: http://evaluacion.ssm.gob.mx/pdf/gpc/eyr/SS-019-08.pdf (accessed on 29 March 2021).
9. Secretaría de Salud (SSA). Norma Oficial Mexicana NOM-035-SSA2-2012, Para la Prevención y Control de Enfermedades en la Perimenopausia y Postmenopausia de la Mujer. Criterios Para Brindar Atención Médica. Available online: https://www.cndh.org.mx/sites/default/files/doc/Programas/VIH/Leyes%20y%20normas%20y%20reglamentos/Norma%20Oficial%20Mexicana/NOM-035-SSA2-2012.pdf (accessed on 29 March 2021).
10. Duffy, O.; Iversen, L.; Aucott, L.; Hannaford, P. Factors associated with resilience or vulnerability to hot flushes and night sweats during the menopausal transition. *Menopause* **2012**, *20*, 383–392. [CrossRef]
11. Ford, K.; Sowers, M.; Crutchfield, M.; Wilson, A.; Jannausch, M. A longitudinal study of the predictors of prevalence and severity of symptoms commonly associated with menopause. *Menopause* **2005**, *12*, 308–317. [CrossRef]
12. Anderson, D.; Chung, H.; Seib, C.; Dobson, A.; Kuh, D.; Brunner, E.; Crawford, S.; Avis, N.; Gold, E.; Greendale, G.; et al. Obesity, smoking, and risk of vasomotor menopausal symptoms: A pooled analysis of eight cohort studies. *Am. J. Obstet. Gynecol.* **2019**, *222*, 478.e1–478.e17. [CrossRef]
13. Thurston, R.; Sowers, M.; Chang, Y.; Sternfeld, B.; Gold, E.; Johnston, J.; Matthews, K. Adiposity and reporting of vasomotor symptoms among midlife women. *Am. J. Epidemiol.* **2008**, *167*, 78–85. [CrossRef]

14. Vega, G.; Hernández, A.; Leo, G.; Vega, J.; Escartin, M.; Luengas, J.; Guerrero, M. Incidencia y factores relacionados con el síndrome climatérico en una población de mujeres mexicanas. *Rev. Chil. Obstet. Ginecol.* **2007**, *72*, 314–320. [CrossRef]
15. Instituto Nacional de Estadística y Geografía e Informática (INEGI). Población Total por Entidad Federativa y Grupo Quinquenal de edad Según Sexo, Serie de Años Censales de 1990 a 2020. Available online: https://www.inegi.org.mx/app/tabulados/interactivos/?pxq=Poblacion_Poblacion_01_e60cd8cf-927f-4b94-823e-972457a12d4b&idrt=123&opc=t (accessed on 2 May 2022).
16. Academy of Nutrition and Dietetics (AND). Abridged Nutrition Care Process Terminology (NCPT). Reference Manual: Standardized Terminology for the Nutrition Care Process. Available online: https://www.eatrightstore.org/product-type/books/abridged-ncpt-reference-manual (accessed on 14 February 2022).
17. World Health Organization (WHO). Obesity: Preventing and Managing the Global Epidemic. Geneva. Available online: https://apps.who.int/iris/handle/10665/63854 (accessed on 21 May 2021).
18. American Diabetes Association (ADA). American Diabetes Association Standards of Medical Care in Diabetes 2019. *Diabetes Care* **2019**, *42*, S1–S193. [CrossRef]
19. Secretaria de Salud (SSA). Norma Oficial Mexicana NOM-030-SSA2-2009, Para la Prevención, Detección, Diagnóstico, Tratamiento y Control de la Hipertensión Arterial Sistémica. Available online: https://www.cndh.org.mx/DocTR/2016/JUR/A70/01/JUR-20170331-NOR21.pdf (accessed on 21 May 2021).
20. Whelton, P.; Carey, R.; Aronow, W.; Casey, D.; Collins, K.; Dennison, C.; DePalma, S.; Gidding, S.; Jamerson, K.; Jones, D.; et al. 2017 ACC/AHA/AAPA/ ABC/ACPM/AGS/APhA/ASH/ASPC/NMA/PCNA Guideline for the prevention, detection, evaluation, and management of high blood pressure in adults: Executive summary: A report of the American College of Cardiology/American Heart Association task force on Clinical Practice Guidelines. Hypertension. *Am. Heart Assoc. J.* **2018**, *71*, 1269–1324. [CrossRef]
21. Giannitsi, S.; Bougiakli, M.; Bechlioulis, A.; Kotsia, A.; Michalis, L.; Naka, K. 6-minute walking test: A useful tool in the management of heart failure patients. *Ther. Adv. Cardiovasc. Dis.* **2019**, *13*, 1753944719870084. [CrossRef]
22. Tijerina, A.; Tur, J.A. Development and validation of a semiquantitative food frequency questionnaire to assess dietary intake in 40-65-year-old Mexican women. *Ann. Nutr. Metab.* **2020**, *76*, 73–82. [CrossRef]
23. Academia Nacional de Medicina (ANM). Guías Alimentarias y de Actividad Física en Contexto de Sobrepeso y Obesidad en la Población Mexicana. Documentos de Postura. Available online: https://www.anmm.org.mx/publicaciones/CAnivANM150/L29_ANM_Guias_alimentarias.pdf (accessed on 8 April 2022).
24. Prevención con Dieta Mediterránea (PREDIMED). Cuestionario de Actividad Física en el Tiempo Libre de Minnesota. Available online: http://www.predimed.es/uploads/8/0/5/1/8051451/activ_fisica.pdf (accessed on 8 April 2022).
25. Enomoto, H.; Terauchi, M.; Odai, T.; Kato, K.; Iizuka, M.; Akiyoshi, M.; Miyasaka, N. Independent association of palpitation with vasomotor symptoms and anxiety in middle-aged women. *Menopause* **2021**, *28*, 741–747. [CrossRef]
26. Ryu, K.; Park, H.; Kwon, D.; Yang, K.; Kim, Y.; Yi, K.; Shin, J.; Hur, J.; Kim, T. Vasomotor symptoms and metabolic syndrome in Korean postmenopausal women. *Menopause* **2015**, *22*, 1239–1245. [CrossRef]
27. Thurston, R.; Chang, Y.; Mancuso, P.; Matthews, K. Adipokines, adiposity, and vasomotor symptoms during the menopause transition: Findings from the Study of Women's Health Across the Nation. *Fertil. Steril.* **2013**, *100*, 793–800.e1. [CrossRef]
28. Avis, N.; Crawford, S.; Greendale, G.; Bromberger, J.; Everson, S.; Gold, E.; Hess, R.; Joffe, H.; Kravitz, H.; Tepper, P.; et al. Duration of menopausal vasomotor symptoms over the menopause transition. *JAMA Intern. Med.* **2015**, *175*, 531–539. [CrossRef]
29. Schiaffino, A.; Rodríguez, M.; Pasarín, M.I.; Regidor, E.; Borrell, C.; Fernández, E. Odds ratio o razón de proporciones? Su utilización en estudios transversales. *Gac. Sanit.* **2003**, *17*, 51. [CrossRef]
30. Strömberg, U. Prevalence odds ratio v prevalence ratio. *Occup. Environ. Med.* **1994**, *51*, 143. [CrossRef]
31. Edwards, H.; Duchesne, A.; Au, A.; Einstein, G. The many menopauses: Searching the cognitive research literature for menopause types. *Menopause* **2019**, *26*, 45–65. [CrossRef] [PubMed]
32. Pelosi, E.; Simonsick, E.; Forabosco, A.; Garcia, J.; Schlessinger, D. Dynamics of the ovarian reserve and impact of genetic and epidemiological factors on age of menopause. *Biol. Reprod.* **2015**, *92*, 130. [CrossRef] [PubMed]
33. Zhang, X.; Liu, L.; Song, F.; Song, Y.; Dai, H. Ages at menarche and menopause, and mortality among postmenopausal women. *Maturitas* **2019**, *130*, 50–56. [CrossRef] [PubMed]
34. Dorador, M.; Orozco, G. Síntomas psicológicos en la transición menopáusica. *Rev. Chil. Obstet. Ginecol.* **2018**, *83*, 228–239. [CrossRef]
35. Khatoon, F.; Sinha, P.; Shahid, S.; Gupta, U. Assessment of menopausal symptoms using modified menopause rating scale (MRS) in women of Northern India. *Int. J. Reprod. Contracept. Obstet. Gynecol.* **2018**, *7*, 947–951. [CrossRef]
36. Zhu, D.; Chung, H.; Dobson, A.; Pandeya, N.; Anderson, D.; Kuh, D.; Hardy, R.; Brunner, E.; Avis, N.; Gold, E.; et al. Vasomotor menopausal symptoms and risk of cardiovascular disease: A pooled analysis of six prospective studies. *Am. J. Obstet. Gynecol.* **2020**, *1*, 1.e1–1.e16. [CrossRef] [PubMed]
37. Coslov, N.; Richardson, M.; Fugate, N. Symptom experience during the late reproductive stage and the menopausal transition: Observations from the Women Living Better survey. *Menopause* **2021**, *28*, 1012–1025. [CrossRef] [PubMed]
38. Ata, B.; Seyhan, A.; Seli, E. Diminished ovarian reserve versus ovarian aging. *Curr. Opin. Obstet. Gynecol.* **2019**, *31*, 139–147. [CrossRef]

39. Gold, E.; Crawford, S.; Shelton, J.; Tepper, P.; Crandall, C.; Greendale, G.; Matthews, K.; Thurston, R.; Avis, N. Longitudinal analysis of changes in weight and waist circumference in relation to incident vasomotor symptoms: The Study of Women's Health Across the Nation (SWAN). *Menopause* **2017**, *24*, 9–26. [CrossRef]
40. Zeleke, B.; Bell, R.; Billah, B.; Davis, S. Vasomotor and sexual symptoms in older Australian women: A cross-sectional study. *Fertil. Steril.* **2015**, *105*, 149–155. [CrossRef]
41. Kapoor, E.; Collazo, M.; Faubion, S. Weight gain in women at midlife: A concise review of the pathophysiology and strategies for management. *Mayo Clin. Proc.* **2017**, *92*, 1552–1558. [CrossRef]
42. Freedman, R. Hot flashes: Behavioral treatments, mechanisms, and relation to sleep. *Am. J. Med.* **2005**, *118*, 124–130. [CrossRef]
43. Costanian, C.; Zangiabadi, S.; Bahous, S.; Deonandan, R.; Tamim, H. Reviewing the evidence on vasomotor symptoms: The role of traditional and non-traditional factors. *Climacteric* **2020**, *23*, 213–223. [CrossRef]
44. Gast, G.; Samsioe, G.; Grobbee, D.; Nilsson, P.; Van der Schouw, Y. Vasomotor symptoms, estradiol levels and cardiovascular risk profile in women. *Maturitas* **2010**, *66*, 285–290. [CrossRef]
45. Herber, G.; Mishra, G.; Van Der, Y.; Brown, W.; Dobson, A. Risk factors for night sweats and hot flushes in midlife: Results from a prospective cohort study. *Menopause* **2013**, *20*, 953–959. [CrossRef]
46. D'Aurea, C.; Poyares, D.; Piovezan, R.; Passos, G.; Tufik, S.; De Mello, M. Objective short sleep duration is associated with the activity of the hypothalamic-pituitary-adrenal axis in insomnia. *Arq. Neuro-Psiquiatr.* **2015**, *73*, 516–519. [CrossRef]
47. Son, Y.; Ubuka, T.; Tsutsui, K. Regulation of stress response on the hypothalamic-pituitary-gonadal axis via gonadotropin-inhibitory hormone. *Front. Neuroendocrinol.* **2022**, *64*, 100953. [CrossRef]
48. Franco, O.; Muka, T.; Colpani, V.; Kunutsor, S.; Chowdhury, S.; Chowdhury, R.; Kavousi, M. Vasomotor symptoms in women and cardiovascular risk markers: Systematic review and meta-analysis. *Maturitas* **2015**, *81*, 353–361. [CrossRef]
49. Alshak, M.; Das, J. StatPearls Publishing: Neuroanatomy, Sympathetic Nervous System. Available online: https://www.ncbi.nlm.nih.gov/books/NBK542195 (accessed on 23 April 2022).
50. Gallicchio, L.; Miller, S.; Kiefer, J.; Greene, T.; Zacur, H.; Flaws, J. Risk factors for hot flashes among women undergoing the menopausal transition: Baseline results from the Midlife Women's Health Study. *Menopause* **2015**, *22*, 1098–1107. [CrossRef]
51. Bergeron, N.; Chiu, S.; Williams, P.; King, S.; Krauss, R. Effects of red meat, white meat, and nonmeat protein sources on atherogenic lipoprotein measures in the context of low compared with high saturated fat intake: A randomized controlled trial. *Am. J. Clin. Nutr.* **2019**, *110*, 24–33. [CrossRef]
52. Cor, G.; Gast, H.; Mishra, G. Fruit, Mediterranean-style, and high-fat and -sugar diets are associated with the risk of night sweats and hot flushes in midlife: Results from a prospective cohort study. *Am. J. Clin. Nutr.* **2013**, *97*, 1092–1099. [CrossRef]
53. Beezhold, B.; Radnitz, C.; McGrath, R.; Feldman, A. Vegans report less bothersome vasomotor and physical menopausal symptoms than omnivores. *Maturitas* **2018**, *112*, 12–17. [CrossRef]
54. Elavsky, S.; McAuley, E. Personality, menopausal symptoms, and physical activity outcomes in middle-aged women. *Personal. Individ. Differ.* **2009**, *46*, 123–128. [CrossRef]
55. Flesaker, M.; Serviente, C.; Troy, L.; Witkowski, S. The role of cardiorespiratory fitness on quality of life in midlife women. *Menopause* **2021**, *28*, 431–438. [CrossRef]
56. Aparicio, V.; Borges, M.; Ruiz, P.; Coll, I.; Acosta, P.; Špacírová, Z.; Soriano, A. Association of objectively measured physical activity and physical fitness with menopause symptoms. The Flamenco Project. *Climacteric* **2017**, *20*, 456–461. [CrossRef]
57. Elavsky, S.; Gonzales, J.; Proctor, D.; Williams, N.; Henderson, V. Effects of physical activity on vasomotor symptoms: Examination using objective and subjective measures. *Menopause* **2012**, *19*, 1095–1103. [CrossRef] [PubMed]
58. Arias, M.; Castro, L.; Barreiro, J.; Cabanas, P. Una revisión sobre los disruptores endocrinos y su posible impacto sobre la salud de los humanos. *Rev. Esp. Endocrinol. Pediátrica* **2020**, *11*, 33–53. [CrossRef]
59. Gold, E.; Block, G.; Crawford, S.; Lachance, L.; FitzGerald, G.; Miracle, H.; Sherman, S. Lifestyle and demographic factors in relation to vasomotor symptoms: Baseline results from the Study of Women's Health Across the Nation. *Am. J. Epidemiol.* **2004**, *159*, 1189–1199. [CrossRef] [PubMed]

MDPI

Article

The Role of a High-Fat, High-Fructose Diet on Letrozole-Induced Polycystic Ovarian Syndrome in Prepubertal Mice

Joanna Maria Pieczyńska [1], Ewa Pruszyńska-Oszmałek [2], Paweł Antoni Kołodziejski [2], Anna Łukomska [3] and Joanna Bajerska [1,*]

[1] Department of Human Nutrition and Dietetics, Poznań University of Life Sciences, 60-637 Poznań, Poland; joanna.pieczynska@up.poznan.pl
[2] Department of Animal Physiology, Biochemistry and Biostructure, Poznań University of Life Sciences, 60-637 Poznań, Poland; ewa.pruszynska@up.poznan.pl (E.P.-O.); pawel.kolodziejski@up.poznan.pl (P.A.K.)
[3] Department of Preclinical Sciences and Infectious Diseases, Poznań University of Life Sciences, 60-637 Poznań, Poland; anna.lukomska@up.poznan.pl
* Correspondence: joanna.bajerska@up.poznan.pl; Tel.: +48-61-8487335

Abstract: This study aims to investigate the effects of a high-fat, high-fructose (HF/HFr) diet on metabolic/endocrine dysregulations associated with letrozole (LET)-induced Polycystic Ovarian Syndrome (PCOS) in prepubertal female mice. Thirty-two prepubertal C57BL/6 mice were randomly divided into four groups of eight and implanted with LET or a placebo, with simultaneous administration of an HF/HFr/standard diet for five weeks. After sacrifice, the liver and blood were collected for selected biochemical analyses. The ovaries were taken for histopathological examination. The LET+HF/HFr group gained significantly more weight than the LET-treated mice. Both the LET+HF/HFr and the placebo-treated mice on the HF/HFr diet developed polycystic ovaries. Moreover the LET+HF/HFr group had significantly elevated testosterone levels, worsened lipid profile and indices of insulin sensitivity. In turn, the HF/HFr diet alone led to similar changes in the LET-treated group, except for the indices of insulin sensitivity. Hepatic steatosis also occurred in both HF/HFr groups. The LET-treated group did not develop endocrine or metabolic abnormalities, but polycystic ovaries were seen. Since the HF/HFr diet can cause substantial metabolic and reproductive dysregulation in both LET-treated and placebo mice, food items rich in simple sugar—particularly fructose—and saturated fat, which have the potential to lead to PCOS progression, should be eliminated from the diet of young females.

Keywords: polycystic ovary syndrome; pre-pubertal mice; high-fat and high-fructose diet; metabolic disorders; endocrine disorders

Citation: Pieczyńska, J.M.; Pruszyńska-Oszmałek, E.; Kołodziejski, P.A.; Łukomska, A.; Bajerska, J. The Role of a High-Fat, High-Fructose Diet on Letrozole-Induced Polycystic Ovarian Syndrome in Prepubertal Mice. *Nutrients* **2022**, *14*, 2478. https://doi.org/10.3390/nu14122478

Academic Editors: Birgit-Christiane Zyriax and Nataliya Makarova

Received: 19 May 2022
Accepted: 13 June 2022
Published: 15 June 2022

1. Introduction

Polycystic ovary syndrome (PCOS) is one of the most common hormonal disorders among women of reproductive age. It significantly impairs their fertility and increases the risk of obesity, type 2 diabetes, hyperlipidemia, and cardiovascular disease [1]. Although the exact cause of PCOS is unknown, recent reviews of the PCOS research have found that genetic susceptibility is associated with PCOS and that environmental factors—such as endocrine disruptors and poor diet—are likely to play an important role in the expression of those genetic traits [2]. PCOS often manifests in the early reproductive years; puberty has been suggested as a critical developmental time period for the development and pathology of PCOS [3]. One factor commonly believed to be a risk factor for the development of adolescent PCOS is excess body weight [4]. With normal body weight, the level of total testosterone physiologically increases and the concentration of sex hormone binding globulin (SHBG) decreases, leading to an increase in the concentration of free testosterone;

in obese girls, these changes are much more pronounced, leading to hyperandrogenemia [5]. This occurs because excessively developed adipose tissue is unable to properly secrete leptin and adiponectin, which are responsible for regulating androgen concentrations [6]. Moreover, excessive body weight is often associated with the development of insulin resistance (IR), where higher insulin concentrations stimulate the ovaries to secrete more androgens [7].

At present, young people are increasingly subjected to significant exposure to diets rich in saturated fat, sugar, and fructose in particular [8]. Soft drinks, energy drinks, fruit juices and nectars, and more generally, free sugars in liquid form represent the main source of fructose consumed by today's young generation, especially obese individuals. Of course, fructose is also contained in fruit, but in minimal quantities compared to the weight of the fruit itself [9]. This type of dietary environment has undoubtedly contributed to the alarming increased prevalence of childhood obesity [10]. The results of the survey by Pathak and Nichter imply that the childhood exposure of girls to a diet high in saturated fat and simple sugars may also cause hormone disturbances during the prepubescent reproductive maturation period, leading to lifelong ovarian dysfunction and the progression of PCOS [11].

Many animal models have been developed with the aim of coming to a fuller understanding of the potential mechanisms underlying PCOS. Prepubertal exposure to the aromatase inhibitor letrozole (LET) has been used for this purpose [12]. Although there are inconclusive data on metabolic disorders resulting from exposure to LET [13–15], polycystic changes in ovaries have occurred and disturbed endocrine parameters have been observed; these include elevated levels of testosterone and luteinizing hormone and reduced levels of estradiol, progesterone, and follicle-stimulating hormone [16,17]. The results of earlier studies indicate that the addition of high-fat diet to the LET model affords good metabolic aberrations, along with ovarian cysts [1,18,19]. However, all these studies were conducted among adult female rodents. Pilot studies employing chronic, mild elevation of androgen in prepubertal rhesus macaques maintained through young adulthood demonstrated that these female monkeys developed metabolic and ovarian dysfunction when they were exposed to a high-fat, calorie-dense diet [20,21]. These findings revealed for the first time that diet can directly modulate the reproductive and metabolic symptoms associated with hyperandrogenemia. A similar effect is caused by high-fat and high-sugar (HF/HS) diets administered in the prepubertal period. However, it has been suggested that polycystic ovaries and hyperandrogenism develop secondarily to the IR induced by the tested diet [22].

Although the increased consumption of highly processed food, rich in simple sugar particularly fructose and saturated fat, has been associated with obesity and metabolic disorders in young people, the effects of this diet on the symptoms of PCOS around the time of puberty are not clear. This study aims to investigate the effects of a HF/HFr diet on metabolic/endocrine dysregulations associated with letrozole-induced PCOS in prepubertal female mice.

2. Materials and Methods

2.1. Experimental Animals and Treatment

Every effort was made to minimize both the number of animals used and their suffering. We calculated the sample size using G*Power software (RRID:SCR_013726); the sample size of the mice was also determined in accordance with Zheng et al. [23]. The effect size was calculated to be 2.05 on the basis of the differences in HOMA-IR between the control HFD (high fat diet) group and the PCOS+HFD group. With an alpha value of 0.05, a sample size of eight mice per group would yield a power of 0.95.

Thirty-two (32) prepubertal C57BL/6 mice (average body weight 13.5 g) with an age of three weeks were involved in the experiment. The animals were purchased from the Mossakowski Institute of Experimental and Clinical Medicine, Polish Academy of Sciences, Warsaw, Poland, and were housed in the vivarium at the Department of Physiology,

Biochemistry and Animal Biostructure, part of the Faculty of Veterinary Medicine and Animal Sciences at Poznań University of Life Sciences. They were allowed to adapt to the laboratory environment for ten days. All animals were housed in standard polycarbonate cages and maintained in a controlled environment, with a temperature of 21 ± 1 °C, humidity of 55–65%, and a twelve-hour light–dark cycle. After acclimatization, at four weeks of age, the mice were randomly assigned to four groups: (1) mice receiving placebo pellet fed a standard diet (n = 8); (2) mice receiving placebo pellet fed the HF/HFr diet (n = 8); (3) mice receiving LET pellet fed a standard diet (n = 8); and (4) mice receiving LET pellet fed the HF/HFr diet (n = 8). Subcutaneous implantation of continuous release letrozole (3 mg, 50 µg/day) or the placebo pellet was performed to induce PCOS or form a control group, respectively. Letrozole was purchased from Innovative Research of America.

Such young animals were used because PCOS should be induced in the prepubertal period [13]. Two groups of mice were fed with a standard laboratory diet (3.8 kcal/g, energy supply ratio: protein 18%, carbohydrate 66%, fat 16%). In turn, the other two groups were fed the HF/HFr diet (4.7 kcal/g, energy supply ratio: protein 17%, carbohydrate 37.5% (mainly fructose), fat 45.5%). The experimental diets were bought from Morawski Animal Feed (Kcynia, Poland).

The animals had unlimited access to water and food throughout the experimental period. Once a week, the animals were weighed with a Sartorius MSE2202S-100-D0 (Germany) precision balance. During the fourth week of the experiment, dietary consumption was assessed by randomly selecting four mice from each group and placing them in a semi-metabolic cage for 3 days. For this purpose, the diet provided and the diet that remained uneaten were weighted, and the difference was calculated to give the weight consumed. The study was approved by the Local Ethical Commission under permission No. 51/2021 and was performed in line with the ARRIVE 2.0 guidelines for animal research [24].

2.2. Sample Collection

The animals were sacrificed after five weeks of the experiment by decapitation. Blood was collected into nonheparinized sample bottles. The blood was centrifuged (3500× g, 15 min, 4 °C) to obtain serum samples, which were subsequently kept frozen at −80 °C until needed for biochemical assays. The liver was collected and frozen in liquid nitrogen.

Immediately after the last blood draw, ovary samples were rapidly removed from the animals. Ovaries from each mouse were fixed in 10% formalin (formaldehyde in saline). The ovaries were stored at 4 °C in 50 mL of 20% sucrose in PBS for 24 h before sectioning. Ovary sections of 4–5 µm thickness were obtained using a cryostat (CM1860 Ag Protect; Leica Biosystems, Warsaw, Poland) and collected on microscopic slides (Menzel-Glaser, SuperFrost Ultra Plus, Thermo Scientific, Budapest, Hungary). Subsequently, the sections were stained with hematoxylin-eosin (Sigma-Aldrich, Madrid, Spain) in line with the standard histological procedures; they were cover-slipped with DPX Mountant for Histology (Sigma-Aldrich, Madrid, Spain). Slides were examined under a light microscope (Leica, DM500, Leica Biosystems) and analyzed with LAS 4.9 software (Leica Biosystems). The ovarian preparations were analyzed in terms of the number of follicles and their diameters, the corpus luteum, and the thickness of the theca layer and the follicular wall.

2.3. Serum Biochemical Analysis

Serum glucose (GLU), triglycerides (TG), total cholesterol (TC), HDL-C, LDL-C, C-reactive protein (CRP), and total antioxidant capacity (TAC) were measured using commercially available colorimetric and enzymatic assays from Pointe Scientific (Lincoln Park, MI, USA). The concentration of non-esterified fatty acids (NEFA) was determined using an enzymatic test from Wako (Oxoid, Dardilly, France). Concentrations of insulin in blood serum were measured using an immunoassay (ELISA) kit obtained from Sunlong Biotech (Hangzhou, Zhejiang, China). The level of testosterone was analyzed using an immunoassay (ELISA) kit from LDN (Nordhorn, Germany). Liver cholesterol and triglycerides were analyzed after lipid extraction using a cholesterol and triglycerides kit (Pointe Scientific,

Lincoln Park, MI, USA), as described by Folch et al. [25]. The total antioxidant capacity was measured using the TCA method with a TBARS Assay Kit. (Cayman Chemical, Ann Arbor, MI, USA). The optical density of the samples was measured using a Synergy 2 microplate reader (Biotek, Winooski, VT, USA).

2.4. Calculation of the HOMA-IR, HOMA-β, and QUICKI Indices

Insulin resistance and β-cell function were evaluated using the Homeostasis Model Assessment Method. The HOMA-IR was calculated using the following formula:

$$\text{HOMA} - \text{IR} = \text{fasting glucose [mmol/L]} \times \text{fasting insulin [}\mu\text{IU/mL]}/22.5$$

HOMA-β was calculated using the following equation: $\text{HOMA} - \beta = \text{FI} \times 20/(\text{FG} - 3.5)$, where FI is fasting insulin (in μU/mL) and FG is fasting glucose (in mmol/L).

The quantitative insulin sensitivity check index (QUICKI) was calculated using the following formula:

$$\text{QUICKI} = 1 \log \text{(fasting glucose [mg/dL])} + \log \text{(fasting insulin [}\mu\text{IU/mL])}$$

Non-HDL was calculated using the following formula: Non-HDL = total cholesterol (mg/dL) − HDL (mg/dL).

2.5. Statistical Analysis

The results were statistically evaluated using Statistica 13.3.0 (TIBCO Software, Palo Alto, CA, USA; 2017). The results are presented in the tables and figures as arithmetic means ± standard deviations (SD), and the data in some figures are presented as medians with boxes and whiskers representing the interquartile range and the 5th–95th percentiles (GraphPad Prism 9.3.1. (471), GraphPad Software, San Diego, CA, USA). One-way analysis of variance (ANOVA) was used to compare the mean values of variables among the groups. Tukey's post hoc test was used to identify the significance of pairwise comparison of mean values among the groups. Values with different letters (a, b) show statistically significant differences ($p < 0.05$, a < b).

3. Results

There were no significant differences in body weight at the beginning of the experiment (placebo-treated mice: 14.6 ± 1.2 g; placebo-treated mice on HF/HFr diet: 15.0 ± 1.5 g; LET-treated mice: 14.7 ± 1.5 g; and LET-treated mice on HF/HFr diet 14.5 ± 0.6 g; data not shown).

Both the placebo and the LET-treated group of mice on the HF/HFr diet showed significantly higher diet consumption than the corresponding controls ($p < 0.05$). Moreover, the LET+HF/HFr group ate significantly more than mice receiving the placebo ($p < 0.05$). On the other hand, placebo-treated mice on the HF/HFr diet ate significantly more ($p < 0.05$) than LET-treated mice (Figure 1A). After 35 days of the experiment, mice from the LET+HF/HFr group had gained significantly more weight (by a factor of 1.3; $p < 0.05$) than the related control. Moreover, this group of mice showed a significantly higher weight gain than placebo-treated mice either on the control diet ($p < 0.05$) or on the HF/HFr diet ($p < 0.05$; Figure 1B).

Both placebo- and LET-treated mice on the HF/HFr diet showed significantly ($p < 0.05$) higher cholesterol concentrations than the corresponding controls. Moreover, the latter group had a significantly ($p < 0.05$) higher cholesterol concentration than the placebo-treated mice. In turn, the placebo-treated mice on the HF/HFr diet had substantially higher cholesterol levels than the LET-treated mice ($p < 0.05$; Table 1). Exactly the same dependencies were observed for the concentrations of HDL cholesterol, non-HDL cholesterol, and LDL cholesterol. TG concentration and TC/HDL ratio fluctuated at the same level in all four groups, and no significant differences were observed here. Substantially higher levels of plasma NEFA were seen in the HF/HFr placebo-treated mice than in the related controls ($p < 0.05$). This parameter in the placebo-treated mice on the HF/HFr diet was also considerably higher ($p < 0.05$) than in the LET-treated group and even than in the LET+HF/HFr group. HF/HFr feeding also disturbed the hepatic lipid metabolism of the experimental mice. The hepatic concentrations of both TC and TG increased remarkably compared to the control ($p < 0.05$) after the placebo mice had fed on the HF/HFr diet. The hepatic concentrations of TG in this group were also significantly higher than in the LET-treated group ($p < 0.05$). Additionally, the hepatic concentrations of TG were also significantly higher ($p < 0.05$) in the LET+HF/HFr group than in the related control. In this group of mice, the hepatic concentrations of both TC and TG were also significantly higher than in placebo mice fed a standard diet (Figure 2A,B). Plasma glucose levels were significantly higher ($p < 0.05$) by a factor of 1.2 in the placebo-treated mice on the HF/HFr diet than in the LET-treated mice. The LET+HF/HFr group of mice had significantly higher values ($p < 0.05$) of the HOMA-IR and QUICKI indices than the related controls. No significant differences in HOMA-β and CRP levels were observed between groups, while the placebo-treated mice fed the standard diet had a significantly higher TAC, by a factor or 4.9, than the LET+HF/HFr group.

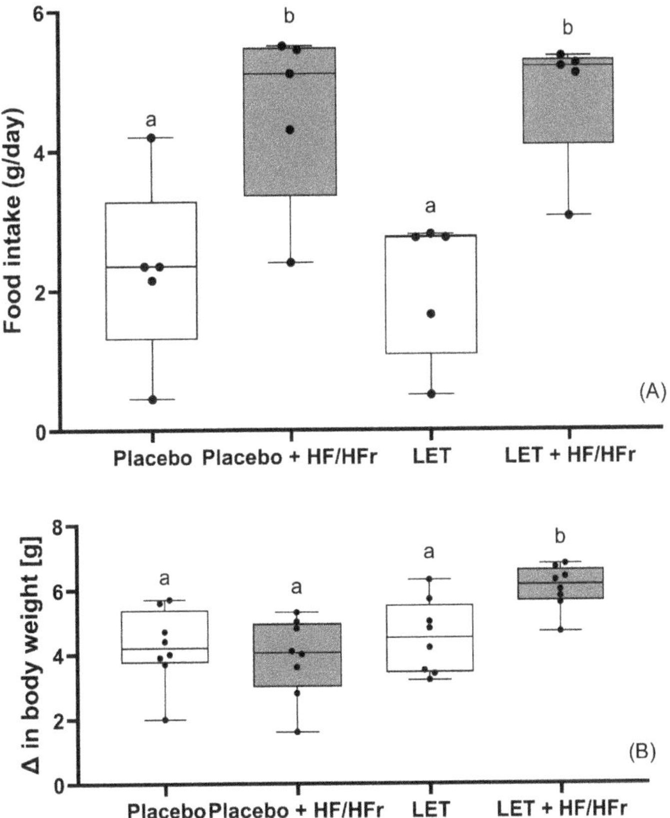

Figure 1. Effects of HF/HFr (high fat/high fructose) diet on food intake (**A**) and changes in body weight (**B**) in placebo or LET-treated (letrozole-treated) mice. The data are presented as medians, with boxes and whiskers representing the interquartile range and the 5th–95th percentiles ($n = 8$ per group), analyzed using one-way ANOVA followed by Tukey's post hoc test. Values with different letters show statistically significant differences ($p < 0.05$).

Table 1. Effects of the HF/HFr diet on metabolic parameters in placebo and LET-treated mice.

Variables	Placebo		Letrozole	
	Control	HF/HFr	Control	HF/HFr
Total cholesterol (mg/dL)	131.9 ± 21.8 [a]	188.4 ± 26.0 [b]	137.6 ± 23.0 [a]	191.7 ± 39.8 [b]
LDL cholesterol (mg/dL)	51.6 ± 5.9 [a]	77.1 ± 4.9 [b]	49.5 ± 2.5 [a]	74.5 ± 6.2 [b]
HDL cholesterol (mg/dL)	30.8 ± 4.1 [a]	40.0 ± 3.2 [b]	28.9 ± 1.8 [a]	38.3 ± 1.6 [b]
Non-HDL cholesterol (mg/dL)	101.2 ± 22.0 [a]	148.4 ± 28.3 [b]	108.7 ± 23.7 [a]	153.4 ± 40.6 [b]
Triglycerides (mg/dL)	141.2 ± 16.4	135.9 ± 21.3	136.2 ± 16.6	128.7 ± 12.9
TC/HDL ratio	4.4 ± 0.9	4.8 ± 1.0	4.8 ± 0.9	5.0 ± 1.2
NEFA (mmol/L)	1.09 ± 0.05 [a]	1.24 ± 0.06 [b]	1.14 ± 0.11 [a]	1.12 ± 0.06 [a]
Glucose (mg/dL)	119.9 ± 14.0 [ab]	140.1 ± 23.0 [b]	113.9 ± 12.0 [a]	136.3 ± 18.4 [ab]
Insulin (mU/L)	2.9 ± 0.8	3.4 ± 1.5	2.6 ± 0.7	3.4 ± 0.7
HOMA-IR	0.9 ± 0.3 [ab]	1.0 ± 0.2 [ab]	0.7 ± 0.2 [a]	1.2 ± 0.3 [b]
HOMA-β	18.5 ± 2.9	17.7 ± 10.9	20.3 ± 9.3	20.0 ± 8.0
QUICKI	0.8 ± 0.1 [ab]	0.7 ± 0.1 [ab]	0.8 ± 0.1 [a]	0.7 ± 0.1 [b]
CRP (mg/L)	25.4 ± 4.2	27.2 ± 1.5	27.5 ± 2.9	29.4 ± 3.0
TAC (μmol/L)	6.5 ± 5.6 [a]	2.8 ± 1.1 [ab]	4.9 ± 3.5 [ab]	1.6 ± 0.5 [b]

Results are expressed as mean ± SD (n = 8 per group). Values with different letters (a,b) show statistically significant differences ($p < 0.05$, Tukey's post hoc test). NEFA: non-esterified fatty acids; TC: total cholesterol; HDL: high-density lipoprotein; LDL: low-density lipoprotein; HOMA-β: homeostasis model assessment of β-cell function; HOMA-IR: homeostasis model assessment of insulin resistance; QUICKI: quantitative insulin sensitivity check index; CRP: C-reactive protein; TAC: total antioxidant capacity; HF/HFr: high fat, high fructose.

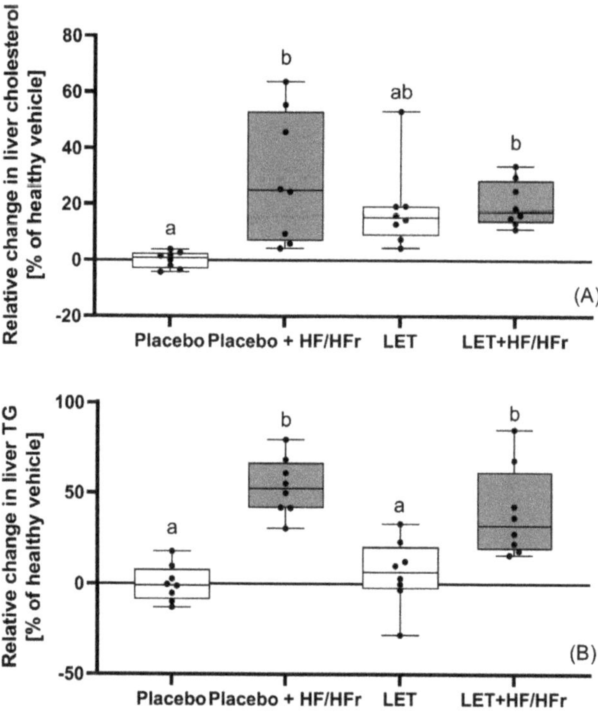

Figure 2. Terminal changes in the liver cholesterol (**A**) and triglycerides (**B**). The data are presented as medians, with boxes and whiskers representing the interquartile range and the 5th–95th percentiles (n = 8 per group), analyzed using one−way ANOVA followed by Tukey's post hoc test. Values with different letters show statistically significant differences ($p < 0.05$).

Serum testosterone concentration was robustly higher by a factor of 1.6 in the LET+HF/HFr group of mice ($p < 0.05$, Figure 3) compared to the control. Moreover, testosterone levels in the placebo mice on the HF/HFr diet were significantly higher ($p < 0.05$) than in the LET-treated group. In the LET-treated group, the LET+ HF/HFr group, and the placebo mice on the HF/HFr diet, an increased number of corpus luteum were observed compared to the placebo-treated mice on the standard diet. Moreover, the placebo mice on the HF/HFr diet had the largest number of ovarian follicles of the four groups. This group also had the thickest granulosa layer in the follicle, while in both the LET-treated and the LED+HF/HFr groups, this layer's thickness was visibly reduced. Furthermore, the LET+HF/HFr group had the greatest follicle diameter and the thickest theca folliculi. Follicular atresia and the presence of cysts were observed not only in both the LET-treated groups but also in the group of placebo-treated mice on the HF/HFr diet. Interestingly, the changes in ovarian morphology in the LET-treated mice were milder than in the LET+HF/HFr group, at least partly due to the smaller number of cysts and lesser degradation of the granular cell layer (Figures 4–9).

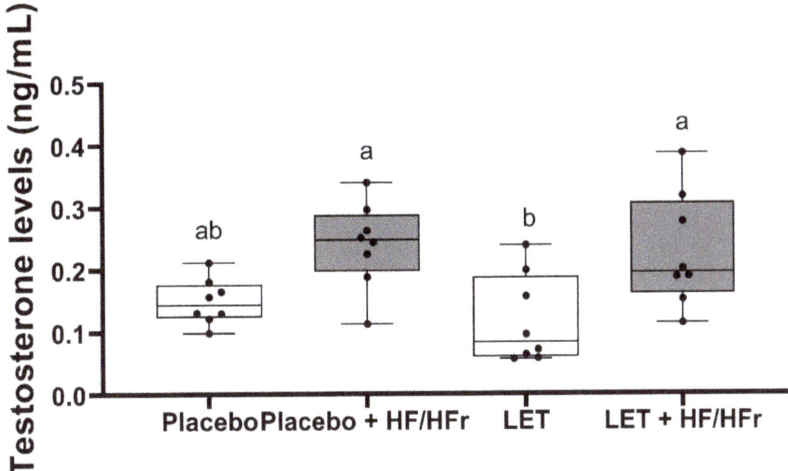

Figure 3. Effects of HF/HFr diet on testosterone levels in placebo or letrozole-treated mice. The data are presented as medians, with boxes and whiskers representing the interquartile range and the 5th–95th percentiles ($n = 8$ per group), analyzed using one-way ANOVA followed by Tukey's post hoc test. Values with different letters show statistically significant differences ($p < 0.05$).

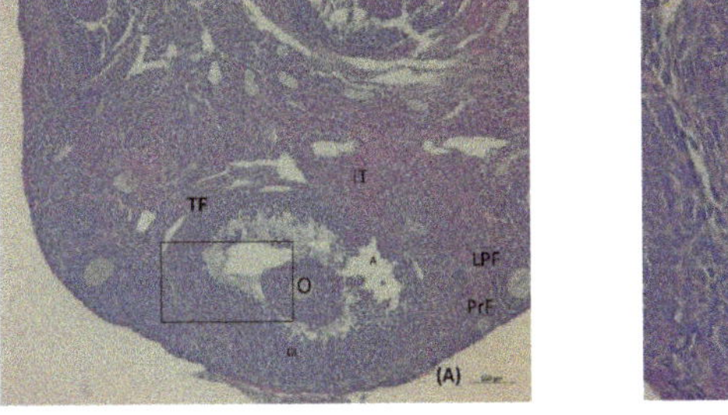

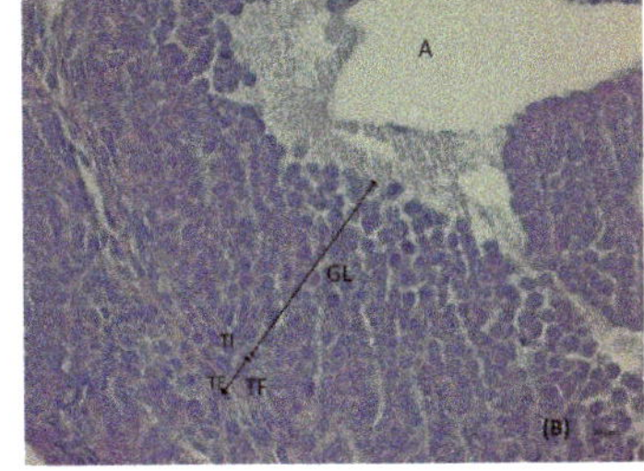

Figure 4. Histological sections of the ovaries of intact mice: tertiary follicle. O: oocyte; GL: granulosa layer; A: antrium; LPF: late primary follicle; PrF: primodial follicle; IT: interstitial tissue; TF: theca folliculi; TE: theca externa; TI: theca interna (H&E; power 100× (**A**)/400× (**B**)).

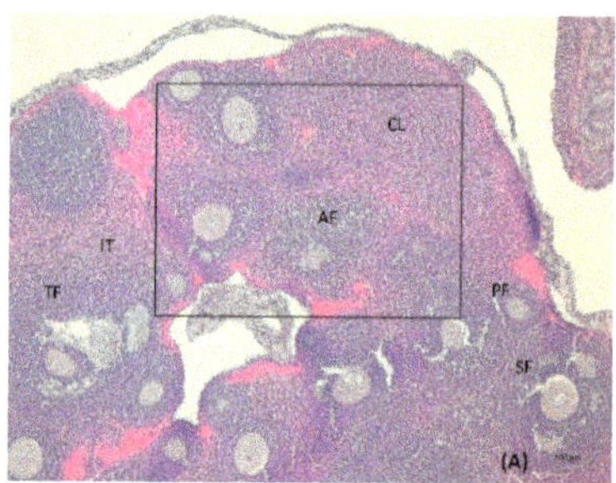

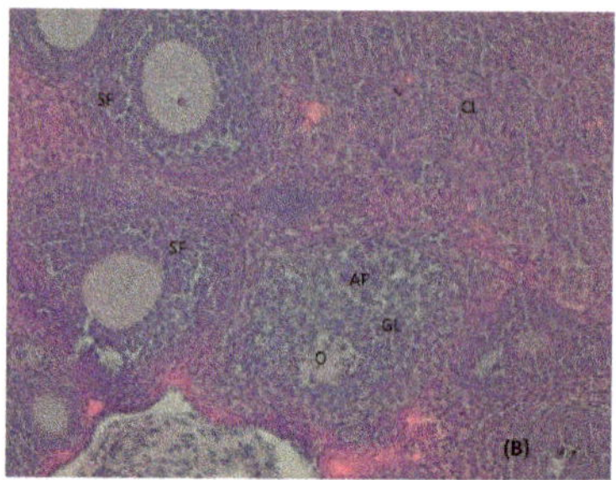

Figure 5. Histological sections of the ovaries with atretic follicles (AF) in intact mice on the HF/HFr diet. TF: tertiary follicle; PF: primary follicle; SF: secondary follicle; CL: corpus luteum; IT: interstitial tissue; O: oocyte; GL: granulosa cells in the process of degradation (H&E; power 100× (**A**)/200× (**B**)).

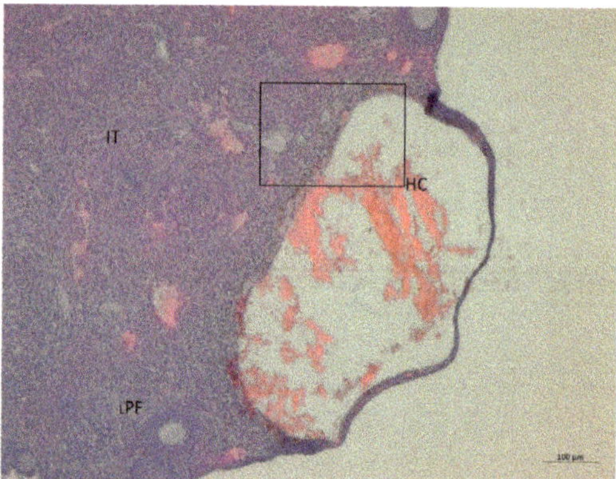

Figure 6. Histological sections of ovaries with hemorrhagic cysts (HC) in LET-treated mice. LPF: late primary follicle; IT: interstitial tissue (H&E; power 100×).

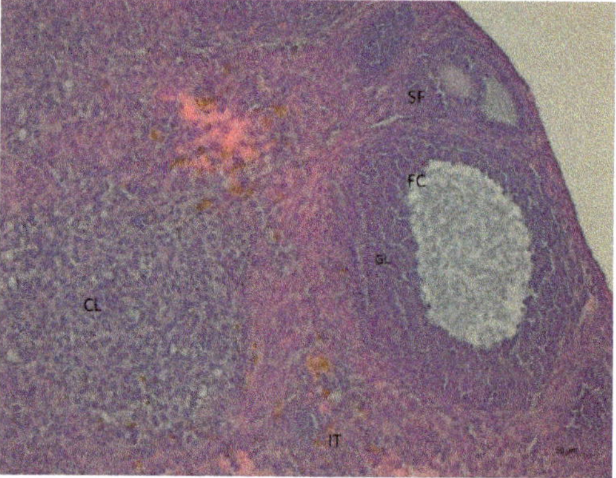

Figure 7. Histological sections of ovaries with follicular cysts (FC) in LET-treated mice. GL: granulosa cells; CL: corpus luteum; IT: interstitial tissue (H&E; power 200×).

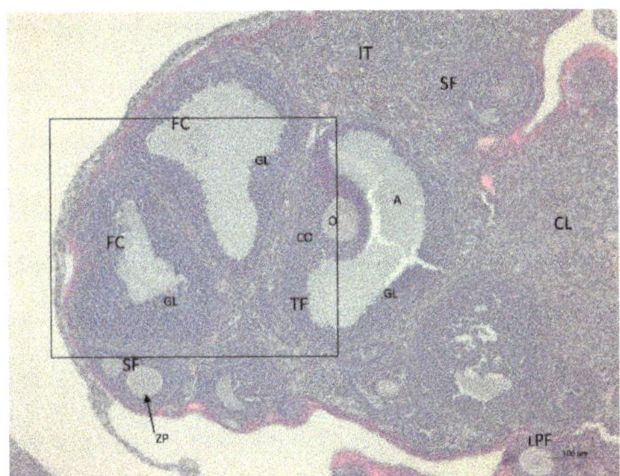

Figure 8. Histological sections of ovaries with follicular cysts (FC) in LET-treated mice on the HF/HFr diet. TF: tertiary follicle; O: oocyte; GL: irregular granulosa layer; CO: cumulus oophorus; A: antrium; LPF: late primary follicle; SF: secondary follicle; ZP: zona pellucida; IT: interstitial tissue (H&E; power 100×).

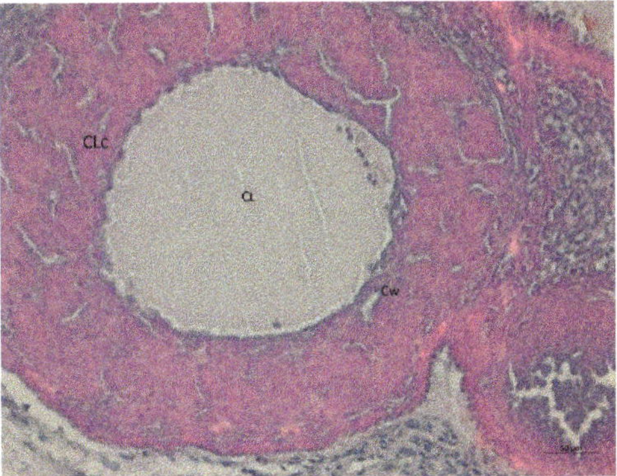

Figure 9. Histological sections of ovaries with corpus luteum cysts (CLc) in LET-treated mice on the HF/HFr diet. Cw: cyst wall; CL: cyst lumen (H&E; power 400×).

4. Discussion

Since PCOS often manifests in the early reproductive years, puberty is considered to be a critical time period for the development of PCOS. Indeed, our study demonstrated that five weeks of HF/HFr feeding initiated at the prepubertal age provoked some reproductive and metabolic features of PCOS in LET-treated female mice. More specifically, in the LET+HF/HFr group of mice, elevated testosterone levels and morphological changes in the ovaries were seen, suggesting PCOS. Indeed as was shown previously, elevated serum testosterone levels are related to endocrine imbalances and contribute to PCOS symptoms such as ovarian dysfunction and irregular ovarian or estrous cycles [26]. The LET+HF/HFr group of mice also had a worsened lipid profile (TC, LDL-C, HDL-C and non-HDL, except of TG) and insulin sensitivity indices (HOMA-IR and QUICKI). Hepatic steatosis also occurred in this group of mice. Our findings are in line with those of previous studies showing that the use of letrozole with a high-fat diet may induce or worsen the symptoms of PCOS [1,25]. More specifically, Xu et al. noted that twelve weeks of administration of LET with a high-fat diet in female Sprague–Dawley rats induced anovulatory cycles and polycystic ovary morphology, body weight gain, elevated testosterone levels, abnormal glucose and lipid metabolism, as well as insulin resistance [27]. Begum et al. indicated that twelve weeks of administration of the LET+HF diet in Wistar female rats induced additional glucose intolerance [1]. It should, however, be highlighted that all PCOS symptoms seen in our study developed as early as week five of the experiment when LET+HF/HFr was administrated; moreover, hepatic steatosis also occurred. Worsened insulin sensitivity indices, as observed in the LET-treated mice on the HF/HFr diet, seem to be associated with the excess body weight of mice. Indeed, obesity is associated with inflammation and the generation of reactive oxygen species that have a potent role in inducing insulin resistance [15]. In line with this, in our LET-treated mice on the HF/HFr diet, we observed substantial decreases in total antioxidant capacity (TAC) compared to placebo-treated mice on a standard diet. This indicates that the occurrence of PCOS is associated with oxidative stress in PCOS women, which may even contribute to the pathogenesis of this disorder [28]. Agreeing with this, a case-control study showed statistically significant decreases in the TAC levels of women with PCOS as compared to the control group [29].

Interestingly, it was seen in our study that LET itself did not lead to the development of any endocrine or metabolic abnormalities in experimental mice, but polycystic ovaries were observed. In contrast, Arroyo et al. and Skarra et al. demonstrated that five weeks of LET treatment resulted in the hallmarks of PCOS, including elevated testosterone and luteinizing hormone (LH) levels, acyclicity, and the appearance of cystic ovarian follicles [13,30]. However, despite the hormonal variations shown in different animal studies, letrozole in general manifests good reproducibility for PCOS-like features in rodents [15] and is believed to cause the lean reproductive phenotype of PCOS [31].

The HF/HFr diet itself may also lead to some features of PCOS. As was observed in our study, five weeks of exposure to the HF/HFr diet significantly elevated serum testosterone of the female mice and also disturbed some lipid parameters (TC, LDL-C, HDL-C and non-HDL, except of TG). Elevated levels of glucose and polycystic changes in ovaries were also observed. However, worsened insulin sensitivity was not observed in this model. Roberts et al. indicated that a high-fat, high-sugar diet given for eleven weeks led to hyperinsulinemia but not to hyperandrogenemia in experimental rats [22]. However, elevated testosterone levels in that study were predictive of a high number of ovarian cysts [22]. In our study, the elevated testosterone levels seen in a group of mice on an HF/HFr diet seem to be the effect of higher levels (though not statistically significant levels) of insulin. Indeed, insulin acts directly through its own receptor in PCO theca cells to increase androgen production [32]. Interestingly, metabolic disturbances in the placebo-treated mice fed the HF/HFr diet were seen even when the body weight of those mice did not increase significantly. It was also surprising that, despite the equally high consumption of the HF/HFr diet, only the LET-treated mice gained significantly more weight, while the body weight of the placebo mice did not differ from that of the other groups. Similar results were obtained by Patel and Shah, but this was associated with a reduction in food intake, which was not observed in our experiment [33]. Huang et al. explained that female rodents are relatively resistant to hyperphagia and weight gain in response to a high-fat diet, in part due to the effects of estrogen, which suppress food intake and increase energy expenditure [34].

Since increased prevalence of NAFLD has been reported in women with PCOS [35], we also assessed the hepatic accumulation of TC and TG in our experimental mice. More specifically, the hepatic accumulation of both TC and TG increased remarkably after the placebo-treated mice were fed the HF/HFr diet, as compared with control. In LET-treated mice on the HF/HFr diet, only the TG level was significantly higher than that of the related control. The accumulation of excess triglycerides in hepatocytes is generally the result of the increased delivery of non-esterified fatty acids (NEFAs), increased synthesis of NEFAs, impaired intracellular catabolism of NEFAs, impaired secretion as triglyceride, or a combination of these abnormalities [36]. In our study, only higher levels of the plasma NEFA in the mice fed a HF/HFr diet compared to placebo-treated mice was seen. Interestingly, despite the visible NAFLD, we did not observe significant differences in serum triglyceride concentration, as hypertriglyceridemia develops secondarily to hepatic steatosis [37] and the period of 5 weeks was likely insufficient for its full appearance [38].

5. Conclusions

Our findings reveal for the first time that HF/HFr feeding given around puberty may directly stimulate reproductive and metabolic symptoms, not only in LET-treated mice but also in placebo-treated mice. These findings indicate that a diet that is highly processed, high in simple sugars (particularly fructose), and high in saturated fats may, if eaten every day, have a great impact on PCOS progression in young females. Food products that are rich in these ingredients should therefore be eliminated from the diets of women with PCOS, especially younger women. Furthermore, the combination of the HF/HFr diet with LET causes visible metabolic disorders, comparable to those found in women with PCOS. Moreover, hepatic steatosis also occurred. Thus, this animal model can be used to the test various options for PCOS treatment. In turn, the model based on letrozole alone was not sufficient to induce the above-mentioned disorders, although it caused visible changes in the morphological structure of the ovaries.

Author Contributions: Conceptualization: J.M.P., J.B.; methodology: P.A.K., E.P.-O., J.M.P.; data curation: J.M.P., P.A.K., E.P.-O.; data analysis: J.M.P., J.B., P.A.K., E.P.-O., A.Ł.; writing the first draft: J.M.P.; review and editing: J.B., P.A.K., E.P.-O., A.Ł. All authors have read and agreed to the published version of the manuscript.

Funding: The APC was funded within the framework of the Polish Ministry of Science and Higher Education's program: "Regional Initiative Excellence" in the years 2019–2022 (No. 005/RID/2018/19)", financing amount 12 000 000 PLN.

Institutional Review Board Statement: The study was approved by the Local Ethical Commission under permission No. 51/2021 and was performed in line with the ARRIVE 2.0 guidelines for animal research [24].

Informed Consent Statement: Not applicable.

Data Availability Statement: The datasets used and analyzed in the current study are available from the corresponding author upon reasonable request.

Conflicts of Interest: The authors declare no conflict of interest.

References

1. Begum, N.; Manipriya, K.; Veeresh, B. Role of high-fat diet on letrozole-induced polycystic ovarian syndrome in rats. *Eur. J. Pharmacol.* **2022**, *917*, 174746. [CrossRef] [PubMed]
2. Merkin, S.S.; Phy, J.L.; Sites, C.K.; Yang, D. Environmental determinants of polycystic ovary syndrome. *Fertil. Steril.* **2016**, *106*, 16–24. [CrossRef] [PubMed]
3. Torres, P.J.; Skarra, D.V.; Ho, B.S.; Sau, L.; Anvar, A.R.; Kelley, S.T.; Thackray, V.G. Letrozole treatment of adult female mice results in a similar reproductive phenotype but distinct changes in metabolism and the gut microbiome compared to pubertal mice. *BMC Microbiol.* **2019**, *19*, 57. [CrossRef] [PubMed]
4. Rosenfield, R.L. Identifying Children at Risk for Polycystic Ovary Syndrome. *J. Clin. Endocrinol. Metab.* **2007**, *92*, 787–796. [CrossRef] [PubMed]
5. Burt Solorzano, C.M.; McCartney, C.R.; Blank, S.K.; Knudsen, K.L.; Marshall, J.C. Hyperandrogenaemia in adolescent girls: Origins of abnormal gonadotropin-releasing hormone secretion. *BJOG Int. J. Obstet. Gynaecol.* **2010**, *117*, 143–149. [CrossRef] [PubMed]
6. Zeng, X.; Xie, Y.-J.; Liu, Y.-T.; Long, S.-L.; Mo, Z.-C. Polycystic ovarian syndrome: Correlation between hyperandrogenism, insulin resistance and obesity. *Clin. Chim. Acta* **2019**, *502*, 214–221. [CrossRef]
7. Rosenfield, R.L.; Ehrmann, D.A. The Pathogenesis of Polycystic Ovary Syndrome (PCOS): The Hypothesis of PCOS as Functional Ovarian Hyperandrogenism Revisited. *Endocr. Rev.* **2016**, *37*, 467–520. [CrossRef]
8. Munt, A.E.; Partridge, S.R.; Allman-Farinelli, M. The barriers and enablers of healthy eating among young adults: A missing piece of the obesity puzzle: A scoping review: Barriers and enablers of healthy eating. *Obes. Rev.* **2017**, *18*, 1–17. [CrossRef]
9. Giussani, M.; Lieti, G.; Orlando, A.; Parati, G.; Genovesi, S. Fructose Intake, Hypertension and Cardiometabolic Risk Factors in Children and Adolescents: From Pathophysiology to Clinical Aspects. A Narrative Review. *Front. Med.* **2022**, *9*, 792949. [CrossRef]
10. Tsan, L.; Décarie-Spain, L.; Noble, E.E.; Kanoski, S.E. Western Diet Consumption During Development: Setting the Stage for Neurocognitive Dysfunction. *Front. Neurosci.* **2021**, *15*, 632312. [CrossRef]
11. Pathak, G.; Nichter, M. Polycystic ovary syndrome in globalizing India: An ecosocial perspective on an emerging lifestyle disease. *Soc. Sci. Med.* **2015**, *146*, 21–28. [CrossRef] [PubMed]

12. Ressler, I.B.; Grayson, B.E.; Ulrich-Lai, Y.M.; Seeley, R.J. Diet-induced obesity exacerbates metabolic and behavioral effects of polycystic ovary syndrome in a rodent model. *Am. J. Physiol. Metab.* **2015**, *308*, E1076–E1084. [CrossRef] [PubMed]
13. Arroyo, P.; Ho, B.S.; Sau, L.; Kelley, S.T.; Thackray, V.G. Letrozole treatment of pubertal female mice results in activational effects on reproduction, metabolism and the gut microbiome. *PLoS ONE* **2019**, *14*, e0223274. [CrossRef] [PubMed]
14. Manneras, L.; Cajander, S.; Holmang, A.; Seleskovic, Z.; Lystig, T.; Lonn, M.; Stener-Victorin, E. A New Rat Model Exhibiting Both Ovarian and Metabolic Characteristics of Polycystic Ovary Syndrome. *Endocrinology* **2007**, *148*, 3781–3791. [CrossRef] [PubMed]
15. Ryu, Y.; Kim, S.W.; Kim, Y.Y.; Ku, S.-Y. Animal Models for Human Polycystic Ovary Syndrome (PCOS) Focused on the Use of Indirect Hormonal Perturbations: A Review of the Literature. *Int. J. Mol. Sci.* **2019**, *20*, 2720. [CrossRef]
16. Kafali, H.; Iriadam, M.; Ozardalı, I.; Demir, N. Letrozole-induced polycystic ovaries in the rat: A new model for cystic ovarian disease. *Arch. Med. Res.* **2004**, *35*, 103–108. [CrossRef]
17. Kauffman, A.S.; Thackray, V.G.; Ryan, G.E.; Tolson, K.; Glidewell-Kenney, C.A.; Semaan, S.J.; Poling, M.C.; Iwata, N.; Breen, K.M.; Duleba, A.J.; et al. A Novel Letrozole Model Recapitulates Both the Reproductive and Metabolic Phenotypes of Polycystic Ovary Syndrome in Female Mice1. *Biol. Reprod.* **2015**, *93*, 69. [CrossRef]
18. Wang, H.-B.; Loh, D.H.; Whittaker, D.S.; Cutler, T.; Howland, D.; Colwell, C.S. Time-Restricted Feeding Improves Circadian Dysfunction as well as Motor Symptoms in the Q175 Mouse Model of Huntington's Disease. *Eneuro* **2018**, *5*. [CrossRef]
19. Huang, L.; Liang, A.; Li, T.; Lei, X.; Chen, X.; Liao, B. Mogroside V Improves Follicular Development and Ovulation in Young-Adult PCOS Rats Induced by Letrozole and High-Fat Diet Through Promoting Glycolysis. *Front. Endocrinol.* **2022**, *13*, 838204. [CrossRef]
20. Varlamov, O.; Chu, M.P.; McGee, W.K.; Cameron, J.L.; O'Rourke, R.; Meyer, K.A.; Bishop, C.; Stouffer, R.L.; Roberts, C. Ovarian Cycle-Specific Regulation of Adipose Tissue Lipid Storage by Testosterone in Female Nonhuman Primates. *Endocrinology* **2013**, *154*, 4126–4135. [CrossRef]
21. McGee, W.K.; Bishop, C.V.; Pohl, C.R.; Chang, R.J.; Marshall, J.C.; Pau, F.K.; Stouffer, R.L.; Cameron, J.L. Effects of hyperandrogenemia and increased adiposity on reproductive and metabolic parameters in young adult female monkeys. *Am. J. Physiol. Metab.* **2014**, *306*, E1292–E1304. [CrossRef] [PubMed]
22. Roberts, J.S.; Perets, R.A.; Sarfert, K.S.; Bowman, J.; Ozark, P.A.; Whitworth, G.B.; Blythe, S.N.; Toporikova, N. High-fat high-sugar diet induces polycystic ovary syndrome in a rodent model. *Biol. Reprod.* **2017**, *96*, 551–562. [CrossRef] [PubMed]
23. Zheng, Y.; Yu, J.; Liang, C.; Li, S.; Wen, X.; Li, Y. Characterization on gut microbiome of PCOS rats and its further design by shifts in high-fat diet and dihydrotestosterone induction in PCOS rats. *Bioprocess Biosyst. Eng.* **2020**, *44*, 953–964. [CrossRef] [PubMed]
24. Percie du Sert, N.; Hurst, V.; Ahluwalia, A.; Alam, S.; Avey, M.T. Baker MThe ARRIVE guidelines 2.0: Updated guidelines for reporting animal research. Boutron I, redaktor. *PLoS Biol.* **2020**, *18*, e3000410.
25. Folch, J.; Lees, M.; Stanley, G.H.S. A simple method for the isolation and purification of total lipides from animal tissues. *J. Biol. Chem.* **1957**, *226*, 497–509. [CrossRef]
26. Eldar-Geva, T.; Margalioth, E.J.; Gal, M.; Ben-Chetrit, A.; Algur, N.; Zylber-Haran, E.; Brooks, B.; Huerta, M.; Spitz, I.M. Serum anti-Mullerian hormone levels during controlled ovarian hyperstimulation in women with polycystic ovaries with and without hyperandrogenism. *Hum. Reprod.* **2005**, *20*, 1814–1819. [CrossRef]
27. Xu, J.; Dun, J.; Yang, J.; Zhang, J.; Lin, Q.; Huang, M.; Ji, F.; Huang, L.; You, X.; Lin, Y. Letrozole Rat Model Mimics Human Polycystic Ovarian Syndrome and Changes in Insulin Signal Pathways. *Med Sci. Monit.* **2020**, *26*, e923073-1. [CrossRef]
28. Lu, J.; Wang, Z.; Cao, J.; Chen, Y.; Dong, Y. A novel and compact review on the role of oxidative stress in female reproduction. *Reprod. Biol. Endocrinol.* **2018**, *16*, 80. [CrossRef]
29. Alipour, B.; Roohelhami, E.; Rashidkhani, B.; Shahrdami, F.; Edalati, S. Evaluating indicators of oxidative stress and their relationship with Insulin Resistance in polycystic ovary syndrome. *Prog. Nutr.* **2019**, *21*, 178–183. [CrossRef]
30. Skarra, D.V.; Hernández-Carretero, A.; Rivera, A.J.; Anvar, A.R.; Thackray, V.G. Hyperandrogenemia Induced by Letrozole Treatment of Pubertal Female Mice Results in Hyperinsulinemia Prior to Weight Gain and Insulin Resistance. *Endocrinology* **2017**, *158*, 2988–3003. [CrossRef]
31. Stener-Victorin, E.; Padmanabhan, V.; Walters, A.K.; Campbell, E.R.; Benrick, A.; Giacobini, P.; Dumesic, A.D.; Abbott, D.H. Animal Models to Understand the Etiology and Pathophysiology of Polycystic Ovary Syndrome. *Endocr. Rev.* **2020**, *41*, bnaa010. [CrossRef] [PubMed]
32. Baptiste, C.G.; Battista, M.-C.; Trottier, A.; Baillargeon, J.-P. Insulin and hyperandrogenism in women with polycystic ovary syndrome. *J. Steroid Biochem. Mol. Biol.* **2009**, *122*, 42–52. [CrossRef] [PubMed]
33. Patel, R.S.; Shah, G. High-fat diet exposure from pre-pubertal age induces polycystic ovary syndrome (PCOS) in rats. *Reproduction* **2018**, *155*, 139–149. [CrossRef] [PubMed]
34. Huang, K.-P.; Ronveaux, C.C.; Knotts, T.A.; Rutkowsky, J.R.; Ramsey, J.J.; Raybould, H.E. Sex differences in response to short-term high fat diet in mice. *Physiol. Behav.* **2020**, *221*, 112894. [CrossRef] [PubMed]
35. Qu, Z.; Zhu, Y.; Jiang, J.; Shi, Y.; Chen, Z. The clinical characteristics and etiological study of nonalcoholic fatty liver disease in Chinese women with PCOS. *Iran. J. Reprod. Med.* **2013**, *11*, 725–732.
36. Basaranoglu, M. Fructose as a key player in the development of fatty liver disease. *World J. Gastroenterol.* **2013**, *19*, 1166. [CrossRef]
37. Fabbrini, E.; Magkos, F. Hepatic Steatosis as a Marker of Metabolic Dysfunction. *Nutrients* **2015**, *7*, 4995–5019. [CrossRef]
38. de Moura e Dias, M.; dos Reis, S.A.; da Conceição, L.L.; Sediyama CMN de, O.; Pereira, S.S.; de Oliveira, L.L. Diet-induced obesity in animal models: Points to consider and influence on metabolic markers. *Diabetol. Metab. Syndr.* **2021**, *13*, 32. [CrossRef]

 nutrients

Review

Dietary Natural Compounds and Vitamins as Potential Cofactors in Uterine Fibroids Growth and Development

Iwona Szydłowska [1,*], Jolanta Nawrocka-Rutkowska [1], Agnieszka Brodowska [1], Aleksandra Marciniak [1], Andrzej Starczewski [1] and Małgorzata Szczuko [2]

[1] Department of Gynecology, Endocrinology and Gynecological Oncology, Pomeranian Medical University, 71-252 Szczecin, Poland; jolanta.nawrocka.rutkowska@pum.edu.pl (J.N.-R.); agabrod@wp.pl (A.B.); o.marciniak@wp.pl (A.M.); andrzejstarcz@o2.pl (A.S.)

[2] Department of Human Nutrition and Metabolomics, Pomeranian Medical University, 71-460 Szczecin, Poland; malgorzata.szczuko@pum.edu.pl

* Correspondence: iwonaszyd@wp.pl; Tel.:+48-91-425-0541

Abstract: An analysis of the literature generated within the past 20 year-span concerning risks of uterine fibroids (UFs) occurrence and dietary factors was carried out. A link between Vitamin D deficiency and UFs formation is strongly indicated, making it a potent compound in leiomyoma therapy. Analogs of the 25-hydroxyvitamin D3, not susceptible to degradation by tissue 24-hydroxylase, appear to be especially promising and tend to show better therapeutic results. Although research on the role of Vitamin A in the formation of fibroids is contradictory, Vitamin A-enriched diet, as well as synthetic retinoid analogues, may be preventative or limit the growth of fibroids. Unambiguous conclusions cannot be drawn regarding Vitamin E and C supplementation, except for alpha-tocopherol. Alpha-tocopherol as a phytoestrogen taking part in the modulation of estrogen receptors (ERs) involved in UF etiology, should be particularly avoided in therapy. A diet enriched in fruits and vegetables, as sources of carotenoids, polyphenols, quercetin, and indole-3-carbinol, constitutes an easily modifiable lifestyle element with beneficial results in patients with UFs. Other natural substances, such as curcumin, can reduce the oxidative stress and protect against inflammation in leiomyoma. Although the exact effect of probiotics on uterine fibroids has not yet been thoroughly evaluated at this point, the protective role of dairy products, i.e., yogurt consumption, has been indicated. Trace elements such as selenium can also contribute to antioxidative and anti-inflammatory properties of a recommended diet. In contrast, heavy metals, endocrine disrupting chemicals, cigarette smoking, and a diet low in antioxidants and fiber were, alongside genetic predispositions, associated with UFs formation.

Keywords: uterine fibroids; diet; green tea; curcumin; vitamin A, C, D, E; selenium; trace elements

Citation: Szydłowska, I.; Nawrocka-Rutkowska, J.; Brodowska, A.; Marciniak, A.; Starczewski, A.; Szczuko, M. Dietary Natural Compounds and Vitamins as Potential Cofactors in Uterine Fibroids Growth and Development. *Nutrients* **2022**, *14*, 734. https://doi.org/10.3390/nu14040734

Academic Editor: Sadia Afrin

Received: 13 January 2022
Accepted: 7 February 2022
Published: 9 February 2022

1. Introduction

Uterine leiomyomas are the most frequent tumors in women at reproductive age. The origin of these benign neoplasms is multifactorial. Genetic, inflammatory, hormonal, and other associated factors play an important role in uterine fibroids (UFs) development. Molecular analysis of this type of tumors points towards the mediator complex subunit 12 (*MED12*) mutations and high mobility group AT-hook 2 (HMGA2). The above genetic variants have different gene expression profiles [1–3]. The diversity of metabolic processes and reduced levels of specific vitamins and other co-factor metabolites enable tumor growth and formation. This process is conditioned by alterations in enzyme function and signaling pathways [4]. *MED12* mutations are associated with the induction of gene expression of wingless-type mouse mammary tumor virus integration site family, member 4 (Wnt4), and activation of β-catenin signaling in UFs [5]. The activation of mammalian target of rapamycin (mTOR) pathway can also be responsible for the growth of fibroids [6]. Oxidative stress has also been indicated as a potential factor [7]. Profibrotic tumors can form a result of the imbalance between connective tissue production and degradation. Oxidative stress is involved in the deposition of extracellular matrix (ECM) components in myoma, such as collagen, proteoglycan, and fibronectin [8]. Local and general inflammatory status may also present as a possible mechanism underlying the onset of myomas [9–11]. There is a growing body of knowledge on the management and conservative treatment of UFs. Dietary components can alter various factors associated with uterine myoma formation and have both promoting and inhibiting effects on tumor formation and growth [12–21].

The most popular, that is hormonal and surgical treatment of myoma, can have many potential side effects and complications. Thus, the trend of finding less invasive and more natural methods of managing leiomyoma appears to be very important in the era of postponement of patients' procreation plans and the growing problem of infertility. In the era of increased interest in organics, environmentalism, and ever changing patient expectations, natural substances represent a promising strategy of action in the prevention, treatment, and management of fibroids and, as such, may provide complementary therapy for this common pathology.

1.1. Nutrients and Their Deficiencies

A recent study by Makwe et al. revealed lower serum levels of vitamin C, vitamin D, and calcium in black women with uterine fibroids [22]. The hypothesis was that these vitamins and minerals played a role in etiopathogenesis, progression, and growth of UFs. In a study by Orta et al., the effect of frequent dairy consumption was highlighted as a factor that might influence the incidence of uterine leiomyoma [13]. Despite no straightforward correlation between dairy consumption and fibroid formation, the authors noticed that a greater intake of yogurt and calcium-rich dairy could inversely reduce the risk of fibroids occurrence. Similar observations were made by Shen et al. and Zhou et al. who claim that dairy consumption may indeed have preventative effects when it comes to UFs formation [15,17]. Wise et al. observed an inverse association between dairy consumption and uterine leiomyomata development in African American women [20,21]. Since the dairy consumption is significantly lower among Black Americans than White Americans, they attributed dietary habits as a possible factor contributing to the frequency of the disease prevalence by race [21]. This may be related to the microflora of dairy products and sources of calcium in food alike. Dairy can be considered not only as a food source supporting the development of normal bacterial flora (probiotics, prebiotics, intestinal passage support), but also as the supplier of compounds and antioxidants.

Fruit and Vegetables are rich in dietary fibers, phytochemicals, vitamins and minerals, and antioxidants. These phytochemicals have well-known anti-inflammatory, antiproliferation, antifibrotic, and anti-vascular properties [23], thus in light of current knowledge, a diet based on a large amount of plant and dairy products seems to be a favorable recommendation. The focus of our study was on natural compounds that protect against the risk of UFs and can be useful in treatment of these tumors. After an extensive review of the available Literature on the effects of dietary components on the occurrence of UFs (past two decades), we hypothesized that foods with proven antioxidant, anti-inflammatory, antiproliferative, and antifibrotic properties might be considered a potential treatment for patients with uterine myomas. All the studies discussed in the review are presented in Supplementary Table S1.

1.2. The Influence of Intestinal Dysbiosis

Dysbiosis, or dysregulation of the gut microbiota, may play a role in the pathogenesis and perpetuation of inflammatory processes [24]. Dairy products, as a source of vitamins and minerals, may potentially reduce the inflammation and tumor growth and thus have a beneficial effect on leiomyoma occurrence and growth. Some of these products, produced by bacterial fermentation of milk, i.e., yogurts, may change intestinal microbiota in consumers. Short chain fatty acids (SCFA) are produced in the gut as bacterial metabolites [25]. Butyric acid (BA), one of SCFAs, can induce differentiation and apoptosis and inhibit proliferation and angiogenesis [26]. Dysbiosis may stimulate the pro-inflammatory cytokines or growth factors. Growth factors and cytokines interact through the estrogen and progesterone action, which plays an important role in uterine leiomyoma growth [27]. A beneficial gut microbial environment can potentially affect the uterus environment and reduce the risk of uterine leiomyoma formation [28]. The gut microbiota covers the entire population of microorganisms: bacteria, fungi, viruses, and protozoa that exist in symbiosis with the human gut. Probiotics are living microorganisms that have a positive effect on gut microbiota. Changes in gut microbiota associated with probiotics intake have been shown to have positive effects in a number of diseases, mainly associated with chronic inflammation and oxidative stress [29–34], but to date, there have not been many studies on probiotic impact on uterine leiomyoma—only an indirect role of intestinal microflora changes resulting from yoghurt consumption and their effect on myoma formation has been observed [13,17,20,21].

2. Vitamins

Particular attention should be paid to carotenoids and Vitamin A derivatives. Carotenoids sourced directly from diet actively decrease reactive oxygen species and thus the oxidative stress response in tissues. Carotenoids are a major source of vitamin A in a diet. Lycopene, for instance, displays antioxidant properties and provitamin A activity, and is contained in many yellow, orange, and red fruits and vegetables [12,14,15,35,36]. In other words, carotenoids may play a role in diminishing the number and size of leiomyomas by its antioxidant effect. They cause suppression of cell proliferation and induce cells differentiation and apoptosis [16]. Vitamins C and E, as antioxidants, protect cell membranes and the DNA from oxidative stress, and vitamin A is essential for cell differentiation and proliferation control and may help reduce fibroid growth [18].

Retinoids, as derivatives of Vitamin A, have structural or functional similarity to vitamin A. Retinoids can be natural or synthetic. The proven effects of retinoids include reduction in inflammation, regulation of cell growth and proliferation, and inhibition of carcinogenesis. Studies confirm that retinoids inhibit the growth of primary cultures of human uterine myomas [37–41]. Tomatoes and tomato-based products are particularly high in lycopene, folate, vitamin C, vitamin A, and flavonoids. These bioactive compounds, as potent antioxidants, can be used with therapeutic effects in leiomyoma treatment. Changes in diet may have more beneficial effects than selective supplementation.

2.1. *Carotenoids, such as Lycopene*

Carotenoids as antioxidants can potentially reduce the risk of uterine leiomyoma. He et al. argued that a consumption of fruit and vegetable products rich in fibers and lycopene, significantly decreased the risk of fibroids in premenopausal women, while in postmenopausal women that correlation was insignificant [14]. This was confirmed in Shen Y. et al. 2016 study [15]. They revealed that a vegetarian diet, with greater intake of fresh fruits (i.e., apples) and vegetables (i.e., cruciferous vegetables, especially cabbage, Chinese cabbage, broccoli, and tomatoes rich in lycopene) was likely to significantly reduce the incidence of UFs. Similarly, Sahin et al. observed in their animal model study that high doses of lycopene supplementation in a form of tomato powder could prevent development and/or cause shrinkage of fibroids [35,36]. The doses of lycopene used by Sahin et al. were however much higher than in a typical human diet. Zhou et al. noticed that the risk of UFs formation could be significantly decreased with increased nut and vegetable consumption (especially legumes, seaweed, and carrots). Their study found no significant correlation with fruit intake (except for kiwi), which was inversely associated with the risk of UFs [17]. In contrast, Wise et al. showed that citrus fruit consumption was inversely associated with fibroid risk among Black American women [16]. Martin et al. suggested a positive but not statistically significant association between b-carotene and UFs [18]. Some studies do not support the hypothesis of an exclusively beneficial role of carotenoids in the risk of myoma. No association was found between UF risk and dietary intake of carotenoids (e.g., α- and β-carotene, lycopene: tomato juice, spaghetti, watermelon, salad greens, carrots, spinach, sweet potatoes, greens) in the 2021 study by Wise et al., confirming previous results published in 2011 [12,16]. Terry et al. observed a similar lack of correlation, stressing that cigarette smoking combined with high β-carotene intake exacerbated the risk of fibroid formation [19]. Czeczuga-Semeniuk analyzed the presence of different types of carotenoids (β-carotene, β-cryptoxanthin, lutein, neoxanthin, violaxanthin, and mutatoxanthin) in tissue of a healthy uterus and various uterine tumors, including myomas [42]. Lutein epoxide and mutatoxanthin were predominant in myomas. This points toward the fact that carotenoids with provitamin A activity may induce fibroid growth.

Dietary intake of carotenoids raises many questions with regard to their effects on fibrotic tumors, especially in female patients who smoke cigarettes, and as such requires further study.

2.2. Vitamin A. Retinoids

A 2020 study by Wise et al. found no association of UF incidence with dietary Vitamin A intake (i.e., salad greens, carrots, spinach, sweet potatoes, eggs, cheese, and cereal products) [12]. These results were generated within a particular racial group of women and as such might be worth exploring in the remaining population. Previously, in 2011, Wise et al. observed an inverse correlation between dietary intake of Vitamin A and the risk of UFs. They noted that it was predominantly conditioned by preformed Vitamin A derived from animal sources (i.e., liver and milk), but not by provitamin A from fruit and vegetable sources (i.e., carrots, sweet potatoes/yams, and collard greens) [16]. These findings may complement the molecular study of Heinonen et al. who showed that levels of vitamin A were specifically reduced in leiomyomas of the *MED12* subtype [4], yet appeared to be contradictory to conclusions drawn by Martin et al. who noted positive, statistically significant, and dose-dependent association between vitamin A and uterine fibroid risk, except for the population of Hispanic women [18].

A possible explanation for the positive association between Vitamin A and the risk of UFs is that exposure to high levels of Vitamin A can activate peroxisome proliferator-activated receptors (PPARs). These nuclear receptors, in combination with retinoid X receptors, can activate gene expression, and thus simultaneously increase the risk of UFs in some women [18]. Retinoids are small molecule derivatives of Vitamin A. Studies have shown that the retinoid pathway is significantly altered in fibroids compared to normal myometrium. Retinol requires conversion to retinoic acid (RA), which requires the activity of specific enzymes. In fibroid fibroblasts, aldehyde dehydrogenase (ALDH1) has been specifically identified as one of them. Studies concluded that alterations in the retinoid pathway could lead to abnormal RA production and signaling, which might be important in fibroid development [38,39,43,44]. Zaitseva reports that transcription factor II (known as NR2F2) and CTNNB1(b-catenin) genes are potentially causal factors in the development of UFs [45]. According to her research, the combination of RA and progesterone regulates NR2F2 expression and thus affects fibroid growth. This suggests that retinoids, by causing these molecular alternations, may be useful in the treatment of UFs. In their in vitro studies, Ben-Sasson et al. and Malik et al. demonstrated decreased cell proliferation, ECM formation, RA metabolism, transforming growth factor beta (TGF- β) regulation, and increased apoptosis in human leiomyoma treated with retinoic acid (ATRA-all-trans-retinoic acid) [40,41]. Broaddus et al. showed, again in vitro, that treatment with 4-(N-hydroxyphenyl)-retinamide (4-HPR) or a-difluoromethylornithine (DFMO) resulted in growth inhibition of primary cultures of human uterine leiomyoma [37]. 4-HPR is a synthetic retinoid analog that, in comparison to other retinoids, has a reduced toxicity potential. It promotes apoptosis and induces growth inhibition through induction of p53, p21, and p16, and modulation of extracellular matrix in human uterine leiomyomas. It was noted that the mechanisms of growth inhibition and apoptosis induction could be independent of binding to nuclear retinoid receptors [37].

The above results suggest that a diet rich in Vitamin A and retinoids can prevent fibroids and inhibit tumor growth. Synthetic retinoid analogues can also be effective.

2.3. Vitamin E

Vitamin E is a potent antioxidant that acts by scavenging lipid hydroperoxyl radicals and so can protect cells from the effects of free radicals [46]. Food sources of Vitamin E include canola oil, olive oil, almonds and peanuts, meat, dairy, leafy greens, and fortified cereals. It is also available in oral supplements. Little data is available on Vitamin E and its effects on UFs. Wise et al. found no associated risks with consumption of diet-derived Vitamin E [16]. Martin et al. highlighted a positive, dose-dependent link between vitamin E and the incidence of UFs; the findings were, however, not statistically significant [18]. Ciebiera et al. showed higher serum concentration levels of α-tocopherol (the most common form of vitamin E) in Caucasian women, which may be an important factor in fibroid development [47].

Vitamin E, despite its antioxidant properties, appears not to demonstrate proven beneficial effects in terms of leiomyoma prevention and management.

2.4. Vitamin D

Vitamin D is obtained mainly from sun exposure (skin synthesis), food (oily fish such as trout, salmon, tuna, mackerel, and fish liver oils), and vitamin supplements. Prohormonal forms of vitamin D require hydroxylation in the liver to 25-hydroxyvitamin D (25(OH)D) and in the kidney to their active form, that is 1,25-dihydroxyvitamin D (1,25(OH)$_2$D$_3$). Vitamin D$_3$ exerts its biological functions by interacting with and activating the nuclear vitamin D receptor (VDR). In vitro studies point toward uterine myoma cells exhibiting lower levels of VDR expression [48,49]. In addition, a negative correlation between decreased levels of vitamin D receptor (VDR) and increased levels of estrogen and progesterone receptors (ER-α, PR-A, PR-B) was observed in myoma tumors [50]. It is estimated that Vitamin D deficiency affects 25–50% (possibly more) of patients [51]. Several studies found that insubstantial levels of Vitamin D can contribute to the development of UFs in African American, Caucasian, and Asian women alike [52–60]. Vitamin D$_3$ deficiency activates fibroid cell growth, exacerbates DNA damage, and reduces DNA repairability; it promotes uncontrolled proliferation and fibrosis, and increases chronic inflammation. In combination, these processes are highly tumorogenic [61,62]. Othman et al. demonstrated that, in comparison to normal myometrium, myoma tissue contains significantly lower concentrations of 1,25(OH)$_2$D$_3$. Additionally, an overexpression of 24-hydroxylase was found in myoma, which may further suppress the anti-tumor effect of 1,25(OH)$_2$D$_3$ and exacerbate vitamin D deficiency in the tissue [63]. A recent study by Ciebiera et al. revealed an inverse correlation between lower 25(OH)D serum concentrations and increased serum transforming growth factor β3 (TGF-β3) concentrations in women affected by fibroids [64]. This growth factor can be associated with increased fibrosis and ECM accumulation in myoma [65,66]. Findings on inverse correlation between serum levels of 25(OH)D and fibroid volume vary, ranging from significant to insignificant association [52,67,68]. No correlation was however observed between 25(OH)D serum levels and number of fibroids [67]. Some data suggest that Vitamin D supplementation reduces leiomyoma cell proliferation and thus prevents leiomyoma growth [56,69–71]. A significant downregulation of ER-α, PR-A, PR-B, and steroid receptor coactivators in human myoma cells may be one of the mechanisms—an effect similar to that observed during the course of hormone therapy with GnRH analogues and ulipristal acetate (UPA) [50,72,73]. Halder et al. and Li et al. point to the antifibrotic activity of Vitamin D [74,75]. Vitamin D$_3$ inhibited TGF-β3-induced protein expression and all TGF-β3-mediated effects involved in the fibrotic processes in leiomyoma. Other studies indicate that increasing Vitamin D levels by one unit can reduce the risk of developing UFs by 4–8% [59,64]. Hajhashemi et al. confirmed a significant decrease in leiomyoma size after 10 weeks of vitamin D administration [76]. A slight, statistically insignificant reduction in fibroid volume after a short-term Vitamin D supplementation was observed by Arjeh et al. [77], Davari Tanha et al. [78], and Suneja et al. [79] in patients with hypovitaminosis D (12, 16, and 8 weeks, respectively). Ciavattini et al. reported a similar effect after 12 months of Vitamin D_3 supplementation [67]. Vitamin D_3 supplementation may inhibit the growth of UFs, reduce fibroid-related symptoms, and reduce the need for surgical or medical treatment for progression of fibroids [67,78,79]. Especially, a long-term course of treatment can have antiproliferative, antifibrotic, and proapoptotic effects in leiomyoma, as demonstrated by Corachán et al. [80], a finding consistent with other studies in vitro [61,74,75]. Beside apoptosis induction, Vitamin D suppresses catechol-O-methyltransferase (COMT) expression and activity in myoma cells—an enzyme that plays a vital role in myoma formation [74]. In fact, physiological concentrations of vitamin D can effectively inhibit the growth of myoma cells [61]. $1,25(OH)_2D_3$ can significantly reduce the expression of ECM-associated proteins and structural actin fibers in human leiomyoma cells, as observed by Halder et al. [48]. This effect was a consequence of previous significant induction of nuclear vitamin D receptor (VDR) expression by $1,25(OH)_2D_3$ in a concentration-dependent manner. In another study, Halder et al. observed a significant reduction in MMP-2 and MMP-9 mRNA levels, as well as a reduction in MMP-2 and MMP-9 protein levels in uterine fibroid cells in a concentration-dependent manner and concluded that through this mechanism, $1,25(OH)_2D_3$ might limit fibroid growth and ECM deposition [81]. Al-Hendy et al. observed that $1,25(OH)_2D_3$ spontaneously induced its own VDR, while significantly downregulating the expression of sex steroid receptors (ERs and PRs) and receptor coactivators, which affected myoma formation and growth; hence, $1,25(OH)_2D_3$ suppressed estrogen-induced proliferation in leiomyoma cells [50]. Cell proliferation and extracellular matrix production in myoma tumors can be affected by Vitamin D as it can suppress tumor-promoting Wnt4/β-catenin expression and reduce activation of mTOR signaling in human UF cells [82]. As observed by Corachán et al., Vitamin D inhibits the Wnt/β-catenin and TGFβ pathways, reducing proliferation and extracellular matrix formation, in different molecular subtypes of uterine myomas (MED12-mutated and wild-type human tumors) [83]. Ali et al. hypothesized that myoma tumor progression might be inhibited by recovering the damaged DNA repair system [84]. They showed in vitro that vitamin D_3 treatment significantly reduces DNA damage, restores the normal DNA damage response, and is accompanied by induction of VDR in fibroid cells [84]. DNA repair in cells exposed to classic DNA damage inducers in UF pathogenesis (endocrine-disrupting chemicals -EDCs) was achieved by a $1,25(OH)_2D_3$ treatment of myoma cells in animal models [85]. Clinical trials show promising results in patients with UFs and hypovitaminosis D treated with Vitamin D_3 supplementation, revealing a significant decrease in tumor size and numbers [76,86]. At the time of the review, search results pointed to a randomized trial (RCT) being conducted in women at reproductive age affected with uterine myoma, aiming to evaluate whether supplementation with Vitamin D_3 could reduce the risk and inhibit the growth of fibroids [87]. Results of the evaluation of Vitamin D_3 effects in this particular group of women can be of value in everyday gynecological practice.

Paricalcitol, an analog of $1,25(OH)_2D_3$, has less calcemic activity and, therefore, appears to be safer in long-term use than $1,25(OH)_2D_3$. Halder et al. study indicates that treatment with paricalcitol has an inhibitory effect on uterine fibroid cell proliferation [69]. On a murine model, both paricalcitol and $1,25(OH)_2D_3$ significantly reduced fibroid size, but paricalcitol was more potent. Porcaro et al. observed a significant reduction in myoma volume and overall improvement in quality of life in patients treated with a combination of Vitamin D, EGCG, and vitamin B6 [88]. This combined supplementation treatment presents as quite an innovative approach to treating leiomyoma with oral supplementation. Shen et al., on the other hand, arrived at contrary conclusions. In their study, Vitamin D supplementation had no effects on the risk of UFs [15].

A study by Güleç et al. determined the association between Vitamin D receptor polymorphisms and the occurrence of uterine myomas [89]. It shows that among the fokl polymorphisms of the vitamin D receptor, the presence of the CC fok1 genotype may be a risk-reducing factor, and the T allele may increase the risk of uterine myomas. A recent study by Fazeli et al. evaluated CYP24A1 gene expression in uterine myoma tissue [90]. CYP24A1 is a mitochondrial enzyme that catalyzes the degradation of $1,25(OH)_2D_3$ to its less active 25-D3 form and regulates the amount of active Vitamin D in tissues. The expression of CYP24A1 in leiomyoma suggests that local degradation of $1,25(OH)_2D_3$ may also have a role in fibroma development.

Overall, Vitamin D_3 may be a promising option in prevention and treatment of UFs. The majority of presented studies consider treatment with Vitamin D_3 as safe and effective.

2.5. Vitamin C

Vitamin C (ascorbic acid) is an antioxidative nutrient that prevents against the effects of oxidative stress and has anti-inflammatory properties [22,46]. It cannot however be biosynthesized by the human organism [91]. The best source of this vitamin is a diet rich in fruits (currants, acerola, cherries, and citrus fruits) and vegetables (tomatoes, peppers, cabbage, broccoli, and spinach). Vitamin C supplements are also available in a variety of forms. Pleiotropic effects of vitamin C (antioxidant, anti-inflammatory, immune system support, cofactor in hormone biosynthesis, and microcirculation protector) may suggest its use in treatment of myomas. Sadly, there is not much data regarding this matter.

Martin et al. noted that the risk of UFs increased with higher Vitamin C levels, but the association was not significant [18]. A study by Heinonen et al. demonstrated dysregulation of Vitamin C metabolism in leiomyoma of the *MED12* subtype—a common type of mutation in UFs [4]. Ascorbic acid was also used as bleeding prevention in surgical myoma treatments, but the conclusions were contradictory [92,93]. Pourmatroud et al. observed that Vitamin C administration can reduce blood loss during abdominal myomectomy; Lee et al. did not confirm that in women undergoing laparoscopic myomectomy.

To conclude, there is very little data on Vitamin C supplementation in terms of its effects on leiomyoma, but higher consumption of fruits and vegetables may reduce the risk of myoma incidence.

2.6. Other Vitamins

Apart from two studies, there is virtually no data on the effects of other vitamins on the formation and management of UFs. The literature showed no clear associations between Vitamin B6, Vitamin B12, folate, and the occurrence of UFs [16,18].

3. The Active Compounds from Plants

3.1. Green Tea—Polyphenols

Green tea is widely known for its antioxidant activity and is extensively consumed, especially in Asian countries. Compared to other beverages, it has a much higher catechin content. Components of green tea include polyphenols (epigallocatechin-3-gallate—EGCG, epigallocatechin—EGC, epicatechin-3-gallate—ECG, epicatechin—EC), flavones, and flavanols (kaempferol, myricetin, quercetin). The average daily intake of EGCG from green tea consumption in the EU ranges from 90 to 300 mg/day, while high-level consumers intake even up to 860 mg EGCG/day. In vitro studies showed that EGCG, consumed in the form of a green tea extract, inhibited proliferation and growth and promoted apoptosis in cultures of human uterine leiomyoma cells in a dose-dependent manner [94,95]. Antiproliferative and gene-modulating effects of EGCG were partially mediated through the effect on catechol-O-methyltransferase (COMT) enzyme activity. Similar EGCG action effects were observed in vivo in animal models [96,97]. Ozercan et al. observed that EGCG extract decreased tumor necrosis factor α (TNFα) levels, a cytokine associated with leiomyoma pathophysiology [97]. It appears that EGCG supplementation, by modulating multiple cellular signaling pathways, reduces tumor size and may be an alternative therapeutic option in treatment of UFs. Roshdy et al. evaluated green tea extract (EGCG) taken orally as safe and effective treatment for symptomatic UFs [98]. EGCG intake at a dose of 800 mg/day resulted not only in a significant reduction in tumor volume but also in a reduction of fibroid-specific symptoms, and many treated patients experienced improved health-related quality of life. However, the European Food Safety Authority (EFSA) notes the potential adverse hepatotoxic effects of green tea catechins at intakes ≥ 800 mg EGCG/day taken as a dietary supplement [99]. Grandi et al. observed that EGCG at a daily dose of 300 mg, when vitamin B6 and vitamin D were added, significantly reduced the volume of intramural (mainly) and subserosal UFs [100]. The 90-day treatment resulted in a significant reduction in the length of menstrual bleeding, but not significant changes in health-related quality of life or an improved comfort of sex-life. Grandi et al. suggest EGCG supplementation as an alternative method of treatment for women in late reproductive age, when hormone therapy is not optional. Similar results were observed by Porcaro et al. for a dose of 150 mg EGCG with 25 μg vitamin D and 5 mg vitamin B6 intake in women at reproductive age with symptomatic myomas [88]. Young women at childbearing age can also benefit from EGCG supplementation. Miriello et al. noted that combined daily supplementation of EGCG (300 mg), vitamin D (50 μg), and vitamin B6 (10 mg) can, with no side effects, reduce myoma volume and related symptoms and improve patients' quality of life [101].

EGCG under normal, physiological conditions is characterized by low stability, poor bioavailability, and high metabolic changes. Therefore, methods are being sought to improve the stability of EGCG as a drug. Ahmed et al. studied the biological properties of pro-drug EGCG analogs (pro-EGCG analogs) in human leiomyoma cell lines [102]. They found that these drugs, with improved stability, bioavailability, and biological activity, exhibited potent antiproliferative, antiangiogenic, proapoptotic, and antifibrotic activities in UFs. Pro-drugs EGCG analogs share the same molecular targets as natural EGCG in inhibiting enzymatic activity and could potentially be more effective than natural EGCG therapeutic agent in a long-term use in women with symptomatic UFs. Contrary results were reported in an observational study by Biro et al. [103]. They found that consumption of green tea extract (GTE) capsules resulted only in a significant improvement in physical quality of life (QoL) score. No changes were observed in myoma size or myoma-related complaints or in global QoL score after GTE supplementation. Shen et al. arrived at the same conclusion: that drinking green tea had no effect on the risk of leiomyoma [15].

To sum up, the effects of UFs treatment with polyphenols can depend on dose, duration of treatment, and patient selection.

3.2. *Curcumin/Turmeric*

Curcumin is one of the three major curcuminoids in turmeric plant (*Curcuma longa*). Numerous studies have highlighted its antioxidant, anti-inflammatory, anti-carcinogenic and immunoregulatory activity at the molecular level [104–106]. By suppressing anti-apoptotic proteins, curcumin can protect against the formation and growth of tumors, including uterine myomas.

Malik et al. demonstrated in vitro that curcumin inhibited the proliferation of uterine leiomyoma cells [107]. The curcumin compound caused upregulation of the apoptotic pathway and inhibited fibronectin production, a component of ECM [107]. Tsuiji et al. noted the effect of curcumin on peroxisome proliferator-activated receptor-gamma (PPARγ) activation [108]. Feng et al. demonstrated in an animal model that Rhizoma Curcumae (RC) and Rhizoma Sparganii (RS), used in traditional Chinese medicine, were effective in preventing and treating UFs in rats [109]. The combination of RC and RS effectively reduced the expression of extracellular matrix component collagen, fibroblast activating protein, and transforming growth factor beta (TGF-β), simultaneously decreasing the expression level of signaling factors (AKT, ERK and MEK) in cell proliferation [109]. Similar effects were observed by Yu et al. [110]. They noted that RC/RS herbs regulated key pathways in UF cell proliferation and ECM formation, such as MAPK, PPAR, Notch, and TGF-β/Smad. Curcumin can thus reduce the oxidative stress and protect against inflammation. This activity is expressed through modulation of proinflammatory cytokines and signaling pathways, including beforementioned peroxisome proliferator-activated receptor gamma (PPAR-γ) [105].

In conclusion, turmeric appears to be a desirable dietary component for women at risk of developing uterine myomas and those already affected by this disease.

3.3. Quercetin and Indole-3-Carbinol

Indole-3-carbinol (I3C) and quercetin, as plant- derived components, were also of interest in a potential treatment of UFs. The former one can be sourced mainly from cruciferous vegetables, the latter is an active compound of onion. A recent in-vitro study revealed anti-fibrotic and anti-migratory effects of quercetin and I3C for uterine leiomyomas, but with no impact on myoma cell proliferation [111].

4. Micro- and Macro Elements

4.1. Selenium

Selenium (Se) plays an integral part of ROS detoxifying selenoenzymes such as glutathione peroxidases and thioredoxin reductases, and that is why its antioxidant function can potentially have an effect on uterine leiomyomas [112]. The expression of selenium-binding protein 1 was found to be decreased in uterine leiomyomas. The role this protein plays in tumorigenesis was highlighted, and selenium intake was indicated for the prevention and treatment of uterine leiomyomas. Tuzcu et al. showed in animal models that dietary selenium supplementation reduces the size of spontaneously occurring leiomyoma [113]. The selenium-rich diet had no effect on the number of tumors.

4.2. Other Trace Elements

There appears to be only very few studies concerning the effects of trace elements on uterine myomas. In Nasiadek et al.'s study, cadmium (Cd) concentration was significantly lower in myoma than in the surrounding muscle. A significant increase in magnesium (Mg) and magnesium/calcium (Mg/Ca) ratio, with simultaneous decrease in iron (Fe) were also found in tumor tissue [114]. This may be associated with dairy consumption. Makwe et al. found that in African women at reproductive age suffering from leiomyoma, serum levels of magnesium (Mg) and phosphorus (P) were not significantly different when compared with healthy women at reproductive age [22]. Johnstone et al. showed that higher blood concentrations of cadmium (Cd) and lead (Pb) combined with higher urinary levels of cobalt (Co) could be positively associated with the higher risk of myoma incidence. They argued that increased exposure to these trace elements may stimulate fibroid growth and that fibroid tumors may act as a tissue reservoir for these elements [115].

In summary, of the dietary trace elements, only selenium has proven efficacy in preventing and treating uterine myomas. Heavy metals that can be transferred to food (i.e., from food containers) increase the risk of myoma and stimulate fibroid growth.

5. Endocrine-Disrupting Chemicals

It may be worth noting that some environmental contaminants can interfere with the beneficial effects of certain foodstuffs. These contaminants are known as endocrine-disrupting chemicals (EDC). By binding to hormone receptors, EDCs can stimulate these receptors and alter the function or production of natural hormones. Induction of both genomic and non-genomic signaling and pro-inflammatory effects of EDCs increase the risk of UFs [116]. Butylated hydroxytoluene (BHT), one of the most widely used antioxidant additives, can improve the stability of fat-soluble vitamins and prevent food spoilage. Results of recent in vitro studies suggest detrimental effects of BHT exposure on uterine myoma progression: increased cell proliferation, colony formation, and ECM accumulation [117]. Polychlorinated biphenyls (PCBs), a class of environmental pollutants found in fruits and vegetables, are also known to be endocrine disruptors. The potential risks of PCBs are associated with the consequences of PCBs binding to estrogen receptors. In their in vitro study, Wang et al. showed that Bisphenol A (BPA) in particular could increase cell proliferation and colony formation, and thus promote tumor growth [118]. Other studies, however, found no correlation between exposure to PCBs and UFs [119,120].

Despite the beneficial effects of food on myoma, the additives contained in food may nullify its positive impact and promote the formation and growth of fibroids.

6. Conclusions

Fibroids are fibrotic tumors with extracellular matrix deposition. Oxidative stress affects the formation and growth of these tumors [8]. The quality of diet can impact the formation and progression of fibroids and, as such, offer an innovative approach to treating these tumors. The micronutrient-uterine fibroids correlation may potentially be modified by ethnicity. Positive and negative nutritional effects on myoma are presented below in Figure 1.

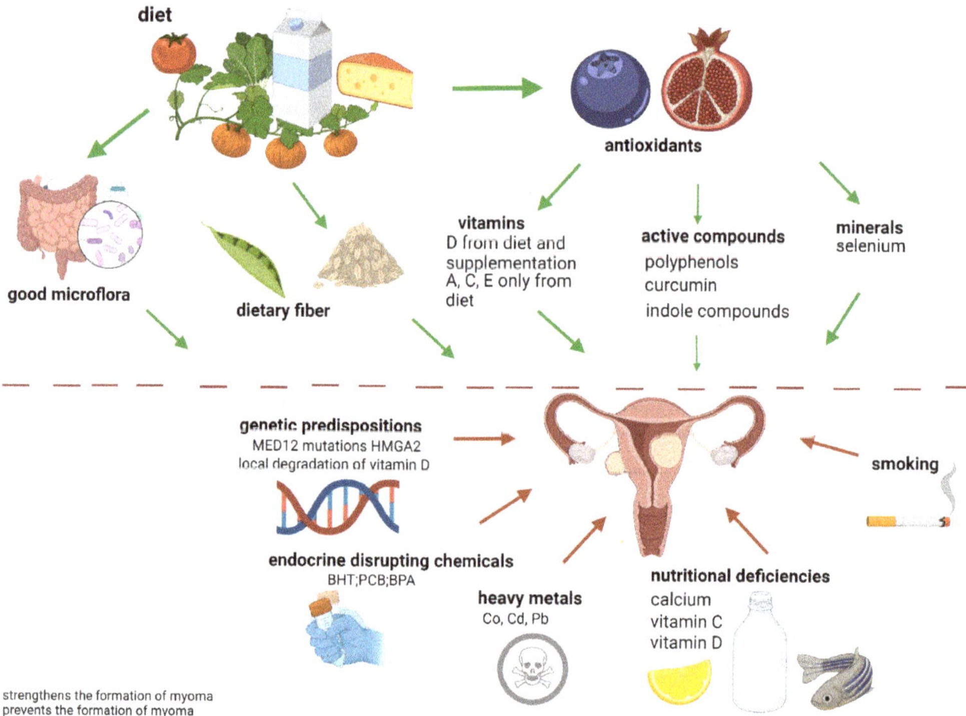

Figure 1. Effects of nutrients and environmental factors on leiomyoma. Legend: MED 12 mutations—mediator complex subunit 12; HMGA2—high mobility group AT-hook 2.; BHT—butylated hydroxytoluene; PCB—polychlorinated biphenyls; BPA—bisphenol A; Co—cobalt; Cd—cadmium; Pb—lead. (Created with BioRender.com https://app.biorender.com/, accessed on 30 November 2021).

The link between Vitamin D deficiency and the risk of UFs was strongly indicated. Vitamin D deficiency can stimulate cell proliferation and leiomyoma growth [61]. Physiological vitamin D concentrations effectively inhibit fibroid cell growth, especially in patients with hypovitaminosis D. The active form of Vitamin D has an affinity for binding to Vitamin D receptors (VDR) expressed in uterine myoma tissue. In fact, Vitamin D_3 appears to be an ideal, inexpensive therapeutic agent for non-invasive treatment of UFs, or at least fibroid growth stabilization, without contraindications and side effects of hormone therapy. So far, the combination of ulipristal acetate and vitamin D_3 treatment for UFs has not been explored extensively in the literature [121,122]. The use of Vitamin D_3 analogues can eliminate the degradation of Vitamin D_3 by tissue 24-hydroxylase, while maintaining the beneficial effects of Vitamin D_3 on UFs [69]. Vitamin D_3 can also promote a healthier composition of the intestinal microbiota, which improves the gut barrier function [123].

The literature findings about the role Vitamin A plays in uterine myoma are contradictory. It has been, however, indicated that treatment with retinoids reduces cell proliferation and ECM formation and increases apoptosis in fibroids [37,40,41]. A diet rich in Vitamin A, as well as the use of synthetic retinoid analogs, can prevent the development of fibroids and limit their growth. A diet rich in carotenoids, such as lycopene, can reduce the incidence of UFs and promote decrease in volume [14–17,35,36].

Based on the available studies, no clear conclusions can be drawn regarding vitamin E and C supplementation. Researchers emphasize the role of alpha-tocopherol, as phytoestrogen, in the modulation of estrogen receptors (ERs) [47]. ERs and estrogens are proven to be involved in the etiology of fibroids.

A daily diet enriched in fruits and vegetables, which are known sources of carotenoids, quercetin, and indole-3-carbinol, may be one of the simplest and modifiable lifestyle changes with beneficial effects in patients affected by leiomyoma. EGCG and green tea extract may also have applications in the prevention and treatment of UFs [15,94,96,98,123]. A combination of natural substances and vitamins i.e., Vitamin D and EGCG, has been evaluated as useful and effective in reducing fibroid-related symptoms and improving quality of life [88,100]. It can be applicable in patients with contraindications to hormonal treatment or those who wish to avoid surgical treatment altogether. Curcumin, another natural substance with proven therapeutic effects on UFs, can reduce oxidative stress and protect against inflammation in leiomyoma. Its action is expressed through modulation of pro-inflammatory cytokines and signaling pathways, including peroxisome proliferator-activated receptor-gamma (PPAR-γ) [105,107–110].

Very few studies considered the effects of probiotics on UFs. However, a protective role of dairy products, including yogurt consumption, has been pointed out [13,15,17,20,21].

To date, there have been only few studies evaluating the potential effects of trace elements on UFs [22,112,113,115]. Selenium, in particular, has shown to have protective properties against oxidative damage and inflammation and has been indicated for supplantation in leiomyoma [112,113].

Using natural compounds in treatment of UFs appears to be a worthwhile endeavor. Natural compounds present as an alternative route in UF treatment, especially in patients with contraindications for hormonal therapy. In women treated conventionally, natural compounds can strengthen therapeutic effects.

7. Methods

Search Strategy

This study is based on an analysis of available studies focusing on correlations between uterine fibroid risk and treatment and diet. The aim of the review was to evaluate data on natural, non-hormonal, effective, and safe therapeutic options for treatment of this disease. An extensive search of PubMed and Medline resources was conducted, aiming at literature published between 2000 and 2021. A combination of the following phrases was used: "uterine fibroids"; "uterine myoma"; "leiomyomata"; "diet", "fruits"; "vegetables"; "plants"; "dairy products"; "curcumin"; "turmeric"; "green tea"; "selenium"; "carotenoids"; "vitamin D"; "Vitamin C"; "Vitamin E"; "Vitamin A"; "dysbiosis"; and "gut microbiota". All retrieved articles (n = 272), collected through the e-search process, were then reviewed by two researchers. Publications were limited to the English language. Literature unrelated to the review theme or replicated in database and conference abstracts were excluded from the analysis. Only full-text studies fulfilling the relevant, above-outlined criteria were included in the review. Approximately 120 publications were evaluated.

Supplementary Materials: The following supporting information can be downloaded at: https://www.mdpi.com/article/10.3390/nu14040734/s1, Table S1: Influence the dietary compounds on the uterine fibroids.

Author Contributions: Conceptualization, I.S.; Data curation: I.S., J.N-R., A.B. and A.M.; Funding acquisition, A.S., I.S. and J.N.-R.; Methodology I.S., J N-R., A.S. and M.S.; Visualization I.S. and M.S.; Writing—original draft preparation: I.S. and M.S.; Writing—review and editing: I.S., J.N.-R., M.S. and A.S.; Project administration: I.S.; Supervision: I.S., A.S. and M.S. All authors have read and agreed to the published version of the manuscript.

Funding: This research received no external funding.

Institutional Review Board Statement: Does not apply to this study.

Informed Consent Statement: Not applicable.

Data Availability Statement: The study did not report any data.

Conflicts of Interest: The authors declare no conflict of interest.

References

1. Galindo, L.J.; Hernández-Beeftink, T.; Salas, A.; Jung, Y.; Reyes, R.; de Oca, F.M.; Hernández, M.; Almeida, T.A. HMGA2 and MED12 alterations frequently co-occur in uterine leiomyomas. *Gynecol. Oncol.* **2018**, *150*, 562–568. [CrossRef] [PubMed]
2. Machado-Lopez, A.; Simón, C.; Mas, A. Molecular and Cellular Insights into the Development of Uterine Fibroids. *Int. J. Mol. Sci.* **2021**, *22*, 8483. [CrossRef] [PubMed]
3. Bertsch, E.; Qiang, W.; Zhang, Q.; Espona-Fiedler, M.; Druschitz, S.A.; Liu, Y.; Mittal, K.; Kong, B.; Kurita, T.; Wei, J.-J. MED12 and HMGA2 mutations: Two independent genetic events in uterine leiomyoma and leiomyosarcoma. *Mod. Pathol.* **2014**, *27*, 1144–1153. [CrossRef] [PubMed]
4. Heinonen, H.-R.; Mehine, M.; Mäkinen, N.; Pasanen, A.; Pitkänen, E.; Karhu, A.; Sarvilinna, N.S.; Sjöberg, J.; Heikinheimo, O.; Bützow, R.; et al. Global metabolomic profiling of uterine leiomyomas. *Br. J. Cancer* **2017**, *117*, 1855–1864. [CrossRef] [PubMed]
5. Markowski, D.N.; Bartnitzke, S.; Löning, T.; Drieschner, N.; Helmke, B.M.; Bullerdiek, J. MED12 mutations in uterine fibroids-their relationship to cytogenetic subgroups. *Int. J. Cancer* **2012**, *131*, 1528–1536. [CrossRef] [PubMed]
6. Crabtree, J.S.; Jelinsky, S.A.; Harris, H.A.; Choe, S.E.; Cotreau, M.M.; Kimberland, M.L.; Wilson, E.; Saraf, K.A.; Liu, W.; McCampbell, A.S.; et al. Comparison of Human and Rat Uterine Leiomyomata: Identification of a Dysregulated Mammalian Target of Rapamycin Pathway. *Cancer Res.* **2009**, *69*, 6171–6178. [CrossRef]
7. Fletcher, N.M.; Abusamaan, M.S.; Memaj, I.; Saed, M.G.; Al-Hendy, A.; Diamond, M.P.; Saed, G.M. Oxidative stress: A key regulator of leiomyoma cell survival. *Fertil. Steril.* **2017**, *107*, 1387–1394.e1. [CrossRef]
8. Fletcher, N.M.; Saed, M.G.; Abu-Soud, H.M.; Al-Hendy, A.; Diamond, M.P.; Saed, G.M. Uterine fibroids are characterized by an impaired antioxidant cellular system: Potential role of hypoxia in the pathophysiology of uterine fibroids. *J. Assist. Reprod. Genet.* **2013**, *30*, 969–974. [CrossRef]
9. Ciavattini, A.; Di Giuseppe, J.; Stortoni, P.; Montik, N.; Giannubilo, S.R.; Litta, P.; Islam, S.; Tranquilli, A.L.; Reis, F.M.; Ciarmela, P. Uterine Fibroids: Pathogenesis and Interactions with Endometrium and Endomyometrial Junction. *Obstet. Gynecol. Int.* **2013**, *2013*, 173184. [CrossRef]
10. Orciani, M.; Caffarini, M.; Biagini, A.; Lucarini, G.; Carpini, G.D.; Berretta, A.; Di Primio, R.; Ciavattini, A. Chronic Inflammation May Enhance Leiomyoma Development by the Involvement of Progenitor Cells. *Stem Cells Int.* **2018**, *2018*, 1716246. [CrossRef]
11. Protic, O.; Toti, P.; Islam, S.; Occhini, R.; Giannubilo, S.R.; Catherino, W.H.; Cinti, S.; Petraglia, F.; Ciavattini, A.; Castellucci, M.; et al. Possible involvement of inflammatory/reparative processes in the development of uterine fibroids. *Cell Tissue Res.* **2016**, *364*, 415–427. [CrossRef] [PubMed]
12. Wise, L.A.; Wesselink, A.K.; Bethea, T.N.; Brasky, T.M.; Wegienka, G.; Harmon, Q.; Block, T.; Baird, D.D. Intake of Lycopene and other Carotenoids and Incidence of Uterine Leiomyomata: A Prospective Ultrasound Study. *J. Acad. Nutr. Diet.* **2021**, *121*, 92–104. [CrossRef] [PubMed]
13. Orta, O.R.; Terry, K.L.; Missmer, S.A.; Harris, H.R. Dairy and related nutrient intake and risk of uterine leiomyoma: A prospective cohort study. *Hum. Reprod.* **2020**, *35*, 453–463. [CrossRef] [PubMed]
14. He, Y.; Zeng, Q.; Dong, S.; Qin, L.; Li, G.; Wang, P. Associations between uterine fibroids and lifestyles including diet, physical activity and stress: A case-control study in China. *Asia Pac. J. Clin. Nutr.* **2013**, *22*, 109–117. [CrossRef] [PubMed]
15. Shen, Y.; Wu, Y.; Lu, Q.; Ren, M. Vegetarian diet and reduced uterine fibroids risk: A case-control study in Nanjing, China. *J. Obstet. Gynaecol. Res.* **2016**, *42*, 87–94. [CrossRef]
16. Wise, L.A.; Radin, R.G.; Palmer, J.R.; Kumanyika, S.K.; Boggs, D.A.; Rosenberg, L. Intake of fruit, vegetables, and carotenoids in relation to risk of uterine leiomyomata. *Am. J. Clin. Nutr.* **2011**, *94*, 1620–1631. [CrossRef] [PubMed]
17. Zhou, M.; Zhai, Y.; Wang, C.; Liu, T.; Tian, S. Association of dietary diversity with uterine fibroids among urban premenopausal women in Shijiazhuang, China: A cross-sectional study. *Asia Pac. J. Clin. Nutr.* **2020**, *29*, 771–781. [CrossRef]

18. Martin, C.L.; Huber, L.R.B.; Thompson, M.E.; Racine, E.F. Serum Micronutrient Concentrations and Risk of Uterine Fibroids. *J. Women's Health* **2011**, *20*, 915–922. [CrossRef]
19. Terry, K.L.; Missmer, S.A.; Hankinson, S.E.; Willett, W.C.; De Vivo, I. Lycopene and other carotenoid intake in relation to risk of uterine leiomyomata. *Am. J. Obstet. Gynecol.* **2008**, *198*, 37.e1–37.e8. [CrossRef]
20. Wise, L.A.; Palmer, J.R.; Ruiz-Narvaez, E.; Reich, D.E.; Rosenberg, L. Is the Observed Association Between Dairy Intake and Fibroids in African Americans Explained by Genetic Ancestry? *Am. J. Epidemiology* **2013**, *178*, 1114–1119. [CrossRef]
21. Wise, L.A.; Radin, R.; Palmer, J.; Kumanyika, S.K.; Rosenberg, L. A Prospective Study of Dairy Intake and Risk of Uterine Leiomyomata. *Am. J. Epidemiology* **2010**, *171*, 221–232. [CrossRef]
22. Makwe, C.C.; Soibi-Harry, A.P.; Rimi, G.S.; Ugwu, O.A.; Ajayi, A.T.; Adesina, T.A.; Okunade, K.S.; Oluwole, A.A.; Anorlu, R.I. Micronutrient and Trace Element Levels in Serum of Women With Uterine Fibroids in Lagos. *Cureus* **2021**, *13*, e18638. [CrossRef] [PubMed]
23. Islam, S.; Akhtar, M.M.; Segars, J.H.; Castellucci, M.; Ciarmela, P. Molecular targets of dietary phytochemicals for possible prevention and therapy of uterine fibroids: Focus on fibrosis. *Crit. Rev. Food Sci. Nutr.* **2017**, *57*, 3583–3600. [CrossRef] [PubMed]
24. Gomaa, E.Z. Human gut microbiota/microbiome in health and diseases: A review. *Antonie Leeuwenhoek* **2020**, *113*, 2019–2040. [CrossRef] [PubMed]
25. Szczuko, M.; Kikut, J.; Maciejewska, D.; Kulpa, D.; Celewicz, Z.; Ziętek, M. The Associations of SCFA with Anthropometric Parameters and Carbohydrate Metabolism in Pregnant Women. *Int. J. Mol. Sci.* **2020**, *21*, 9212. [CrossRef]
26. Chen, D.; Jin, D.; Huang, S.; Wu, J.; Xu, M.; Liu, T.; Dong, W.; Liu, X.; Wang, S.; Zhong, W.; et al. Clostridium butyricum, a butyrate-producing probiotic, inhibits intestinal tumor development through modulating Wnt signaling and gut microbiota. *Cancer Lett.* **2020**, *469*, 456–467. [CrossRef]
27. Zannotti, A.; Greco, S.; Pellegrino, P.; Giantomassi, F.; Carpini, G.D.; Goteri, G.; Ciavattini, A.; Ciarmela, P. Macrophages and Immune Responses in Uterine Fibroids. *Cells* **2021**, *10*, 982. [CrossRef]
28. Chen, C.; Song, X.; Chunwei, Z.; Zhong, H.; Dai, J.; Lan, Z.; Li, F.; Yu, X.; Feng, Q.; Wang, Z.; et al. The microbiota continuum along the female reproductive tract and its relation to uterine-related diseases. *Nat. Commun.* **2017**, *8*, 875. [CrossRef]
29. Torun, A.; Hupalowska, A.; Trzonkowski, P.; Kierkus, J.; Pyrzynska, B. Intestinal Microbiota in Common Chronic Inflammatory Disorders Affecting Children. *Front. Immunol.* **2021**, *12*, 642166. [CrossRef]
30. De Luca, F.; Shoenfeld, Y. The microbiome in autoimmune diseases. *Clin. Exp. Immunol.* **2019**, *195*, 74–85. [CrossRef]
31. Wieërs, G.; Belkhir, L.; Enaud, R.; Leclercq, S.; De Foy, J.-M.P.; Dequenne, I.; De Timary, P.; Cani, P.D. How Probiotics Affect the Microbiota. *Front. Cell. Infect. Microbiol.* **2020**, *9*, 454. [CrossRef] [PubMed]
32. Eslami, M.; Bahar, A.; Keikha, M.; Karbalaei, M.; Kobyliak, N.; Yousefi, B. Probiotics function and modulation of the immune system in allergic diseases. *Allergol. Immunopathol.* **2020**, *48*, 771–788. [CrossRef] [PubMed]
33. Ballini, A.; Santacroce, L.; Cantore, S.; Bottalico, L.; Dipalma, G.; Topi, S.; Saini, R.; De Vito, D.; Inchingolo, F. Probiotics Efficacy on Oxidative Stress Values in Inflammatory Bowel Disease: A Randomized Double-Blinded Placebo-Controlled Pilot Study. *Endocrine, Metab. Immune Disord. Drug Targets* **2019**, *19*, 373–381. [CrossRef] [PubMed]
34. Rezazadeh, L.; Alipour, B.; Jafarabadi, M.A.; Behrooz, M.; Gargari, B.P. Daily consumption effects of probiotic yogurt containing Lactobacillus acidophilus La5 and Bifidobacterium lactis Bb12 on oxidative stress in metabolic syndrome patients. *Clin. Nutr. ESPEN* **2021**, *41*, 136–142. [CrossRef] [PubMed]
35. Sahin, K.; Ozercan, R.; Onderci, M.; Sahin, N.; Gursu, M.F.; Khachik, F.; Sarkar, F.H.; Munkarah, A.; Ali-Fehmi, R.; Kmak, D.; et al. Lycopene Supplementation Prevents the Development of Spontaneous Smooth Muscle Tumors of the Oviduct in Japanese Quail. *Nutr. Cancer* **2004**, *50*, 181–189. [CrossRef]
36. Sahin, K.; Ozercan, R.; Onderci, M.; Sahin, N.; Khachik, F.; Seren, S.; Kucuk, O. Dietary Tomato Powder Supplementation in the Prevention of Leiomyoma of the Oviduct in the Japanese Quail. *Nutr. Cancer* **2007**, *59*, 70–75. [CrossRef] [PubMed]
37. Broaddus, R.R.; Xie, S.; Hsu, C.-J.; Wang, J.; Zhang, S.; Zou, C. The chemopreventive agents 4-HPR and DFMO inhibit growth and induce apoptosis in uterine leiomyomas. *Am. J. Obstet. Gynecol.* **2004**, *190*, 686–692. [CrossRef] [PubMed]
38. Zaitseva, M.; Vollenhoven, B.J.; Rogers, P.A. Retinoids regulate genes involved in retinoic acid synthesis and transport in human myometrial and fibroid smooth muscle cells. *Hum. Reprod.* **2008**, *23*, 1076–1086. [CrossRef]
39. Lattuada, D.; Vigano, P.; Mangioni, S.; Sassone, J.; Di Francesco, S.; Vignali, M.; Di Blasio, A.M. Accumulation of Retinoid X Receptor-α in Uterine Leiomyomas Is Associated with a Delayed Ligand-Dependent Proteasome-Mediated Degradation and an Alteration of Its Transcriptional Activity. *Mol. Endocrinol.* **2007**, *21*, 602–612. [CrossRef]
40. Ben-Sasson, H.; Ben-Meir, A.; Shushan, A.; Karra, L.; Rojansky, N.; Klein, B.Y.; Levitzki, R.; Ben-Bassat, H. All-trans-retinoic acid mediates changes in PI3K and retinoic acid signaling proteins of leiomyomas. *Fertil. Steril.* **2011**, *95*, 2080–2086. [CrossRef]
41. Malik, M.; Webb, J.; Catherino, W.H. Retinoic acid treatment of human leiomyoma cells transformed the cell phenotype to one strongly resembling myometrial cells. *Clin. Endocrinol.* **2008**, *69*, 462–470. [CrossRef] [PubMed]
42. Czeczuga-Semeniuk, E.; Wołczyński, S. Dietary carotenoids in normal and pathological tissues of corpus uteri. *Folia Histochem. Cytobiol.* **2008**, *46*, 283–290. [CrossRef] [PubMed]
43. Zaitseva, M.; Vollenhoven, B.J.; Rogers, P.A.W. Retinoic acid pathway genes show significantly altered expression in uterine fibroids when compared with normal myometrium. *Mol. Hum. Reprod.* **2007**, *13*, 577–585. [CrossRef]
44. Catherino, W.H.; Malik, M. Uterine leiomyomas express a molecular pattern that lowers retinoic acid exposure. *Fertil. Steril.* **2007**, *87*, 1388–1398. [CrossRef] [PubMed]

45. Zaitseva, M.; Holdsworth-Carson, S.J.; Waldrip, L.; Nevzorova, J.; Martelotto, L.; Vollenhoven, B.J.; Rogers, P.A.W. Aberrant expression and regulation of NR2F2 and CTNNB1 in uterine fibroids. *Reproduction* **2013**, *146*, 91–102. [CrossRef] [PubMed]
46. Traber, M.G.; Stevens, J.F. Vitamins C and E: Beneficial effects from a mechanistic perspective. *Free Radic. Biol. Med.* **2011**, *51*, 1000–1013. [CrossRef] [PubMed]
47. Ciebiera, M.; Szymańska-Majchrzak, J.; Sentkowska, A.; Kilian, K.; Rogulski, Z.; Nowicka, G.; Jakiel, G.; Tomaszewski, P.; Włodarczyk, M. Alpha-Tocopherol Serum Levels Are Increased in Caucasian Women with Uterine Fibroids: A Pilot Study. *BioMed Res. Int.* **2018**, *2018*, 6793726. [CrossRef]
48. Halder, S.K.; Osteen, K.G.; Al-Hendy, A. 1,25-Dihydroxyvitamin D3 Reduces Extracellular Matrix-Associated Protein Expression in Human Uterine Fibroid Cells. *Biol. Reprod.* **2013**, *89*, 150. [CrossRef]
49. Lima, M.S.O.; da Silva, B.B.; de Medeiros, M.L.; dos Santos, A.R.; Brazil, E.D.D.N.; Filho, W.M.N.E.; Ibiapina, J.O.; Brito, A.G.A.; Costa, P.V.L. Evaluation of vitamin D receptor expression in uterine leiomyoma and nonneoplastic myometrial tissue: A cross-sectional controlled study. *Reprod. Biol. Endocrinol.* **2021**, *19*, 67. [CrossRef]
50. Al-Hendy, A.; Diamond, M.P.; El-Sohemy, A.; Halder, S.K. 1,25-dihydroxyvitamin D3 regulates expression of sex steroid receptors in human uterine fibroid cells. *J. Clin. Endocrinol. Metab.* **2015**, *100*, E572–E582. [CrossRef]
51. Kennel, K.A.; Drake, M.T.; Hurley, D.L. Vitamin D Deficiency in Adults: When to Test and How to Treat. *Mayo Clin. Proc.* **2010**, *85*, 752–758. [CrossRef] [PubMed]
52. Al-Hendy, A.; Halder, S.K.; Allah, A.S.A.; Roshdy, E.; Rajaratnam, V.; Sabry, M. Serum vitamin D3 level inversely correlates with uterine fibroid volume in different ethnic groups: A cross-sectional observational study. *Int. J. Women's Health* **2013**, *5*, 93–100. [CrossRef] [PubMed]
53. Baird, D.D.; Hill, M.C.; Schectman, J.M.; Hollis, B.W. Vitamin D and the Risk of Uterine Fibroids. *Epidemiology* **2013**, *24*, 447–453. [CrossRef] [PubMed]
54. Paffoni, A.; Somigliana, E.; Vigano, P.; Benaglia, L.; Cardellicchio, L.; Pagliardini, L.; Papaleo, E.; Candiani, M.; Fedele, L. Vitamin D Status in Women With Uterine Leiomyomas. *J. Clin. Endocrinol. Metab.* **2013**, *98*, E1374–E1378. [CrossRef] [PubMed]
55. Li, S.; Chen, B.; Sheng, B.; Wang, J.; Zhu, X. The associations between serum vitamin D, calcium and uterine fibroids in Chinese women: A case-controlled study. *J. Int. Med Res.* **2020**, *48*, 300060520923492. [CrossRef]
56. Brakta, S.; Diamond, J.S.; Al-Hendy, A.; Diamond, M.; Halder, S.K. Role of vitamin D in uterine fibroid biology. *Fertil. Steril.* **2015**, *104*, 698–706. [CrossRef]
57. Wise, L.A.; Ruiz-Narváez, E.A.; Haddad, S.A.; Rosenberg, L.; Palmer, J.R. Polymorphisms in vitamin D–related genes and risk of uterine leiomyomata. *Fertil. Steril.* **2014**, *102*, 503–510.e1. [CrossRef]
58. Singh, V.; Barik, A.; Imam, N. Vitamin D3 Level in Women with Uterine Fibroid: An Observational Study in Eastern Indian Population. *J. Obstet. Gynecol. India* **2019**, *69*, 161–165. [CrossRef]
59. Tunau, K.A.; Garba, J.A.; Panti, A.A.; Shehu, C.E.; Adamu, A.N.; AbdulRahman, M.B.; Ahmad, M.K. Low plasma vitamin D as a predictor of uterine fibroids in a nigerian population. *Niger. Postgrad. Med J.* **2021**, *28*, 181–186. [CrossRef]
60. Xu, F.; Li, F.; Li, L.; Lin, D.; Hu, H.; Shi, Q. Vitamin D as a risk factor for the presence of asymptomatic uterine fibroids in premenopausal Han Chinese women. *Fertil. Steril.* **2021**, *115*, 1288–1293. [CrossRef]
61. Bläuer, M.; Rovio, P.H.; Ylikomi, T.; Heinonen, P.K. Vitamin D inhibits myometrial and leiomyoma cell proliferation in vitro. *Fertil. Steril.* **2009**, *91*, 1919–1925. [CrossRef] [PubMed]
62. ElHusseini, H.; Elkafas, H.; Abdelaziz, M.; Halder, S.; Atabiekov, I.; Eziba, N.; Ismail, N.; El Andaloussi, A.; Al-Hendy, A. Diet-induced vitamin D deficiency triggers inflammation and DNA damage profile in murine myometrium. *Int. J. Women's Health* **2018**, *10*, 503–514. [CrossRef] [PubMed]
63. Othman, E.R.; Ahmed, E.; Sayed, A.A.; Hussein, M.; Abdelaal, I.I.; Fetih, A.N.; Abou-Taleb, H.A.; Yousef, A.-E.A. Human uterine leiomyoma contains low levels of 1, 25 dihdroxyvitamin D3, and shows dysregulated expression of vitamin D metabolizing enzymes. *Eur. J. Obstet. Gynecol. Reprod. Biol.* **2018**, *229*, 117–122. [CrossRef] [PubMed]
64. Ciebiera, M.; Włodarczyk, M.; Słabuszewska-Jóźwiak, A.; Nowicka, G.; Jakiel, G. Influence of vitamin D and transforming growth factor β3 serum concentrations, obesity, and family history on the risk for uterine fibroids. *Fertil. Steril.* **2016**, *106*, 1787–1792. [CrossRef]
65. Islam, S.; Ciavattini, A.; Petraglia, F.; Castellucci, M.; Ciarmela, P. Extracellular matrix in uterine leiomyoma pathogenesis: A potential target for future therapeutics. *Hum. Reprod. Updat.* **2018**, *24*, 59–85. [CrossRef]
66. Arici, A.; Sozen, I. Transforming growth factor-β3 is expressed at high levels in leiomyoma where it stimulates fibronectin expression and cell proliferation. *Fertil. Steril.* **2000**, *73*, 1006–1011. [CrossRef]
67. Ciavattini, A.; Carpini, G.D.; Serri, M.; Vignini, A.; Sabbatinelli, J.; Tozzi, A.; Aggiusti, A.; Clemente, N. Hypovitaminosis D and "small burden" uterine fibroids: Opportunity for a vitamin D supplementation. *Medicine* **2016**, *95*, e5698. [CrossRef]
68. Kaplan, Z.A.O.; Taşçi, Y.; Topçu, H.O.; Erkaya, S. 25-Hydroxy vitamin D levels in premenopausal Turkish women with uterine leiomyoma. *Gynecol. Endocrinol.* **2018**, *34*, 261–264. [CrossRef]
69. Halder, S.K.; Sharan, C.; Al-Hendy, O.; Al-Hendy, A. Paricalcitol, a Vitamin D Receptor Activator, Inhibits Tumor Formation in a Murine Model of Uterine Fibroids. *Reprod. Sci.* **2014**, *21*, 1108–1119. [CrossRef]
70. Halder, S.K.; Sharan, C.; Al-Hendy, A. 1,25-Dihydroxyvitamin D3 Treatment Shrinks Uterine Leiomyoma Tumors in the Eker Rat Model1. *Biol. Reprod.* **2012**, *86*, 116. [CrossRef]

71. Corachán, A.; Ferrero, H.; Aguilar, A.; Garcia, N.; Monleon, J.; Faus, A.; Cervelló, I.; Pellicer, A. Inhibition of tumor cell proliferation in human uterine leiomyomas by vitamin D via Wnt/β-catenin pathway. *Fertil. Steril.* **2019**, *111*, 397–407. [CrossRef] [PubMed]
72. Khan, K.N.; Fujishita, A.; Koshiba, A.; Ogawa, K.; Mori, T.; Ogi, H.; Itoh, K.; Teramukai, S.; Kitawaki, J. Expression profiles of E/P receptors and fibrosis in GnRHa-treated and -untreated women with different uterine leiomyomas. *PLoS ONE* **2020**, *15*, e0242246. [CrossRef] [PubMed]
73. Szydłowska, I.; Grabowska, M.; Nawrocka-Rutkowska, J.; Piasecka, M.; Starczewski, A. Markers of Cellular Proliferation, Apoptosis, Estrogen/Progesterone Receptor Expression and Fibrosis in Selective Progesterone Receptor Modulator (Ulipristal Acetate)-Treated Uterine Fibroids. *J. Clin. Med.* **2021**, *10*, 562. [CrossRef]
74. Sharan, C.; Halder, S.K.; Thota, C.; Jaleel, T.; Nair, S.; Al-Hendy, A. Vitamin D inhibits proliferation of human uterine leiomyoma cells via catechol-O-methyltransferase. *Fertil. Steril.* **2011**, *95*, 247–253. [CrossRef] [PubMed]
75. Halder, S.K.; Goodwin, J.S.; Al-Hendy, A. 1,25-Dihydroxyvitamin D3 Reduces TGF-β3-Induced Fibrosis-Related Gene Expression in Human Uterine Leiomyoma Cells. *J. Clin. Endocrinol. Metab.* **2011**, *96*, E754–E762. [CrossRef]
76. Hajhashemi, M.; Ansari, M.; Haghollahi, F.; Eslami, B. The effect of vitamin D supplementation on the size of uterine leiomyoma in women with vitamin D deficiency. *Casp. J. Intern. Med.* **2019**, *10*, 125–131. [CrossRef]
77. Arjeh, S.; Darsareh, F.; Asl, Z.A.; Kutenaei, M.A. Effect of oral consumption of vitamin D on uterine fibroids: A randomized clinical trial. *Complement. Ther. Clin. Pr.* **2020**, *39*, 101159. [CrossRef]
78. Tanha, F.D.; Feizabad, E.; Farahani, M.V.; Amuzegar, H.; Moradi, B.; Sadeh, S.S. The Effect of Vitamin D Deficiency on Overgrowth of Uterine Fibroids: A Blinded Randomized Clinical Tria. *Int. J. Fertil. Steril.* **2021**, *15*, 95–100. [CrossRef]
79. Suneja, A.; Faridi, F.; Bhatt, S.; Guleria, K.; Mehndiratta, M.; Sharma, R. Effect of Vitamin D3 supplementation on symptomatic uterine leiomyoma in women with Hypovitaminosis D. *J. Midlife Health* **2021**, *12*, 53–60. [CrossRef]
80. Corachán, A.; Ferrero, H.; Escrig, J.; Monleon, J.; Faus, A.; Cervelló, I.; Pellicer, A. Long-term vitamin D treatment decreases human uterine leiomyoma size in a xenograft animal model. *Fertil. Steril.* **2020**, *113*, 205–216.e4. [CrossRef]
81. Halder, S.K.; Osteen, K.G.; Al-Hendy, A. Vitamin D3 inhibits expression and activities of matrix metalloproteinase-2 and -9 in human uterine fibroid cells. *Hum. Reprod.* **2013**, *28*, 2407–2416. [CrossRef] [PubMed]
82. Al-Hendy, A.; Diamond, M.P.; Boyer, T.G.; Halder, S.K. Vitamin D3 Inhibits Wnt/β-Catenin and mTOR Signaling Pathways in Human Uterine Fibroid Cells. *J. Clin. Endocrinol. Metab.* **2016**, *101*, 1542–1551. [CrossRef] [PubMed]
83. Corachán, A.; Trejo, M.G.; Carbajo-García, M.C.; Monleón, J.; Escrig, J.; Faus, A.; Pellicer, A.; Cervelló, I.; Ferrero, H. Vitamin D as an effective treatment in human uterine leiomyomas independent of mediator complex subunit 12 mutation. *Fertil. Steril.* **2021**, *115*, 512–521. [CrossRef] [PubMed]
84. Ali, M.; Shahin, S.M.; Sabri, N.A.; Al-Hendy, A.; Yang, Q. Hypovitaminosis D exacerbates the DNA damage load in human uterine fibroids, which is ameliorated by vitamin D3 treatment. *Acta Pharmacol. Sin.* **2019**, *40*, 957–970. [CrossRef]
85. Elkafas, H.; Ali, M.; Elmorsy, E.; Kamel, R.; Thompson, W.E.; Badary, O.; Al-Hendy, A.; Yang, Q.; Yang, Q. Vitamin D3 Ameliorates DNA Damage Caused by Developmental Exposure to Endocrine Disruptors in the Uterine Myometrial Stem Cells of Eker Rats. *Cells* **2020**, *9*, 1459. [CrossRef]
86. Xess, S.; Sahu, J. A study to correlate association between vitamin D with fibroid and its supplementation in the progression of the disease. *Int. J. Reprod. Contracept. Obstet. Gynecol.* **2020**, *9*, 1477–1481. [CrossRef]
87. Sheng, B.; Song, Y.; Liu, Y.; Jiang, C.; Zhu, X. Association between vitamin D and uterine fibroids: A study protocol of an open-label, randomised controlled trial. *BMJ Open* **2020**, *10*, e038709. [CrossRef]
88. Porcaro, G.; Santamaria, A.; Giordano, D.; Angelozzi, P. Vitamin D plus epigallocatechin gallate: A novel promising approach for uterine myomas. *Eur. Rev. Med. Pharmacol. Sci.* **2020**, *24*, 3344–3351.
89. Yılmaz, S.G.; Gul, T.; Attar, R.; Yıldırım, G.; Isbir, T. Association between fok1 polymorphism of vitamin D receptor gene with uterine leiomyoma in Turkish populations. *J. Turk. Ger. Gynecol. Assoc.* **2018**, *19*, 128–131. [CrossRef]
90. Fazeli, E.; Piltan, S.; Gholami, M.; Akbari, M.; Falahati, Z.; Yassaee, F.; Sadeghi, H.; Mirfakhraie, R. CYP24A1 expression analysis in uterine leiomyoma regarding MED12 mutation profile. *Arch. Gynecol. Obstet.* **2021**, *303*, 787–792. [CrossRef]
91. Man, A.M.S.-D.; Elbers, P.; Straaten, H.M.O.-V. Vitamin C: Should we supplement? *Curr. Opin. Crit. Care* **2018**, *24*, 248–255. [CrossRef]
92. Lee, B.; Kim, K.; Cho, H.Y.; Yang, E.J.; Suh, D.H.; No, J.H.; Lee, J.R.; Hwang, J.W.; Do, S.H.; Kim, Y.B. Effect of intravenous ascorbic acid infusion on blood loss during laparoscopic myomectomy: A randomized, double-blind, placebo-controlled trial. *Eur. J. Obstet. Gynecol. Reprod. Biol.* **2016**, *199*, 187–191. [CrossRef] [PubMed]
93. Pourmatroud, E.; Hormozi, L.; Hemadi, M.; Golshahi, R. Intravenous ascorbic acid (vitamin C) administration in myomectomy: A prospective, randomized, clinical trial. *Arch. Gynecol. Obstet.* **2012**, *285*, 111–115. [CrossRef]
94. Zhang, D.; Al-Hendy, M.; Richard-Davis, G.; Montgomery-Rice, V.; Rajaratnam, V.; Al-Hendy, A. Antiproliferative and proapoptotic effects of epigallocatechin gallate on human leiomyoma cells. *Fertil. Steril.* **2010**, *94*, 1887–1893. [CrossRef] [PubMed]
95. Zhang, N.; Rajaratnam, V.; Al-Hendy, O.; Halder, S.; Al-Hendy, A. Green Tea Extract Inhibition of Human Leiomyoma Cell Proliferation Is Mediated via Catechol- O -Methyltransferase. *Gynecol. Obstet. Investig.* **2014**, *78*, 109–118. [CrossRef]
96. Zhang, D.; Al-Hendy, M.; Richard-Davis, G.; Montgomery-Rice, V.; Sharan, C.; Rajaratnam, V.; Khurana, A.; Al-Hendy, A. Green tea extract inhibits proliferation of uterine leiomyoma cells in vitro and in nude mice. *Am. J. Obstet. Gynecol.* **2010**, *202*, 289.e1–289.e9. [CrossRef]

97. Ozercan, I.H.; Sahin, N.; Akdemir, F.; Onderci, M.; Seren, S.; Sahin, K.; Kucuk, O. Chemoprevention of fibroid tumors by [−]-epigallocatechin-3-gallate in quail. *Nutr. Res.* **2008**, *28*, 92–97. [CrossRef]
98. Al-Hendy, A.; Roshdy, E.; Rajaratnam, V.; Maitra, S.; Sabry, M.; Allah, A.S.A. Treatment of symptomatic uterine fibroids with green tea extract: A pilot randomized controlled clinical study. *Int. J. Women's Health* **2013**, *5*, 477–486. [CrossRef]
99. EFSA Panel on Food Additives and Nutrient Sources added to Food (ANS); Younes, M.; Aggett, P.; Aguilar, F.; Crebelli, R.; Dusemund, B.; Filipič, M.; Frutos, M.J.; Galtier, P.; Gott, D.; et al. Scientific opinion on the safety of green tea catechins. *EFSA J.* **2018**, *16*, e05239. [CrossRef]
100. Grandi, G.; Del Savio, M.C.; Melotti, C.; Feliciello, L.; Facchinetti, F. Vitamin D and green tea extracts for the treatment of uterine fibroids in late reproductive life: A pilot, prospective, daily-diary based study. *Gynecol. Endocrinol.* **2021**, *38*, 63–67. [CrossRef]
101. Miriello, D.; Galanti, F.; Cignini, P.; Antonaci, D.; Schiavi, M.C.; Rago, R. Uterine fibroids treatment: Do we have new valid alternative? Experiencing the combination of vitamin D plus epigallocatechin gallate in childbearing age affected women. *Eur. Rev. Med. Pharmacol. Sci.* **2021**, *25*, 2843–2851. [CrossRef]
102. Ahmed, R.S.I.; Liu, G.; Renzetti, A.; Farshi, P.; Yang, H.; Soave, C.; Saed, G.; El-Ghoneimy, A.A.; El Banna, H.; Foldes, R.; et al. Biological and Mechanistic Characterization of Novel Prodrugs of Green Tea Polyphenol Epigallocatechin Gallate Analogs in Human Leiomyoma Cell Lines. *J. Cell. Biochem.* **2016**, *117*, 2357–2369. [CrossRef]
103. Biro, R.; Richter, R.; Ortiz, M.; Sehouli, J.; David, M. Effects of epigallocatechin gallate-enriched green tea extract capsules in uterine myomas: Results of an observational study. *Arch. Gynecol. Obstet.* **2021**, *303*, 1235–1243. [CrossRef] [PubMed]
104. Hewlings, S.J.; Kalman, D.S. Curcumin: A Review of Its Effects on Human Health. *Foods* **2017**, *6*, 92. [CrossRef] [PubMed]
105. Xu, X.-Y.; Meng, X.; Li, S.; Gan, R.-Y.; Li, Y.; Li, H.-B. Bioactivity, Health Benefits, and Related Molecular Mechanisms of Curcumin: Current Progress, Challenges, and Perspectives. *Nutrients* **2018**, *10*, 1553. [CrossRef] [PubMed]
106. McCubrey, J.A.; Lertpiriyapong, K.; Steelman, L.S.; Abrams, S.L.; Yang, L.V.; Murata, R.M.; Rosalen, P.L.; Scalisi, A.; Neri, L.M.; Cocco, L.; et al. Effects of resveratrol, curcumin, berberine and other nutraceuticals on aging, cancer development, cancer stem cells and microRNAs. *Aging* **2017**, *9*, 1477–1536. [CrossRef]
107. Malik, M.; Mendoza, M.; Payson, M.; Catherino, W.H. Curcumin, a nutritional supplement with antineoplastic activity, enhances leiomyoma cell apoptosis and decreases fibronectin expression. *Fertil. Steril.* **2009**, *91*, 2177–2184. [CrossRef]
108. Tsuiji, K.; Takeda, T.; Li, B.; Wakabayashi, A.; Kondo, A.; Kimura, T.; Yaegashi, N. Inhibitory effect of curcumin on uterine leiomyoma cell proliferation. *Gynecol. Endocrinol.* **2011**, *27*, 512–517. [CrossRef]
109. Feng, Y.; Zhao, Y.; Li, Y.; Peng, T.; Kuang, Y.; Shi, X.; Wang, G.; Peng, F.; Yu, C. Inhibition of Fibroblast Activation in Uterine Leiomyoma by Components of Rhizoma Curcumae and Rhizoma Sparganii. *Front. Public Health* **2021**, *9*, 650022. [CrossRef]
110. Yu, C.H.; Zhao, J.S.; Zhao, H.; Peng, T.; Shen, D.C.; Xu, Q.X.; Li, Y.; Webb, R.C.; Wang, M.H.; Shi, X.M.; et al. Transcriptional profiling of uterine leiomyoma rats treated by a traditional herb pair, Curcumae rhizoma and Sparganii rhizoma. *Braz. J. Med Biol. Res.* **2019**, *52*, e8132. [CrossRef]
111. Greco, S.; Islam, S.; Zannotti, A.; Carpini, G.D.; Giannubilo, S.R.; Ciavattini, A.; Petraglia, F.; Ciarmela, P. Quercetin and indole 3-carbinol inhibit extracellular matrix expression in human primary uterine leiomyoma cells. *Reprod. Biomed. Online* **2020**, *40*, 593–602. [CrossRef] [PubMed]
112. Zhang, P.; Zhang, C.; Wang, X.; Liu, F.; Sung, C.J.; Quddus, M.R.; Lawrence, W.D. The expression of selenium-binding protein 1 is decreased in uterine leiomyoma. *Diagn. Pathol.* **2010**, *5*, 80. [CrossRef] [PubMed]
113. Tuzcu, M.; Sahin, N.; Özercan, I.; Seren, S.; Sahin, K.; Kucuk, O. The Effects of Selenium Supplementation on the Spontaneously Occurring Fibroid Tumors of Oviduct, 8-Hydroxy-2′-Deoxyguanosine Levels, and Heat Shock Protein 70 Response in Japanese Quail. *Nutr. Cancer* **2010**, *62*, 495–500. [CrossRef] [PubMed]
114. Nasiadek, M.; Krawczyk, T.; Sapota, A. Tissue levels of cadmium and trace elements in patients with myoma and uterine cancer. *Hum. Exp. Toxicol.* **2005**, *24*, 623–630. [CrossRef]
115. Johnstone, E.B.; Louis, G.M.B.; Parsons, P.J.; Steuerwald, A.J.; Palmer, C.D.; Chen, Z.; Sun, L.; Hammoud, A.O.; Dorais, J.; Peterson, C.M. Increased urinary cobalt and whole blood concentrations of cadmium and lead in women with uterine leiomyomata: Findings from the ENDO Study. *Reprod. Toxicol.* **2014**, *49*, 27–32. [CrossRef]
116. Elkafas, H.; Badary, O.A.; Elmorsy, E.; Kamel, R.; Yang, Q.; Al-Hendy, A. Endocrine-Disrupting Chemicals and Vitamin D Deficiency in the Pathogenesis of Uterine Fibroids. *J. Adv. Pharm. Res.* **2021**, *5*, 248–263. [CrossRef]
117. Chiang, Y.-F.; Chen, H.-Y.; Ali, M.; Shieh, T.-M.; Huang, Y.-J.; Wang, K.-L.; Chang, H.-Y.; Huang, T.-C.; Hong, Y.-H.; Hsia, S.-M. The Role of Cell Proliferation and Extracellular Matrix Accumulation Induced by Food Additive Butylated Hydroxytoluene in Uterine Leiomyoma. *Nutrients* **2021**, *13*, 3074. [CrossRef]
118. Wang, K.-H.; Kao, A.-P.; Chang, C.-C.; Lin, T.-C.; Kuo, T.-C. Bisphenol A at environmentally relevant doses induces cyclooxygenase-2 expression and promotes invasion of human mesenchymal stem cells derived from uterine myoma tissue. *Taiwan. J. Obstet. Gynecol.* **2013**, *52*, 246–252. [CrossRef]
119. Neblett, M.F., 2nd; Curtis, S.W.; Gerkowicz, S.A.; Spencer, J.B.; Terrell, M.L.; Jiang, V.S.; Marder, M.E.; Barr, D.B.; Marcus, M.; Smith, A.K. Examining Reproductive Health Outcomes in Females Exposed to Polychlorinated Biphenyl and Polybrominated Biphenyl. *Sci. Rep.* **2020**, *10*, 3314–3319. [CrossRef]
120. Wesselink, A.K.; Henn, B.C.; Fruh, V.; Orta, O.R.; Weuve, J.; Hauser, R.; Williams, P.L.; McClean, M.D.; Sjodin, A.; Bethea, T.N.; et al. A Prospective Ultrasound Study of Plasma Polychlorinated Biphenyl Concentrations and Incidence of Uterine Leiomyomata. *Epidemiology* **2021**, *32*, 259–267. [CrossRef]

121. Ali, M.; Shahin, S.M.; Sabri, N.A.; Al-Hendy, A.; Yang, Q. 1,25 Dihydroxyvitamin D3 Enhances the Antifibroid Effects of Ulipristal Acetate in Human Uterine Fibroids. *Reprod. Sci.* **2019**, *26*, 812–828. [CrossRef] [PubMed]
122. Ciebiera, M.; Męczekalski, B.; Lukaszuk, K.; Jakiel, G. Potential synergism between ulipristal acetate and vitamin D3 in uterine fibroid pharmacotherapy—2 case studies. *Gynecol. Endocrinol.* **2019**, *35*, 473–477. [CrossRef] [PubMed]
123. Afrin, S.; AlAshqar, A.; El Sabeh, M.; Miyashita-Ishiwata, M.; Reschke, L.; Brennan, J.T.; Fader, A.; Borahay, M.A. Diet and Nutrition in Gynecological Disorders: A Focus on Clinical Studies. *Nutrients* **2021**, *13*, 1747. [CrossRef] [PubMed]

MDPI
St. Alban-Anlage 66
4052 Basel
Switzerland
www.mdpi.com

Nutrients Editorial Office
E-mail: nutrients@mdpi.com
www.mdpi.com/journal/nutrients

www.ingramcontent.com/pod-product-compliance
Lightning Source LLC
LaVergne TN
LVHW070921170726
843515LV00005B/833

* 9 7 8 3 0 3 6 5 9 5 8 4 9 *